MW00715316

THEORETICAL PERSPECTIVES IN ENVIRONMENT-BEHAVIOR RESEARCH

**Underlying Assumptions,
Research Problems, and Methodologies**

THEORETICAL PERSPECTIVES IN ENVIRONMENT-BEHAVIOR RESEARCH

Underlying Assumptions, Research Problems, and Methodologies

Edited by

Seymour Wapner
Clark University
Worcester, Massachusetts

Jack Demick
Suffolk University
Boston, Massachusetts and
University of Massachusetts Medical School
Worcester, Massachusetts

Takiji Yamamoto
Research Institute of Human and Health Science
Yokohama, Japan

Hirofumi Minami
Kyushu University
Fukuoka, Japan

Kluwer Academic / Plenum Publishers
New York, Boston, Dordrecht, London, Moscow

Library of Congress Cataloging-in-Publication Data

Theoretical perspectives in environment-behavior research : underlying assumptions,
research problems, and methodologies / edited by Seymour Wapner ... [et al.].
    p. cm.
  Includes bibliographical references and index.
  ISBN 0-306-46192-7
    1. Environmental psychology. I. Wapner, Seymour, 1917-

  BF353 .T48 1999
  155.9--dc21
                                                                99-048266

ISBN: 0-306-46192-7

© 2000 Kluwer Academic / Plenum Publishers, New York
233 Spring Street, New York, N.Y. 10013

10 9 8 7 6 5 4 3 2 1

A C.I.P. record for this book is available from the Library of Congress.

Printed in the United States of America

CONTRIBUTORS

Irwin Altman, Psychology Department, University of Utah, Salt Lake City, UT 84112

Robert B. Bechtel, Department of Psychology, University of Arizona, Tucson, Arizona 85721

Marino Bonaiuto, Università degli Studi di Roma "La Sapienza", Dipartimento di Psicologia dei Processi di Sviluppo e Socializzazione, 00185 Rome (Italy)

Mirilia Bonnes, Università degli Studi di Roma "La Sapienza", Dipartimento di Psicologia dei Processi di Sviluppo e Socializzazione, 00185 Rome (Italy)

David Canter, Centre for Investigative Psychology, Eleanor Rathbone Building, The University of Liverpool, Liverpool L69 7ZA United Kingdom

Herb Childress, Jay Farbstein & Associates, Inc., San Luis Obispo, CA 93401

Arza Churchman, Faculty of Architecture and Town Planning, Technion-Israel Institute of Technology, Haifa Israel

Jack Demick, Department of Psychology, Suffolk University, Boston, Massachusetts 02114, and Center for Adoption Research and Policy, University of Massachusetts, Worcester, Massachusetts 01605

Gary W. Evans, Department of Design & Environmental Analysis, Cornell University, Ithaca, NY 14853-4401

Masami Kobayashi, Department of Global Environmental Engineering, Kyoto University Yoshida-Honmachi, Sakyo-ku, Kyoto, 606-8501, Japan

Brian R. Little, Department of Psychology, Carleton University, Ottawa, ON K1S5B6, Canada

William Michelson, Department of Sociology, University of Toronto, Toronto, Ontario, Canada M5T 1P9, and Faculty of Arts and Sciences, University of Toronto, Toronto, Ontario, Canada M5S 3G3

Hirofumi Minami, Department of Urban Design, Planning and Disaster Management, Graduate School of Human-Environment Studies, Kyushu University, 6-19-1, Hakozaki, Fukuoka, 812 Japan

Ken Miura, Department of Global Environmental Engineering, Kyoto University Yoshida-Honmachi, Sakyo-ku, Kyoto, 606-8501, Japan

Yasushi Nagasawa, Department of Architecture, Graduate School of Engineering, The University of Tokyo, 7-3-1 Hongo, Bunkyo-ku, Tokyo 113 Japan

Ryuzo Ohno, Department of Built Environment, Interdisciplinary Graduate School of Science and Engineering, Tokyo Institute of Technology, Yokohama, 226-8502, Japan

George Rand, UCLA Department of Architecture and Urban Design, Los Angeles, CA 90024

Amos Rapoport, Department of Architecture, University of Wisconsin-Milwaukee, P.O. Box 413, Milwaukee, WI 53201

Leanne G. Rivlin, City University of New York Graduate School, New York, NY 10016

Susan Saegert, Graduate Center CUNY, Environmental Psychology, New York, NY 10036

David Seamon, Department of Architecture, Kansas State University, Manhattan, KS 66506-2901

Daniel Stokols, School of Social Ecology, Room 206, Social Ecology I Building, University of California, Irvine, Irvine, California 92697

Takashi Takahashi, Graduate School of Science and Technology, Niigata University 2-8050 Igarashi, Niigata 950-21 Japan

Seymour Wapner, Heinz Werner Institute for Developmental Analysis, Clark University, Worcester, Massachusetts 01610-1477

Carol M. Werner, Psychology Department, University of Utah, Salt Lake City, UT 84112

Takiji Yamamoto, Research Institute of Human Health Science, 10-51-215, Ichibaue-machi, Tsurumi-ku, Yokohama, 230 Japan

PREFACE

This volume had its origins in the *Handbook of Japan-United States Environment-Behavior Research* edited by Wapner, Demick, Yamamoto, and Takahashi (1997). In that volume, Altman's chapter entitled "Environment and Behavior Studies: A Discipline? Not a Discipline? Becoming a Discipline?" raised questions about the status of environment-behavior research. To advance research, he stressed the necessity of "understanding who we are, why we do what we do, and the connections and gaps between scholars and practitioners adopting diverse perspectives" (p. 423). Altman's focus on confronting fundamental philosophical assumptions was so elegantly elaborated in his chapter with Rogoff (Altman & Rogoff, 1987) and resonates with the work of Wapner and his associates (e.g., Wapner, 1981, 1987; Wapner & Demick, 1998; Wapner, Demick, Yamamoto, & Takahashi, 1997), insofar as they have systematically focused on the interrelations among theory, problem, and method, and the need for all researchers to make their hidden assumptions overt.

Given this background, the precipitating events that led to the systematic attempt to produce this volume were two symposia: one at the 1997 Man-Environment Behavior Association (MERA) meetings (organized by the coeditors) in Tokyo and another invited symposium at the 1998 Congress of the International Association of Applied Psychology in San Francisco (organized by S. Wapner). In both of these symposia, all contributors were asked to focus on the philosophical and theoretical assumptions underlying their environment-behavior research, and the linkage of those assumptions to the research problems they choose to study and the methodologies that they employ. These participants and other well-known investigators in the environment-behavior field contributed chapters to this volume.

We hope that by making overt the assumptions and their linkages to research problems and methods, the investigators and their investigations described here will effectively advance the research and practice of the field of environment-behavior research.

We express our thanks to the many people who helped produce this volume, in particular Joyce Lee, who served as editorial assistant.

REFERENCES

Altman, I. (1997). Environment and behavior studies: A discipline? Not a discipline? Becoming a discipline? In S. Wapner, J. Demick, T. Yamamoto, & T. Takahashi (Eds.), *Handbook of Japan-United States environment-behavior research* (pp. 423–434). NY: Plenum.

Altman, I., & Rogoff, B. (1987). World views in psychology: Trait, interactional, organismic and transactional perspectives. In D. Stokols & I. Altman (Eds.), *Handbook of environmental psychology* (pp. 7–40). New York: Wiley.

Wapner, S. (1981). Transactions of persons-in-environments: Some critical transitions. *Journal of Environmental Psychology, 1*, 223–239.

Wapner, S. (1987). A holistic, developmental, systems-oriented environmental psychology: Some beginnings. In D. Stokols & I. Altman (Eds.), *Handbook of environmental psychology* (pp. 1433–1465). NY: Wiley.

Wapner, S., & Demick, J. (1998). Developmental analysis: A holistic, developmental, systems-oriented perspective. In R. M. Lerner (Ed.), *Theoretical models of human development. Vol. 4 Handbook of child psychology* (5th ed., Editor-in-chief: William Damon). New York: Wiley.

Wapner, S., Demick, J., Yamamoto, T., & Takahashi, T. (Eds.). (1997). *Handbook of Japan-US environment-behavior research: Toward a transactional approach.* NY: Plenum.

CONTENTS

INTRODUCTION

Seymour Wapner, Jack Demick, Takiji Yamamoto,
and Hirofumi Minami

The papers in this volume represent an attempt to have authors with diverse perspectives involved in environment-behavior research and practice present the assumptions underlying their work and the linkage of those assumptions to the research problems they choose to study and the methodologies they employ.

The approaches to environment-behavior research are quite diverse including natural science to human science orientations and cover a variety of world views and spheres of activity that range from aesthetic and theoretical to the practical concerns of everyday life. A brief overview follows.

Wapner and Demick, in Chapter 2, broadly characterize their holistic, developmental, systems-oriented perspective in terms of the following assumptions: organicism (holism); transactionalism (focus on part-whole relations); person-in-environment as unit of analysis of experience and action; levels of organization of person and of environment; constructivism; multiple intentionality; and the orthogenetic principle of development. Moreover, they illustrate the linkage of assumptions and methodologies to research problems by describing a programmatic research area of critical person-in-environment transitions on experience (cognitive, affective, valuative) and action. In this context, they give their perspective on such questions, as for example: whether the researcher holds a natural science view, a human science view, or both; how a particular unit of analysis shapes research problems; and the implications for methodology of the method-oriented versus the problem-oriented researcher.

In Chapter 3, Werner and Altman adhere to Altman's (1995) view that uncovering the underlying assumptions of environment-behavior researchers will make for significant advances in moving environment-behavior study to becoming an integrated discipline. They set the stage for verbalizing such assumptions by drawing on Altman and Rogoff's (1987) description of world views or world hypotheses in conceptualizing environment-behavior research. Their focus is on the philosophy of science, unit of analysis, and temporal qualities. They provide a series of examples—in particular through a project on factors that serve to reduce unnecessary use of toxic chemicals—to show how assumptions are shaped differently depending on the researcher's world view.

The problem of critical person-in-environment transitions is pursued by

Theoretical Perspectives in Environment-Behavior Research, edited by Wapner *et al.*
Kluwer Academic / Plenum Publishers, New York, 2000.

Kobayashi and Miura (Chapter 4) with respect to natural disasters in Japan, Indonesia, and the Philippines. Survivors' place making was traced by focusing on positive and negative means used in each construction phase, including shelter, temporary housing, and rebuilt housing. Based on personalization for each residential phase, the role of the environment in making the transition is viewed as a model concerning transition after a natural disaster. The empirical-theoretical analysis served as the basis for practical action with commercial adoption of a plan for a Disaster Reconstruction Group House.

In Chapter 5, Rivlin makes overt her assumptions relevant to the nature of the environment historical perspectives, the multidisciplinary nature of environment and behavior studies, and the complexity of environment-behavior issues. These assumptions and contemporary factors affecting the selection of real life, in contrast to laboratory staged, research problems were effective in developing designs that addressed the time dimensions of the problem and the value of multiple methods and triangulation in research. Examples are given of the research underpinned by the multiple, multimodal perspective that speaks to the effectiveness of the approach; nevertheless, cautions are expressed concerning its use, such as the generalizability of the research across geographies, cultures, and physical sites.

The move away from laboratory staged projects is evident in Bechtel's (Chapter 6) ecological psychology approach, which centers around the behavior setting as the fundamental unit of analysis of human behavior. In keeping with the perspective, ordinary, more laboratory-oriented psychological studies are questioned as being too much outside observable daily behavior. Research in ecological psychology shows the influence of such variables as the size of social unit on human behavior. Problems and limitations of the research methods are explored.

There is further focus on the interpersonal and sociocultural features of the environment in Bonaiuto and Bonnes' (Chapter 7) presentation. They consider two social-psychological approaches to environment-behavior research, namely, social identity theory and the discursive approach. Their emphasis on the role played by sociocultural factors affecting people's environmental perceptions, evaluations, and actions is underlined in their main assumptions and illustrated by three research examples. Social identity theory stresses that people perceive and evaluate their own environment according to the same principles regulating their relations to other significant features of their social worlds (such as social groups and categories to which they belong). The discursive approach stresses the nature and features of environmental issues that are constructed by different social actors through different discursive selections and framings.

A social ecological model is presented by Little (Chapter 8) in which personal projects are the units of analysis. Four assumptive themes undergird both the theoretical and methodological developments in this research program. The *constructivist* theme emphasizes the need for methods that are personally salient, reflexive, and evocative. The *contextualist* theme underscores the importance of making assessments in an ecologically representative and temporally sensitive manner, and of creating social indicators with our assessment instruments. The *conative* theme emphasizes the importance of using modular, middle-level units of analysis that are examined systematically. The *conciliency* theme enjoins researchers and practitioners to seek connections by employing methodologies that are conjointly focused on both individual and group level phenomena, that are integrative rather than sectoral, and that provide for direct applicability in understanding and enhancing human well-being.

The person aspect of the person-in-environment unit of analysis is also stressed in Churchman's (Chapter 9) consideration of "women and the environment." This topic encompasses far more than the transactions of a particular user group with the physical world of neighborhoods, communities, and settlements. It also addresses fundamental questions about the nature of our environments, the nature of our professions, and the way we study, educate, do research, design, and plan. Much of the motivation for the emergence of the field, its philosophical underpinnings, and it methodological arguments are based on criticisms of the value-laden, male-oriented assumptions of "traditional" theory, research, and practice. Her chapter identifies the various assumptions—both those identified with traditional approaches and with the alternative approaches proposed. Which of these assumptions have been, or should be questioned, and which have not are analyzed. The chapter concludes by presenting the research evidence available with regard to major aspects of the topic of women and the environment.

In Chapter 10, Rapoport argues that environment-behavior studies must be seen as a science, which implies the need for explanatory theory. He describes his approach toward this goal and then deals with such issues as the nature of science, explanatory theory, and the attributes of "good" theory. Finally, Rapoport outlines his work in environment-behavior studies to illustrate how an explanatory theory of environment-behavior relations might be developed.

The natural science perspective stressed by Rapoport also appears to underpin Michelson's (Chapter 11) long-term interest in and research on the relationships of built environments to everyday life. Unforeseen obstacles to scientific advancement have required attention to assumptions, logic, and specifications, including such factors as: (a) the logic underpinning concerns about the kind of theory, levels of theory, directionality, strength of theoretical relationships and the complexity of theoretical schemes; (b) the need for measures that assess built environments and the outcomes hypothesized to occur from them independently but yet together; (c) the integration of built environments with non-contextual factors in the explanation of behavior; and (d) the absolute value of environment-behavior relationships in the analysis of human life and behavior.

Ohno's work on perception (Chapter 12) represents an example in keeping with a natural science perspective insofar as he re-examines a hypothetical model of environmental perception underlying his own development, which emphasizes the role of ambient vision in environmental perception. Two fundamental theories and their underlying assumptions are discussed: one concerns parallel processing and the other concerns Gibson's theory. The manifestation of the theoretical perspectives in empirical studies are reviewed.

The focus by Rapoport and others on scientific theory may be viewed in contrast to the human science approach as evident in Seamon's (Chapter 13) phenomenological approach and Childress' (Chapter 14) narrative approach. Seamon examines the phenomenological approach to environment-behavior research by underlining two major assumptions of the approach, namely, that people and environment compose an indivisible whole and that phenomenological methodology can be described in terms of a "radical empiricism." Three types of phenomenological method are described: (a) first person methodological research; (b) existential-phenomenological research; and (c) hermaneutic-phenomenological research. Finally, such issues as a phenomenological definition of reliability and the value of phenomenological research for architecture and environmental design are considered.

Childress, in Chapter 14, describes his narrative approach simply as the process of living among people about whom you would like to know more, watching, listening, and talking things over, and then reporting what you saw and heard to some third party. Some features central to his approach are that: the wholeness of person and place is best conveyed through the narrative; his interest is in questions for which objectivity is not possible; the individual and his or her circumstances are central; his research will change his participants and himself; his research will always be incomplete and arguable; and the power of the humanities to change minds is as important as the power of the sciences to change conditions.

Canter (Chapter 15), in focusing on psychological research making assumptions about the phenomena studied, speaks to the need for understanding the everyday experience of people in natural occurring contexts, which require methodologies that accurately reflect such experience. In this way, Canter sees the tension between producing studies that have scientific, theoretical merit as opposed to studies that have applied relevance.

Demick, Wapner, Yamamoto, and Minami (Chapter 16), building on the person-in-environment unit of analysis described by Wapner and Demick (Chapter 2), focus on the sociocultural aspect of the environment by considering how assumptions of the holistic, developmental, systems-oriented perspective determine problems and methods relevant to cross-cultural psychology. To show linkages between assumptions with problems for environment-behavior research and methodologies, they consider such research problems as: body/self experience (Japan vs. USA); values mothers hold for handicapped and non-handicapped preschoolers in the home and school context (Japan vs. Puerto Rico vs. Mainland USA); necessity, amenity, and luxury in physical, interpersonal, and sociocultural environments (Japan vs. Italy vs. Russia vs. USA); adaptation to new environments, for example, sojourner behavior (Japan vs. USA) and migration (Puerto Rico/Mainland USA); and compliance with mandatory automobile safety belt legislation (Japan vs. USA/Massachusetts vs. Italy).

The perspectives of architects working on environment-behavior research is evident in the presentations by Nagasawa (Chapter 17) and Takahashi (Chapter 18). Nagasawa notes that despite extensive efforts made by hospital architects, planners, and researchers to develop better hospital functioning and healthier environments for patients, modern hospitals have not yet reached that mark and people avoid them whenever possible. This is regarded, in part, as a consequence of the extensive gap between patients' expectations and the actual physical environment and operational systems within hospitals that have been uncovered by the studies of Nagasawa and his group. A review of the history of hospital buildings for healing is informative in identifying the increasingly dominant role that expensive medical technologies play and how planners may have overlooked important patient issues while focusing on the economic and functional needs of hospitals. Case studies are outlined and the impact of methodological choices are shown to relate to the nature, usefulness, and extent of findings. A change from a system-oriented approach to a more concept-oriented approach is posited as an important step in a wider search for better "healing environments."

Takahashi (Chapter 18) notes that, with the appearance of the modern movement in architecture, theories and methods of design, such as functionalism, were introduced. The functional approach is based on architectural determinism and a problem-solving (engineering) strategy. There is, however, currently a paradigm shift required in environmental design and architectural education. Consideration should be given to the notion of problem-finding and "sympathetic methodologies." It is assumed that such new methods will better enable sustainable environment-behavior situations that can be enhanced by cross-cultural discussions.

Minami and Yamamoto (Chapter 19) are concerned with urban conditions. They examine basic assumptions underlying major concepts in the field of environmental psychology such as crowding, personal space, and principles of urban planning from a socio-cultural perspective. Crowding, which is regarded as a negative state of living conditions in the west, might be regarded as a rather positive aspect of intimate community living in Asian countries. Communal space, rather than personal space, might be more important in Japanese city life. A case study of an urban renewal project in a Japanese city that adopts western models is presented to examine cultural-appropriateness in the concept of current environmental psychology and urban design disciplines. Two sets of cultural assumptions underlying environment-behavior research and urban planning are delineated and discussed. Life-world design as a communicative and participatory approach in the understanding and creation of a humane environment is proposed.

Evans and Saegert's (Chapter 20) central theme is that the accumulation of stressful experiences leads to increased risk for adverse outcomes as the adaptive capabilities of the organism are increasingly challenged. They point out that, if this proposition is correct, then an important omission in environmental stress research has been the examination of crowding, noise, and other stressors in isolation of the naturalistic contexts in which they normally occur. They examine residential crowding from an ecological perspective. They show that stressors accompanying inner city poverty interact with crowding to magnify their harmful influences on children and their families.

Stokols (Chapter 21) focuses on the difficulties involved in developing useful and robust environment-behavior theories in the current sociocultural conditions of accelerating societal change, and of technological and informational complexity. He assumes that three conceptual strategies might address these theoretical challenges more effectively: 1. elucidation of high-impact features of the sociophysical environment that disproportionately affect individuals' behavior and well-being; 2. examination of the mediational and moderating processes by which physical and social qualities of environments mutually influence participants' activities and health; and 3. identification of high-impact settings within people's daily activity systems that exert the greatest impact on their coping capacities and resources.

Stokol's quest for developing robust and useful environment-behavior theories is complemented by Rand's (Chapter 22) assumption that an interdisciplinary, intercultural perspective is critical in the study of person-environment relations. He develops his point of view by applying Pasto's (1964) notion of "space frame" (pervasive spatial scheme) to a wide variety of micro and macro phenomena (e.g., literature, science fiction, poetry) by recognizing that seeing is not purely visual insofar as it involves not only a "geometric-technical mode" but also an "expressive-physiognomic" mode. The focus is on the linkage between both aspects of visual experience. Special emphasis is given to the physiognomic, expressive mode of perception that is related to feeling and access to the unconscious. The "space frame" notion serves to account for many aspects of person-environment relations in a manner compatible with art and architectural history.

Finally, in Chapter 23, Wapner, Demick, Yamamoto, and Minami compare the various contributions of this volume against the backdrop of their holistic, developmental, systems-oriented approach to person-in-environment functioning across the life span. While the contributions may share more similarities than dissimilarities, the authors hope that the uncovering of theoretical assumptions will lead to an even greater accelerated growth of environment-behavior research in the 21st century.

2

ASSUMPTIONS, METHODS, AND RESEARCH PROBLEMS OF THE HOLISTIC, DEVELOPMENTAL, SYSTEMS-ORIENTED PERSPECTIVE

Seymour Wapner* and Jack Demick**

*Heinz Werner Institute for Developmental Analysis
Clark University
Worcester, Massachusetts 01610-1477
**Department of Psychology
Suffolk University
Boston, Massachusetts 02114
and Center for Adoption Research and Policy
University of Massachusetts
Worcester, Massachusetts 01605

1. INTRODUCTION

Beginning with the basic assumption of perspectivism that is grounded in the philosophical position of interpretationism or constructivism (Lavine, 1950 a,b) and Pepper's (1942) notion that understanding the world is based on a metaphoric description of world hypotheses and root metaphors, we shall, in keeping with the theme of this volume, attempt to look inward so-to-speak and uncover the basic assumptions of our holistic, developmental, systems-oriented perspective that shapes the nature of the research problems and methods we use in analyzing environment-behavior relationships. We shall attempt to characterize the main assumptions and illustrate how they mold methodology and the formulation of research problems.

2. ASSUMPTIONS

2.1. Organicism (Holism) and Transactionalism

The perspective is grounded in the world views of *organicism* and *transactionalism* (Altman & Rogoff, 1987). Organicism, synthetic putting together of parts into a

Theoretical Perspectives in Environment-Behavior Research, edited by Wapner *et al.*
Kluwer Academic / Plenum Publishers, New York, 2000.

unified whole, is implied by the *holistic* assumption that a change in one part of a system impacts all other parts, that is, the totality of the system. Transactionalism, which conceptualizes the person and environment as parts of a whole, includes *action* and *experience* with respect to various aspects of the environment, where experience, includes cognitive, that is, sensorimotor, perceptual and conceptual experience, as well as affective and valuative experience. An example of research linked to this assumption is the analysis of the work environment by Sundstrom (1987) who found that relations among levels of organization–physical/biological (temperature, air quality), psychological (feelings about work), and sociocultural (e.g., interpersonal relations)—contribute to the overall effectiveness of an organization. This assumption impacts the choice of our own research problems (e.g., critical person-in-environment transitions across the life span) as well as the methods employed, including experimental, phenomenological, holistic, and ecologically oriented studies with multiple assessments of diverse aspects experience and action.

2.2. Person-in-Environment System as Unit of Analysis/Levels of Organization

In keeping with the transactionalist assumption is the proposition that the *person-in-environment system* is the *unit to be analyzed*. Consider the relevant features of the person and environment categorized in terms of levels of organization, namely, physical/biological, psychological, and sociocultural. Within the psychological level, cognitive processes may be further categorized with respect to sensorimotor, perceptual and conceptual levels. Considering the person-in-environment as unit of analysis, persons are assumed to operate at three levels: *physical/biological* (e.g., physical status), *psychological/intrapersonal* (e.g.,cognitive, affective, evaluative processes), and *sociocultural* (e.g., role). In contrast to other positions, environments are characterized at three levels: *physical* (natural and man-made objects), living organisms (e.g., spouse, pets), and *sociocultural* (e.g., rules, laws of society).

Various researchers have included some of these levels of functioning, for example, Barker's (1968) work on behavior settings, Stokols and Shumaker's (1981) analysis of people in places, Rapoport's (1982) linkage between environment and culture, and Altman's (1976) analysis of the role of the person and environment in his treatment of the integration of social and environmental psychology. However, there remains a need to include three levels of both the person and the environment within the analyses of a single environment-behavior research problem.

2.3. Constructivism/Multiple Intentionality

Human beings are assumed to be spontaneously active, striving agents with capacity to construct, construe, and experience their environments in various ways as well as to act on that experience. For example, the person may adopt different intentions focusing on the environment (e.g., a building or the city out-there), focusing on the self (e.g., I am tall), and focusing on the relation between self and environment (e.g., I know my way around Tokyo). Implicit here is the directedness notion that the person can be directed toward self *qua* object or toward the object out there as well as the distinction between the *experienced*, *behavioral*, or *psychological* environment versus the *physical* environment.

There is much evidence to show that the cognitive-developmental status of the person (e.g., child vs. adult) plays a role in the way the environment is perceived and organized. This was demonstrated by Muchow and Muchow (cf. Werner, 1940/57) when making the distinction between life space and geographic space and demonstrating that a Hamburg dock or department store is construed differently by a child and an adult who have different views of the functional meaning of the place.

This assumption is predominant in Seamon's (1987, ch.13, this volume) phenomenological examination of intentionality in environmental experience. Wohlwill's (1983) characterization of ambience as focal or background is in keeping with the intentionality assumption.

Moreover, intentionality makes for a focus on time or space. Ongoing human experience is spatio-temporal in nature. Both temporal qualities and spatial qualities represent abstraction from ongoing human functioning that is spatio-temporal in character. As we see it, human beings can structure their experience of spatial features as if independent of temporality as well as structure their experience of temporal features as if independent of space. Further, we believe a distinction should be made between general change and developmental change. Developmental change is directed toward some end state; in contrast, no particular end state is specified for general change.

2.4. Development

In contrast to other approaches, the *developmental* aspect of the perspective is not restricted to ontogenesis. A developmental ideal or telos is postulated as characterized by the *orthogenetic principle* (Werner & Kaplan, 1956; Kaplan, 1959; Werner, 1940/57). The orthogenetic principle states, in formal organizational terms, that with development there is a shift from a de-differentiated state toward differentiated and isolated, differentiated and conflict states, and the ideal of a differentiated and hierarchically integrated person-in-environment system state. This developmental ideal implies *control* over self-world relationships, *discrete* (rather than syncretic) mental functioning, *articulate* (rather than diffuse) structures, adoption of *stable* (rather than labile), *flexible* (rather than rigid) modes of coping, with means *subordinated* to ends. Such a state of organization is assumed to involve greater salience of positive affective states, diminution of isolation, anonymity, helplessness, depersonalization, and entrapment; coordination of long-term and short-term goals, and movement toward an integrated unity of overt and covert actions (Wapner, 1987; Wapner & Demick, 1998)

Developmental ordering in terms of the formal, organizational features of the orthogenetic principle holds for such changes as occurs in:

2.4.1. Ontogenesis. (e.g., child's social and cognitive development, etc.) and aging (e.g., cognitive developmental changes in old age);

2.4.2. Microgenesis. (e.g., development of ideas, spatial organizations);

2.4.3. Pathogenesis. (e.g., neuropathogenesis-brain injury; psychopathogenesis-psychosis);

2.4.4. Phylogenesis. (e.g., adaptation manifest by members of different species);

2.4.5. Ethnogenesis. (e.g., changes during history of mankind); and

2.4.6. Conditions of Functioning. (e.g., fatigue vs normal conditions).

It has been demonstrated that children's organization of a large scale urban environment, as reflected in sketch maps, shows increasing differentiation and hierarchic integration with increase in age (e.g., Moore, 1976). Relatively little research has been done with other developmentally-ordered changes such as, microgenesis, etc. as noted above.

An example of a microgenetic study was conducted by Schouela, Steinberg, Leveton, and Wapner (1980). They found:

"Changes in the cognitive organization of a relatively circumscribed new environment (small university campus) over the course of a six month period following arrival in the locale were assessed by sketch map representations. Quantitative and qualitative analyses of the maps produced by the subjects, as well as their responses to a follow up inquiry, provided evidence that: (a) the cognitive organization of the new environment becomes increasingly differentiated and integrated over the course of exposure to the locale, and (b) the process of cognitive organization is based on the use of a personally salient location as an anchor point in relation to which the other parts of the environment are established and articulated." (p. 1.)

Another study dealt with both ontogenetic and microgenetic changes in nonverbal representation of spatial relations. American children–4 1/2, 6 1/2, 9 and 10 years of age–drew sketch maps of a new unfamiliar environment, a town in Holland, over the course of their 9-month visit. Age comparisons as well as microgenetic changes for a given child in keeping with the orthogenetic principle were found. There was progressive differentiation into distinct objects with increasing integration of spatial relations among them (Wapner, Kaplan, & Ciottone, 1981).

3. MODES OF ANALYSIS

3.1. Structural and Dynamic Analyses

The *structural* mode of analysis deals with organization or part-whole relations, for example, how parts, such as physical character of home and intra-psychological experience, fit with each other. Another example based on the influence of a part on the whole is the study, which demonstrated that location in a relatively "bad" versus "good" part of a city impacted experience of the city as a whole (Demick, Hoffman, & Wapner, 1985).

The *dynamic* analysis deals with long- and short-term goals or ends, and the means and instrumentalities to achieve those goals. For example, consider the goal of learning the spatial arrangement of a physical environment such as the University of Tokyo. What instrumentalities can be used to achieve this goal? Can one build a spatial organization through use of an anchor point (first place of entry), which is systematically added to and reorganized with further exposure (see Schouela et al., 1980, above).

4. PROBLEM FORMULATION AND METHODOLOGY

As noted earlier our research focuses on describing relations among and within parts making up the person-in-environment system as well as with specifying conditions that make for changes in the organization of these relations. Methodology for conducting such research is linked to the assumption that both the *"natural science"* and *"human science"* perspectives are appropriate depending on the nature of the problem. Some research problems involving fewer and less complex variables can utilize experimental methods of natural science. In contrast, however, more complicated problems involving variables difficult to control must rely on and utilize more descriptive, phenomenological methods. There is concern not only with *description* of experience and action, but also with the processes underlying *change* in experience. Thus, generalizations are derived from both nomothetic (representative samples) as well as idiographic (sample as prototype) procedures. For example, experimental methods are appropriate for the study of certain aspects of a phenomenon (using examples from our own work, factors such as cognitive style as well as conditions such as planning vs. no planning that affect adaptation to life transitions), while other methods (e.g., phenomenological, narrative) provide access to different aspects of a phenomenon (e.g., experiential changes following such transitions).

Furthermore, there is greater focus on *process* than on achievement (Werner, 1937). The research is *problem-* rather than *means-oriented*, that is, whereas in the former, the method dictates the range of problems that can be studied, the latter gives priority to the phenomenon being studied (Maslow, 1946). Moreover, both *cross-sectional* and *longitudinal* designs are employed. The former involve attempts to characterize an atemporal relation between abstracted variables or a set of relationships by contrasting groups or conditions at a given point in time; the latter involve changes over time.

5. PARADIGMATIC RESEARCH PROBLEMS

5.1. Programmatic Studies of Critical Person-in-Environment Transitions

In attempting to illustrate the linkage between assumptions and research problems as well as the heuristic value of the holistic, developmental, systems-oriented perspective, it is necessary to consider our research program on critical person-in-environment transitions that has been in operation for more than two decades. The paradigmatic problem of critical person-in-environment transitions was chosen because it is linked to all of the assumptions of the perspective, in particular the holistic and developmental assumptions. A potent perturbation to any part of the person-in-environment system at one or more levels of person and environment is expected to impact the system as a whole. With respect to development, powerful changes in the person, the environment, or both may make for *developmental regression* as described by the orthogenetic principle. It is further assumed that, given appropriate conditions, there may be a further change of *developmental progression* in the functioning of the person-in-environment system.

Examination of Table 1 reveals a number of sites and examples where a powerful perturbation may initiate a critical person-in-environment transition (see Wapner & Demick, 1998, for a description of some of these studies).

Table 1. Sites and Examples of Perturbations to Person-in-Environment System which May Initiate Critical Transitions

PERSON (x ENVIRONMENT)	ENVIRONMENT (x PERSON)
PHYSICAL (BIOLOGICAL)	*PHYSICAL*
Age (e.g., onset of puberty, menopause, death)	*Objects* (e.g., *acquisition or loss of cherished*
Pregnancy	*possessions*)
Disability	*Disaster* (e.g., *onset of flood, hurricane, earthquake,*
Illness	*tornado, volcanic eruption, nuclear war*)
– Addiction (e.g., onset and *termination of alcoholism,*	*Relocation* (e.g., *psychiatric community, nursing home,*
obesity, drug addiction)	*rural*, urban, *transfer to new college, migration*)
– *Chronic* (e.g., *onset of diabetes, rheumatoid arthritis*)	Urban Change (e.g., decline, *renewal*)
– *Acute* (e.g., onset and treatment of cancer, *AIDS*)	Rural Change (e.g., industrialization)
PSYCHOLOGICAL	*LIVING ORGANISMS*
Body Experience (e.g., *increase or decrease in size of*	*Peer Relations* (e.g., *making or dissolving a friendship*
body, onset of experience of positive or negative	*or social network, falling in or out of love*)
body evaluation, acquisition or loss of cherished	Family (e.g., change in extended family, immediate
possessions)	family, parents, relatives)
Self Experience (e.g., self-concept and experience of	Neighbors
control, dignity, identity, power, security as in *onset*	Co-workers
of or recovery from mental illness, changing role in	Teachers
social network)	Animals (pets)
SOCIO-CULTURAL	*SOCIO-CULTURAL*
Role	
– *Work* (e.g., becoming employed, temporarily	Economics (e.g., new technology, job opportunity)
employed, *unemployed, retired*)	*Educational* (e.g., *nursery school*, kindergarten,
– Financial (e.g., becoming rich as in winning lottery,	elementary school, *high school, college, sojourn to*
becoming poor as in stock market crash)	*university abroad*, graduate, or *professional school*)
– Educational (e.g., professor, students administrator)	*Legal* (e.g., abortion legislation, driving age, *automobile*
– Marital (e.g., being married, *divorced*, widowed,	*seat belt legislation*, child abuse, retirement
parenthood, adoption)	legislation, euthanasia)
– *Religious* (e.g., becoming priest, minister, rabbi, nun,	Mores (e.g., attitude toward sex)
Jesuit, "Born Again" Christian, conversion)	Political (e.g., social, country, prison, defection)
– Political (e.g., becoming a refugee, undercover agent,	Religious (e.g., oppression, change in policy re female
war veteran, Holocaust survivor, survivor of	ministers and rabbis, celibacy of priests)
terrorism, elected official)	*Organizational (Industry) Leadership*
– Cultural (e.g., becoming a celebrity, member of a	
cult group)	
– Ethnicity (e.g., becoming aware, proud of, ashamed	
of background)	
– Gender (e.g., changing sexual orientation, from	
justice to caring orientation)	

[1] Italicized items indicate published studies or studies in progress.

Let us briefly describe some of these studies on critical person-in-environment transitions.

5.1.1. Person: Physical/biological. How does a person, who through an accident is unable to walk, experience the physical arrangements of access to a building? Given children who are handicapped (e.g., orthopedic disability, blindness, deafness, cerebral palsy, at risk with respect to an educational handicap), how do they experience and act with respect to their home and school environment (cf. Quirk & Wapner, 1995)? Relevant here is a study, utilizing the holistic and intentionality assumptions, that deals with changes in ambulatory status from prefracture to 3 and 6 months post-fracture.

"... participants who had greater previous experience with illness and who had positive expectations for recovery were likely to have less negative change in ambulation from prefracture to 3 months and better overall ambulation at 3 months. . . . The findings also revealed that patients who perceived their problem as directed toward the environment out-there, that is, caused by the environment, showed greater improvement in ambulation at 3 and 6 months relative to those who were directed toward self, that is, who perceived the fracture as an internal or organic problem." (Quirk & Wapner, 1995, p. 97)

5.1.2. Person: Intra-psychological. Individuals differing with respect to nature of mental illness–schizophrenics compared with anti-social personalities–responded differentially prior to and following relocation of a psychiatric therapeutic community (Demick & Wapner, 1980). Significant differences were found between these two groups of patients over time with respect to experience of self (body *qua* object), of environment (location of functional spatial areas, hospital rules), and of self-environment relations (hours spent in different locales; relationships with others).

5.1.3. Person: Sociocultural. What impact does change in role, for example, becoming a parent, have on the manner in which one experiences and acts with respect to the physical and interpersonal features of the world out-there? Pertinent here is a Ph.D. dissertation titled "Transition to parenthood: The experience and action of first-time parents" that is being conducted by William Clark (in progress). "The addition of the first child to a childless family is assumed to be a critical transition which is not replicated by the transition with subsequent children. The transition to parenthood for first time parents precipitates changes in the family person-in-environment system which has impact on each parent and their individual and joint relations to the environment, for example, their interpersonal relationship with significant others, such as extended family, friends, and colleagues (work and community)" (p. 4.) In short, the changes are expected to impact all six aspects of the person-in-environment system. For example, with respect to the physical aspect of the environment: there is furnishing or decorating a room for the baby; giving up personal space in the home, for example, a home office; use of a variety of child safety aids for home and automobile, etc.

5.1.4. Environment: Physical. Utilizing in-depth interviews concerning transactions (experience and action) with people involved in Hurricane Andrew, it was found that with the onset of the hurricane, the person's transactions were regressed developmentally (e.g., afraid to go to sleep, hollering, clinging to parents, nightmares); with time, they appeared to return to functioning at a more advanced developmental level (e.g., hurricane play behavior, encouraging others to be prepared) (Chea & Wapner, 1995).

5.1.5. Environment: Living Organisms. King (1995) analyzed similarities and differences in experience and action of men and women involved in dissolution of a love relationship. She found various precipitating events or triggers making for the loss, including changes in the physical environment (e.g., going away to college, distant location), the interpersonal environment (e.g., unfaithfulness, physical violence), and psychological experience (e.g., loss of attraction, personal problems). There were also differences in modes of action taken to alleviate the loss.

5.1.6. Environment: Sociocultural

5.1.6.1. Educational Transitions. The educational sociocultural context provides many examples of critical person-in-environment transitions that impact the individual at various stages in the life cycle. Consider some examples: 1. A pilot study (Ciottone & Quirk, 1985) on entry into nursery school pointed to the importance on adaptation of the transitional objects that children brought with them and of anchor points or the beginning base of operations, whether they be social (teacher) or physical (some outstanding object near entry) (Ellefsen, 1987). 2. A study of transition from an elementary school to a large junior high school in Japan (Yamamoto & Ishii, 1995) showing the greater increase in social network of students from the a large school as compared with those from a small school. 3. The process of development of the cognitive organization of entering a relatively circumscribed new, small university environment (Schouela, Steinberg, Leveton, & Wapner, 1980) that utilized sketch maps on six occasions following entry and after the sixth occasion where participants were asked to provide reasons for their choice of starting point and how they went about producing the map. There was clear cut evidence of the use of a personally salient anchor point or base of operations for the formation of a cognitive organization of the new environment. 4. A parallel study on the development of the interpersonal environment on entry into college utilized the Psychological Distance Map (PDM) for people (Wapner, 1978). Initials of persons in small circles are placed on a piece of paper at varying distances from a small circle labeled "me" at the center of an otherwise blank page. Analysis of the entries indicated that over a 6 month period, the number of entries from the home environment decreased while the number of entries from the new environment increased markedly; there was also a slow increase in home to university connections (also see Minami, 1985). 5. Wofsey, Rierden, and Wapner (1979) found striking evidence that, compared with seniors who have no fully articulated plans following graduation, those who have highly articulated plans (e.g., admitted to graduate school, having a job in a far away city, etc.), construe the base college environment in which they are still located as reflecting greater self-world distancing (see also Apter, 1976). 6. Quirk, Ciottone, Letendre, and Wapner (1986, 1987) studied entering the first year of medical school and the shift from pre-clinical to clinical training following the second year of training using intensive retrospective, introspective, and prospective interviews. They found that "During the pre-clinical transition there is a significant decline in perceived health, class standing, perceived quality of instruction, and satisfaction with social activities. Concurrently, students experience an increase in the level of felt stress" (Quirk et al., 1987, p. 418). To cope with this transition as well as with the shift from non-clinical to clinical activities, such strategies are employed as use of interpersonal support systems, structuring of time, and lowering of personal aspirations regarding academic and clinical performance.

In addition to these studies on educational critical transitions, it is appropriate in illustrating the sociocultural environmental context to include studies that involve transitions to a new country (sojourn and migration to a new country) and experiential and action changes due to introduction of a law involving use of automobile safety belts. Since cross-cultural research is involved in transition to a new country, this is briefly mentioned here and more fully described in Demick, Wapner, Yamamoto, and Minami (Chapter 16). Since the study on effect of legislative changes in automobile safety belt usage was first considered as an example studying the impact of a change in legislation on experience and action, below we shall more fully describe the study on usage of automobile safety belts as an example of a sociocultural/environmental critical transition. Thus, our person-in-environment unit of analysis directs us to research problems that involve the combination of critical person-in-environment transitions and cross-cultural research.

5.1.6.2. Transitions to a New Country. The studies on sojourner experience (Japan, US) and migration (Japan, US, Puerto Rico) are more fully described in Demick, Wapner, Yamamoto, and Minami (Chapter 16) (Also see Wapner, Demick, Inoue, Ishii, & Yamamoto, 1986; Wapner, Fujimoto, Imamichi, Inoue, & Towes, 1997; Lucca-Irizarry, Wapner, & Pacheco, 1981; Pacheco & Lucca-Irizarry, 1985). Here, we may simply note that these two studies not only utilize the assumption of multiple intentionality and its correlate of multiple worlds (persons are assumed to live in different, yet related experiential worlds such as family, work, school, recreation, etc.), but also that of adaptation (optimal relations between self and environment that may involve change in either or both of these components).

5.1.6.3. Compliance with Mandatory Automobile Seat Belt Legislation (Japan, Italy, US). The work on critical transitions concerning automobile safety belt usage derived originally from the attempts to obtain reliable evidence concerning experience and action following the introduction of a new societal law. Automobile safety belt use was specifically chosen because at the time of the study drivers' use could be accurately observed. A law concerning automobile safety belt use that was to be introduced in Massachusetts and Japan served as the occasion for conducting the studies on this problem.

Demick, Inoue, Wapner, Ishii, Minami, Nishiyama, and Yamamoto (1992) examined cultural differences in automobile safety belt usage prior to and following the initiation of mandatory safety belt legislation in two independent studies, one in Massachusetts (United States) and the second in Hiroshima (Japan). Comparison of the pattern of findings indicated cultural differences over time. In both sociocultural contexts, there was an increase in drivers' use of safety belts on the highway and in the city from the test occasion prior to passage of the law to the test occasion immediately following legislation. However, whereas the usage rates began to level off and remained constant or increased further in Hiroshima, they continued to decrease steadily in Massachusetts. Further, a significant number of the Massachusetts participants voiced their concern that mandatory safety belt legislation was an invasion of privacy/infringement on human rights, which ultimately resulted in repeal of the legislation and further decrease in safety belt use. No parallel phenomenon was manifest among the Japanese. A subsequent pilot study in Italy (Bertini & Wapner, 1992) revealed even sharper rates of decline in seat belt usage following mandatory legislation in Italy versus in the United States.

6. RELATIONS BETWEEN EXPERIENCE AND ACTION

The cultural differences in impact of legislation on automobile safety belt usage opened another important problem, namely, the relations between experience and action. Here, the concern is the processes underlying the translation of experience and intentionality into action: "Do I, in fact, do what I know I should do or what I want to do?" Our most extensive work in this area has been preventative, as described above, in the sense of taking self-protective action such as use of automobile safety belts.

Here, we have found it useful to employ Turvey's (1977) musical instrument metaphor, used in understanding neurological mechanisms. This metaphor distinguishes between "tuning" and "activating" inputs. As Gallistel (1980) stated: "Consider for example a piano or guitar. Turning the tuning pegs to tighten or loosen the strings does not produce music, but it profoundly alters the music that is produced. The tightness of each string is a parameter of a piano or guitar. Signals that adjust these parameters are called parameter adjusting or tuning inputs to distinguish them from activating inputs like key-strikes and string-plucking" (p. 364). In our analysis, tuning corresponds to the category of general factors preparing the individual for some action (e.g., knowledge of safety belt effectiveness, anxiety about driving, prior experience in an accident). Specific precursors or triggers for initiating the concrete behavior of "buckling up" in the particular context of the automobile (e.g., imagination of an accident, desire to serve as a role model for children in the car) are analogous to striking a key, which directly leads to the action of automobile safety belt usage. Thus, our assumptions and our problems operate synergistically, that is, assumptions lead to new problems and vice versa.

In addition to the cross-cultural study on frequency of use of automobile safety belts, research was conducted on the basis for such action. A phenomenologically oriented, structural description of the use of automobile safety belts (Rioux & Wapner, 1986) was obtained with three groups of people: those who were self-proclaimed as non-users, variable users, and committed users. In general, *committed users* initiated and maintained usage through imagining accidents, fear of personal injury, memories of accidents involving others, etc. *Nonusers* distanced themselves from a potential accident by viewing themselves as in control and minimizing risk of personal injury. *Variable users* were context oriented. The decision to wear a safety belt depended on such factors as weather conditions, size of car, with whom they were driving, and familiarity with area.

A central open question is what can be done to move persons so that they *do what they want to do*. Here, we have attempted to assess the nature of the triggers or precipitating events that have preceded their action in accordance with their wishes. It is possible that, if a new group of people struggling with the goal of "doing what they want to do" are made aware of these triggers, these people may in turn change their behavior in keeping with this goal. Another approach, experimental in form, comes from a study on AIDS where a personalized treatment was used to decrease the psychological distance between the subject and the threat of HIV/AIDS (Clark, 1995; Clark, Wapner, & Quirk, 1995; Quirk & Wapner, 1995). This treatment led to reports of a significantly greater frequency of practicing safe sex. An open question remains whether, for example, decreasing the psychological distance between the driver of an automobile and an accident (e.g., show the driver a constructed video of him or her in an auto involved in an accident) would increase use of automobile safety belts.

7. CONCLUSIONS

Our holistic, developmental, systems-oriented perspective to person-in environment functioning has heuristic value in opening a broad variety of problems for study as evident in the programmatic research described in the areas of critical person-in-environment transitions and cross-cultural problems. We believe that this approach can effectively contribute to many aspects of research and praxis in various fields of psychology (see Wapner & Demick, 1998, for a survey of its applicability to diverse areas of psychology) as well as to related disciplines that are concerned with moving human beings toward optimal functioning in their everyday environments. Central to such a belief are those features of the approach as: its focus on formal, organizational aspects of experience and action; its longstanding commitment to the complementarity of normative explication (description) and causal explanation (conditions under which cause-effect relations occur); its transactional emphasis of the person-in-environment system as the unit of analysis; its holistic and broadly defined developmental emphasis; and its deep concern to make explicit the assumptions and their linkage to problems of research and methodologies employed.

We believe that the making explicit of our assumptions together with those implicit in the research of a variety of other investigators will serve as significant steps in uncovering common and diverse ground of those working in the environment-behavior field. Such steps will hopefully move us toward fulfilling Altman's (1997) belief that understanding of "who we are, why we do what we do, and the connections between scholars and practitioners adopting diverse perspectives" will further "our future understanding of people and environments" (p. 424).

REFERENCES

Altman, I. (1976). Environmental psychology and social psychology. *Personality and Social Psychology Bulletin, 2,* 96–113.

Altman, I. (1997). Environment and behavior studies: A discipline? Not a discipline? Becoming a discipline? In S. Wapner, J. Demick, T. Yamamoto, & T. Takahashi (Eds.), *Handbook of Japan-United States environment-behavior research* (pp. 423–434). NY: Plenum.

Altman, I., & Rogoff, B. (1987). World views in psychology: Trait, interactional, organismic and transactional perspectives. In D. Stokols & I. Altman (Eds.) *Handbook of environmental psychology* (pp. 7–40). New York: Wiley.

Apter, D. (1976). *Modes of coping with conflict in the presently inhabited environment as a function of variation in plans to move to a new environment.* Unpublished Master's thesis, Clark University

Barker, R. G. (1968). *Ecological psychology: Concept and methods for studying the environment of human behavior.* Stanford, CA: Stanford University Press.

Bertini, G., & Wapner, S. (1992). *Automobile seat belt use prior to and following legislation.* Unpublished manuscript, Clark University, Worcester, MA.

Chea, W. E., & Wapner, S. (1995). Retrospections of Bahamians concerning the impact of hurricane Andrew. In J. L. Nasar, P. Grannis, & K. Hanyu (Eds.), *Proceedings of the twenty-sixth annual conference of the Environmental Design Research Association* (pp. 87–92). Oklahoma: EDRA.

Ciottone, R. A., & Quirk, M. (1985). *The integration of two worlds: Home and school.* Paper presented at the annual meeting of the Eastern Psychological Association, Boston, MA.

Clark, E. F. (1995). *Women's self-reported experience and action in relation to protection against sexual transmission of HIV: A randomized case comparison study of three interventions.* Doctoral dissertation, Clark University, Worcester, MA.

Clark W. (In progress). *The transition to parenthood: The experience and actions of first time parents.* Doctoral Dissertation, Clark University, Worcester, MA.

Clark, E., Wapner, S., & Quirk, M. (1995, Aug). *Interventions in protecting women against sexual transmission of HIV.* Presented at annual meeting of the American Psychological Association, New York.

Demick, J., Hoffman, A., & Wapner, S. (1985). Residential context and environmental change as determinants of urban experience. *Children's Environments Quarterly, 2*(3), 44–54.

Demick, J., Inoue, W., Wapner, S., Ishii, S., Minami, H., Nishiyama, S., & Yamamoto, T. (1992). Cultural differences in impact of governmental legislation: Automobile safety belt use. *Journal of Cross-Cultural Psychology, 23*(4), 468–487.

Demick, J., & Wapner, S. (1980). Effect of environmental relocation upon members of a psychiatric therapeutic community. *Journal of Abnormal Psychology, 89,* 444–452.

Ellefsen, Karen F. (1987). *Entry into nursery school: Children's transactions as a function of experience and age.* Unpublished master's thesis, Clark University, Worcester, MA.

Gallistel, C. R. (1980). *The organization of action: A new synthesis.* Hillsdale, NJ: Erlbaum.

Kaplan, B. (1959). The study of language in psychiatry. In S. Arieti (Ed.), *American handbook of psychiatry* (Vol. 3, pp. 659–668). New York: Basic Books.

King, K. (1995). Women's experience and action following the dissolution of a love relationship. Unpublished manuscript, Clark University.

Lavine, T. (1950a). Knowledge as interpretation : An historical survey. *Philosophy and Phenomenological research, 10,* 526–540.

Lavine, T. (1950b). Knowledge as interpretation: An historical survey. *Philosophy and Phenomenological research, 11,* 80–103.

Lucca-Irizarry, N., Wapner, S., & Pacheco, A. M. (1981). Adolescent return migration to Puerto Rico: Self-identity and bilingualism. *Agenda: A Journal of Hispanic Issues, 11,* 15–17, 33.

Maslow, A. H. (1946). Problem-centering vs. means-centering in science. *Philosophy of Science, 13,* 326–331.

Minami, H. (1985). *Establishment and transformation of personal networks during the first year of college: A developmental analysis.* Unpublished doctoral dissertation, Clark University, Worcester, MA.

Moore, G. T. (1976). Theory and research on development of environmental knowing. In G. T. Moore & R. G. Golledge (Eds.), *Environmental knowing: Theories, research, and methods* (pp. 138–164). Stroudsburg, PA: Dowden, Hutchinson & Ross.

Pacheco, A. M., Lucca, N., & Wapner, S. (1985). The assessment of interpersonal relations among Puerto Rican migrant adolescents. In R. Diaz-Guerrero (Ed.), *Cross-cultural and national studies in social psychology* (pp. 169–176). N. Holland: Elsevier Science Publishers B.V.

Pepper, S. C. (1942). *World hypotheses.* Berkeley: University of California Press.

Quirk, M., Ciottone, R., Letendre, D., & Wapner, S. (1987). *Critical person-in-environment transitions in medical education.* Presented at the 21st International IAAP Congress, Jerusalem, Israel.

Quirk, M., Ciottone, R., Letendre, D., & Wapner, S. (1987). Critical person-in-environment transitions in medical education. *Medical Teacher, 9*(4), 415–423.

Quirk, M., & Wapner, S. (1995). Environmental psychology and health. *Environment and Behavior, 27,* 90–99.

Rapoport, A. (1982). *The meaning of the built environment.* Beverly Hills, CA: Sage.

Rioux, S., & Wapner, S. (1986). Commitment to use of automobile seat belts: An experiential analysis. *Journal of Environmental Psychology, 6,* 189–204.

Rokeach, M. (1973). *The nature of human values.* New York: Free Press.

Schouela, D. A., Steinberg, L. M., Leveton, L. B., & Wapner, S. (1980). Development of the cognitive organization of an environment. *Canadian Journal of Behavioural Science, 12,* 1–16.

Seamon, D. (1987). Phenomenology and the environment. *Journal of Environmental Psychology, 7,* 367–377.

Stokols, D., & Shumaker, S. A. (1981). People in places: A transactional view of settings. In J. Harvey (Ed.), *Cognition, social behavior, and the environment* (pp. 441–488). Hillsdale, NJ: Erlbaum.

Sundstrom, E. (1987). Work environments: Offices and factories. In D. Stokols & I. Altman (Eds.), *Handbook of environmental psychology* (pp. 733–782). New York: Wiley.

Turvey, M. T. (1977). Preliminaries to a theory of action with reference to vision. In R. Shaw & J. Bransford (Eds.), *Perceiving, acting, and knowing.* Hillsdale, NJ: Erlbaum.

Valsiner, J. (1998, July). *Culture in the mind: Historical nature of human ontogeny.* Presented at the meetings of the International Society for the Study of Behavioural Development. Bern, Switzerland.

Wapner, S. (1978). Some critical person-environment transitions. *Hiroshima Forum for Psychology, 5,* 3–20.

Wapner, S. (1987). A holistic, developmental, systems-oriented environmental psychology: Some beginnings. In D. Stokols & I. Altman (Eds.), *Handbook of environmental psychology* (pp. 1433–1465). NY: Wiley.

Wapner, S., & Demick, J. (1998). Developmental analysis: A holistic, developmental, systems-oriented perspective. In R. M. Lerner (Ed.), *Theoretical models of human development. Vol. 4 Handbook of child psychology* (5th ed., Editor-in-chief: William Damon) (pp. 761–805). New York: Wiley.

Wapner, S., Demick, J., Inoue, W., Ishii, S., & Yamamoto, T. (1986). Relations between experience and action: Automobile seat belt usage in Japan and the United States. In W. H. Ittelson, M. Asai, & M. Carr (Eds.), *Proceedings of the 2nd USA/Japan seminar on environment and behavior* (pp. 279–295). Tucson, AZ: Department of Psychology, University of Arizona.

Wapner, S., Fujimoto, J., Imamichi, T., Inoue, Y., & Toews, K. (1997). Sojourn in a New Culture: Japanese Students in American Universities and American Students in Japanese Universities. In S. Wapner, J. Demick, T. Yamamoto, & T. Takahashi (Eds.), *Handbook of Japan-US environment-behavior research: Toward a transactional approach.* NY: Plenum.

Wapner, S., Kaplan, B., & Ciottone, R. (1981). Self-world relationships in critical environmental transitions: Childhood and beyond. In L. Liben, A. Patterson, & N. Newcombe (Eds.), *Spatial representation and behavior across the life span* (pp. 251–282). NY: Academic Press.

Wapner, S., McFarland, J. H., & Werner, H. (1963). Effect of visual spatial context on perception of one's own body. *British Journal of Psychology, 54,* 41–49.

Werner, H. (1937). Process and achievement: A basic problem of education and developmental psychology. *Harvard Educational Review, 7,* 353–368.

Werner, H. (1940/57). *Comparative psychology of mental development.* New York: International Universities Press. [Originally published in German, 1926, in English, 1940].

Werner, H. (1957). The concept of development from a comparative and organismic point of view. In D. B. Harris (Ed.), *The concept of development: An issue in the study of human behavior.* Univ. of Minnesota Press.

Werner, H., & Kaplan, B. (1956). The developmental approach to cognition: Its relevance to the psychological interpretation of anthropological and ethnolinguistic data. *American Anthropologist, 58,* 866–880.

Wohlwill, J. F. (1983). Physical and social environment as factors in development. In D. Magnuson & U. L. Allen (Eds.), *Personality development as person-environment interaction* (pp. 111–129). New York: Academic.

Wofsey, E., Rierdan, J., & Wapner, S. (1979). Planning to move: Effects on representing the currently inhabited environment. *Environment and Behavior, ll,* 3–32.

Yamamoto, T., & Ishii, S. (1995). Developmental and environmental psychology: A microgenetic developmental approach to transition from a small elementary school to a big junior high school. *Environment and Behavior, 27,* 33–42.

3

HUMANS AND NATURE

Insights from a Transactional View

Carol M. Werner and Irwin Altman

Psychology Department
University of Utah
Salt Lake City, UT 84112

1. INTRODUCTION

As part of our graduate student recruitment process, we routinely send copies of key articles to prospective students. It gives them an opportunity to see current work and anticipate what they might do as graduate advisees. Over the years, some students have taken a rather dim view of our work on transactional world views and philosophical underpinnings of research, describing it variously as "not my thing," "a real snoozer," and the like. We also find this lack of interest in philosophical issues to be true of some of our professional colleagues. For understandable reasons, they are usually more interested in formulating specific research questions, designing research paradigms that will be robust, probing their data and results, learning new statistical techniques that will help them find outcomes that are interesting (and publishable), and rounding out the Discussion sections of their papers with ideas for specific future research projects. It is the tangibles of research, not the philosophical underpinnings, that are immediately rewarding and relevant to researchers' everyday interests and careers.

At the same time, the lack of interest in philosophical issues surprises us, because we have found articulating research assumptions to be quite liberating. By putting traditional assumptions in perspective, we recognize they are just one of several ways of doing research. We feel comfortable trying out alternative approaches and exploring new ways of thinking about and studying phenomena (e.g., Oxley, Haggard, Werner, & Altman, 1985). Indeed, being aware of alternative ways of knowing has helped us see limitations in traditional psychological approaches. In our own work, it helps us see where we have been, where we could go, as well as enabling us to see what we have

Theoretical Perspectives in Environment-Behavior Research, edited by Wapner *et al.*
Kluwer Academic / Plenum Publishers, New York, 2000.

overlooked. We believe that our work as individual investigators as well as the progress of the field are dependent upon understanding the philosophical assumptions that undergird all research.

The purposes of this paper are to lay out various research assumptions, show how they inform practical research decisions, and suggest how they relate to the totality of topics encompassed by "environment and behavior research". In accord with the theme of the book, we focus on our own research and the decisions we make with respect to assumptions and guiding issues.

2. ENVIRONMENT AND BEHAVIOR: RESEARCH TOPICS, WORLD VIEWS, AND ASSUMPTIONS

We define the environment and behavior field as it is defined in most environmental psychology textbooks: how humans relate to and use, influence and are influenced by, nature and the built environment. Like most environment and behavior researchers, we approach this theme through an over-arching interest in quality of life as reflected in such issues as: how environmental design enhances wayfinding and privacy processes; how design can create well-functioning homes and neighborhoods; how city forms support community relations; how people use a variety of work, recreational, and social environments; and so on. In addition and in contrast to many environment and behavior researchers, one of us (CW) is part of an emerging group interested in quality of life for the biosphere (of which humans are a part), and ask such questions as: How do humans relate to nature affectively, cognitively, and behaviorally? How can backyards provide wildlife habitat? How can we lead low-impact lifestyles? How can nature and humans coexist? What individual and group processes lead to decisions to conserve wetlands and open spaces versus paving them over as highways and parking lots?

Research on every single environment and behavior issue is necessarily informed by implicit or explicit world views and underlying philosophical assumptions. Elsewhere, we have provided detailed descriptions of four world views and how they have guided or might guide environment and behavior research (Altman & Rogoff, 1987; Werner, Altman, Oxley, & Haggard, 1987; Werner & Altman, 1998). They include "trait" (behavior driven by internal qualities), "interactional" (traits and situational cues influence behavior), "organismic" (systems model in which separate elements exert mutual and reverberating influences), and "transactional" views (holistic model emphasizing the flow and patterning of events).

The world views offer an opportunity to illuminate different ways of thinking about phenomena and different underlying assumptions. For example, organismic and transactional approaches both aim to understand complex, holistic, and dynamic systems. In many ways, one could expect similar research projects from these perspectives. Researchers can focus productively on micro-level processes without thinking about whether the system has teleological goals (one difference between the two). However, the organicist ideas of independent elements and teleological goals can also create a very different vision of phenomena and very different research questions and expectations, compared to the transactionalist description of mutually defining aspects and short- and long-term goals. In fact, it is not uncommon to find research projects that contain combinations of world views and assumptions. To some extent this is because researchers are not aware of the world views, and to some extent it is because they intuitively draw on different assumptions as they notice aspects of phenomena that favor different ways of thinking.

Table 1 emphasizes three issues that differentiate among various world views:

1) unit of analysis; 2) time and temporal qualities; and 3) philosophy of science as reflected in such issues as the objectivity/subjectivity of the researcher, level or scale of analysis, the search for unique versus general principles of behavior, and form of determinism. We consider each separately and then show how they have informed a current research project.

2.1. Unit of Analysis

One question is whether a phenomenon should be conceived as a collection of separate entities or as a holistic unity in which the various aspects are mutually defining and inseparable. Most current social science research is based on "interactional" or "organismic" views in which separate elements bounce against and move or influence each other. In this way of thinking, change only occurs when elements influence one another. An alternative—which has guided much of our own work—is the transactional world view in which phenomena are seen holistically, composed of mutually defining aspects that change with experience. If one adopts the interactional view, one defines people and situation by separate characteristics, whereas if one adopts a transactional view, one defines them by their unifying processes, and the changing relations among them.

A good example of this distinction is how one proposes to study landscaping around the home. From an interactional perspective, one would focus on particular qualities of yards that are separate from individuals, such as the size, layout, number, and kinds of plants, as well as microclimates and general environmental setting. These qualities are different from those one would use to describe the residents. Research questions ask how the qualities in one domain influence or are related to qualities in the other. These might include questions such as what qualities of yards are related to housing satisfaction for different family types, who makes decisions about the yard's upkeep and appearance, is access to the out of doors related to health outcomes, and so on.

In contrast, the transactional view emphasizes the dynamic unity between people and setting. In this approach, one focuses on psychological processes that can be used to define both the yard and the residents, and one assumes that both yard and residents are changed through their transactions. For example, "identity expression" is a concept that can be used equally well to describe a setting and the people who live there. People use their homes and yards to express their identities as unique individuals as well as their identities as members of groups and the broader society. Thus the yard and family are inseparable and mutually defining. Research questions based on these processes include: How does this yard reflect the family's self-expression processes? Do families with different individual or communal identities select different kinds of yards and change the landscaping in ways that support their self-expressive goals? How do individual and collective styles change across the lifespan, and how do yards mirror these changing styles? How do different cultures express identity, and how do yards reflect these expressive styles? Thus, how you define the unit of analysis—holistic vs. composed

Table 1. General Comparison of Trait, Interactional, Organismic, and Transactional World Views

| | | Selected Goals and Philosophy of Science | | | |
	Unit of Analysis	Time and Change	Causation	Observers	Other
Trait	Person, psychological qualities of persons	Usually assume stability; change infrequent in present operation; change often occurs according to preestablished teleological mechanisms and developmental stages	Emphasizes *material causes*, i.e., cause internal to phenomena	Observers are separate, objective, and detached from phenomena; equivalent observations by different observers	Focus on trait and seek universal laws of psychological functioning according to few principles associated with person qualities; study predictions and manifestations of trait in various psychological domains
Interactional	Psychological qualities of person and social or physical environment treated as separate entities with interaction between parts	Change result from interaction of separate person and environment entities; change sometimes occurs in accord with underlying regulatory mechanisms, e.g., homeostasis; time and change not intrinsic to phenomena	Emphasizes *efficient causes*, i.e., antecedent-consequent relations	Observers are separate, objective, and detached from phenomena; equivalent observations by different observers	Focus on elements and relations between elements; seek laws of relations between variables and parts of system; understand system by prediction and control and by cumulating additive information about relations between elements

	Holistic entities	Change/Stability	Causes	Observers	Focus
Organismic	Holistic entities composed of separate person and environment components whose relations and interactions yield qualities of the whole that are "more than the sum of the parts"	Change results from interaction of person and environment entities. Change usually occurs in accord with underlying regulatory mechanisms, e.g., homeostasis and long-range directional teleological mechanisms. Change irrelevant once ideal state is reached; assumes that system stability is goal	Emphasizes *final causes*, i.e., teleology. "pull" toward ideal state. Systems model in which changes reverberate throughout system	Observers are separate, objective, and detached from phenomena; equivalent observations by different observers	Focus on principles that govern the whole; emphasize unity of knowledge; principles of holistic systems and hierarchy of subsystems; identify principles and laws of whole system
Transactional or Contextual	Holistic entities composed of "aspects," not separate parts or elements; aspects are mutually defining; temporal qualities are intrinsic features of wholes	Stability/change are intrinsic and defining features of psychological phenomena; change occurs continuously; directions of change emergent and not preestablished	Emphasizes *formal causes*, i.e., description and understanding of patterns, shapes, and form of phenomena	Relative: Observers are aspects of phenomena; observers in different "locations": (physical and psychological) yield different information about phenomena	Focus on event, i.e., confluence of people, space, and time; describe and understand patterning and form of events; openness to seeking general principles, but primary interest in accounting for event; pragmatic application of principles and laws as appropriate to situation; openness to emergent explanatory principles; prediction acceptable but not necessary

Note. Adapted from Altman and Rogoff, 1987. Used with permission.

of independent elements—guides the questions you pose and how you study them. (We invite readers to try out the interesting exercise of addressing a single research question from both an interactional and transactional perspective.)

As an aspect of our work, we focus on dialectic processes as a means for linking people and settings, although an extensive review of dialectical philosophy is outside the scope of the present chapter (see Altman & Chemers, 1980, for dialectical analyses of homes and cities; Brown, Werner, & Altman, 1998 for dialectics in relationships; and Werner & Altman, 1998, for dialectical analyses of children in cities). Individual and community identity is one of these dialectics. As noted above, people use landscaping to express cultural values (yards pretty much look alike), while simultaneously using landscaping to express unique tastes and values (variations provide a sense of uniqueness to individual homes). Being unique and fitting in involve competing processes, so there is a constant dialectical tension, a tension that is played out in part through changes in yard appearance, upkeep, home decorating, and related displays.

2.2. Time/Temporal Qualities

Time is one of the least explicated of research assumptions but one of the most complex and interesting. One issue is whether time is viewed as internal or external to events. That is, does the researcher assume that phenomena contain natural, unfolding temporal qualities, or does the researcher apply external temporal markers or time-lines to phenomena? A second issue is whether the researcher takes a short- or long-term view of phenomena. For example, do we use interventions designed to effect short-term changes in behavior, or do we attempt to identify those that lead to long-term, internalized change? A third issue is how time and temporal qualities are conceived. Is the emphasis on linear or cyclical qualities of events, or some combination of these? Does the researcher assume that events move forward in an ever-changing *linear* way? Or, is it assumed that events recur in a *cyclical* pattern, such as the daily rhythms by which people use their yards, and the annual rhythms of seasonal climate changes? Or does the research imply "spiraling" time, in which recurrent events are of interest, but these events are expected to change somewhat with each reenactment. Often, the researcher makes implicit assumptions about time rather than making conscious decisions. Our view is that an awareness of these assumptions leads to better decisions about how to measure and study temporal processes and a better understanding of phenomena.

2.3. Philosophy of Science

The final set of assumptions emphasizes rules of evidence—the criteria for collecting and evaluating data. These judgments vary, depending upon one's philosophical stance on the issues of a) whether the researcher can be objective or subjective, b) the proper level or scale of analysis, c) whether the goal of research is to identify unique and/or general principles of behavior, and d) what form of determinism should be studied.

Our own transactional approach to philosophy of science has been flexible rather than restrictive. With respect to the researcher's objectivity/subjectivity, we assume that every researcher brings expectations and biases to the observations and that these should be acknowledged, described, and understood, rather than denied or ignored. In some cases, it may be worthwhile to develop research techniques for reducing the impact of unwanted factors on the phenomenon being studied. But, a transactional perspective highlights the point that there are several potentially valuable observers of phenomena, including participants and others directly or indirectly involved in an event. The insights of multiple perspectives, rather than a single "objective" observer are valued in a transactional world view. Thus, a resident's, neighbor's, outsider's, and researcher's analyses of plastic pink flamingos as yard decor are all potentially useful. The very questions that are posed and judgments about their worth, the criteria for valid measures, the choices of settings for conducting observations, all contain implicit values, expectations, and assumptions that should be acknowledged.

Some debate whether the goal of scientific study is to understand general principles or to describe unique and nonreplicable events. We have adopted the transactional orientation that both are valuable, as long as one knows which is being studied. Similarly, we favor analyses at multiple levels of scale, from single settings to entire complex communities, depending upon what is of interest.

The final philosophy of science issue is one's preferred form of determinism or view of causality. We have focused on two of Aristotle's four forms of determinism, "efficient" and "formal" cause. *Efficient cause* refers to traditional "antecedent-consequent" or "cause and effect" notions, the form of determinism that has dominated Western social science. Indeed, the common definition of science as "the study of the prediction and control of events" presumes efficient causal relationships. On the other hand, Aristotle's *formal cause* describes the pattern and form of an event. Elements are not seen as pushing or causing one another in an antecedent-consequent fashion, but rather as fitting together as aspects of total holistic unities.

[One tries] . . . to identify relationships among component parts and processes—but none of the components is "caused" by the prior occurrence of another component; and even more important, none of the components "causes" the action or act of which they are components (Ginsburg, 1980, p. 307).

In the transactional world view it is assumed that psychological events unfold in a purposeful and goal-directed fashion but that goals can be short- and long-term and can change with time and circumstances.

Fisher (1982) suggested combining formal with efficient cause such that formal cause is used to understand the meaningful patterns by which phenomena operate, and efficient cause is used to understand how outside influences change the nature of those patterns. For example, families usually develop a daily or weekly routine by which they care for and beautify their yards, interact as a family and take time to enjoy nature. None of these aspects "causes" any other, but the total set forms a coherent pattern. Understanding these patterns and studying their organization is the domain of formal cause; how outside influences such as unusual weather or out of town guests disrupt the pattern would be the domain of efficient cause.

2.4. Summary

The selection of unit of analysis, particular approaches to time and one's particular philosophy of science are arbitrary but strongly influential aspects of the research process. No perspective is better than the others, and all can be used fruitfully to understand environment behavior phenomena.

The next section illustrates how questions about guiding assumptions have informed the development and implementation of a research project designed by CW[1] and a group of undergraduate and graduate students to change people's behaviors with respect to hazardous household products. We begin with a brief description of the project's theory and method, and then analyze the project in more detail, explaining our research decisions in terms of the philosophical issues just described. The study was a large project with multiple goals and purposes, and we drew on a variety of world views and philosophical assumptions, depending upon particular research goals and questions.

3. RESEARCH EXAMPLE: CHANGING HOW PEOPLE RELATE TO NATURE AT HOME

3.1. Project Overview

The purpose of the project is to teach residents about proper use and disposal of household hazardous waste, as well as ways of using alternatives instead of toxics. Thus, at its heart, the project highlights different views of humans' relationship to nature—whether we should dominate and destroy with toxic products or try to live in harmony with nature (Werner, Brown, & Altman, 1997). The project is a pilot, and our goal is to prepare an instruction manual for use by communities nationwide. To do the actual teaching, members of the research team found or developed educational materials about cleaners and pesticides for the home and yard. The content to be taught was: a) use nontoxic alternatives first and toxic products only as a last resort; b) share leftovers with friends so that products are used up while still effective; and c) make new purchases jointly, with friends, thereby reducing the amount of leftovers. A wide array of behaviors is implicated in this message content (e.g., using baking soda instead of cleansers; lemon and water instead of ammonia-based window cleaners; vinegar, baking soda, and boiling water instead of lye-based drain cleaners; cedar chips or lavender instead of moth crystals; hand-weeding instead of spraying; picking off insects by hand instead of spraying them; etc.). We did not stress one behavior over another, but rather let participants decide which they found most interesting and useful.

Three research domains have been woven together in this project, attitude and behavior change, behavioral self-regulation, and self-expression in the home. The primary theoretical framework is Sansone and colleagues' theory of behavioral self-regulation (Sansone & Harakiewicz, 1996; Sansone, Weir, Harpster, & Morgan, 1992). Sansone suggests that people persist at tasks because the actions or experiences of doing the tasks are positive. If a task is naturally boring or difficult, they may stop. A major premise is that *if people have a reason to persist at a boring task, they will figure out a way to make it phenomenally interesting*. That self-generated phenomenal interest will sustain the behavior over the long term; it will also feed back to and strengthen the initial reason to persist.

---

[1] This research project was supported by an EPA-NSF Partnership grant to CW, #R825827-01-0.

Sansone's model can easily be seen as consistent with a transactional world view. With respect to unit of analysis, individuals are connected to the physical environment through actions and through cognitive and phenomenal experiences (see Werner, Altman, & Oxley, 1985, for psychological processes that connect people with place). With respect to temporal qualities, time is clearly integral to self-regulation. Sansone and colleagues posit a dynamic and fluid process by which an originally boring task is slowly transformed to be more interesting. This transformation can occur as individuals see new aspects to the behavior, bring in outside strategies for raising interest, or simply reconstrue the activity to be more interesting. Furthermore, the phenomenal experiences feed back to the original "reason to persist" and undermine or strengthen it, depending upon how positive the experiences are. Thus, Sansone describes transactional processes that transform the individual, the individual's ways of behaving, and the individual's ways of viewing the behavior in a constant interplay of actions and reactions.

Werner and Makela (in press) reinterpreted this theory from an attitude and behavior change perspective. They suggested that Sansone's "reason to persist" was essentially an attitude (or set of attitudes), and that one way in which attitudes guide behavior is through the mediator of phenomenal experience. Werner and Makela found that people who had the strongest attitudes towards recycling (strongest reasons to persist) were most likely to describe positive phenomenal experiences associated with recycling. Residents said they enjoyed crushing cans, used recycling as an opportunity to learn about the waste stream, and used recycling as an opportunity to work together as a family, among other positive experiences. In addition, consistent with Sansone's theory, people who described positive experiences were more likely to maintain the behavior, that is, to recycle on a regular basis.

In adapting Sansone's framework to the current project, the research team focused on her "reason to persist" as an opportunity to use persuasive information and attitude change. We hoped that giving people strong reasons to persist at trying nontoxic alternatives would set in motion the phenomenal and cognitive changes she had documented in her research. Thus, if the new attitudes were strong enough, they would get people to try the new behaviors and also be the impetus for the development of positive phenomenal experiences. We did not systematically encourage people to think of ways to make nontoxics interesting or fun to use. Self-persuasion and autonomy notions suggest that change is more enduring if self-initiated, so we let the transformation processes emerge more naturally (Deci, 1992; Deci & Ryan, 1987; Sansone & Harackiewicz, 1996).

In addition to self-regulation and persuasion, the third theoretical component is our view of homes as places in which people reflect their individual and group identities (Gauvain, Altman, & Fahim, 1983; Werner, Altman, & Oxley, 1985). People care very much about the appearance of their homes and yards; they care about the image they project through their homes, and how significant others regard their homes (Altman & Chemers, 1980; Brown & Werner, 1985; Werner, Peterson-Lewis, & Brown, 1989). As a society, many in the U.S. have come to revere the chemically maintained yard (Bormann, 1993; Jenkins, 1994). We reasoned it would be easier to change people's behaviors with respect to use of chemicals if we could simultaneously change their individual and collective images of the ideal home. We reasoned that such a change in values would be facilitated if individuals realized they were not alone—that their friends and neighbors shared their new vision of what a yard should be: a place for wildlife rather than a place devoid of wildlife.

The research team began by contacting approximately 300 church leaders and asking them to identify a member of the church or synagogue who could work with us on a "fewer toxics/safer homes" education program. We said that we wanted to center the program in churches because it provided a safe context for sharing leftover chemicals (compared to accepting chemicals from a stranger or acquaintance in the neighborhood). Our organizational structure was simple. Six members of the research team worked as "contacts" in different parts of a mixed rural-suburban county. Their job was to meet with a church leader or designated "liaison," show him or her the information materials, provide suggestions for getting started on the toxics reduction program, and give encouragement and advice for increasing or maintaining the group's involvement.

The basic persuasive message was that alternatives were just as easy and effective for most problems, plus they were safer to use around families and safer for the ground water. We established the credibility of our program by noting it was sponsored by the County Health Department and the Poison Control Center. Consistent with a cornerstone of Sansone's theory (and current thinking in intrinsic motivation, Deci, 1992; Deci & Ryan, 1987; Sansone & Harackiewicz, 1996), we focused on encouraging people to develop internal reasons for changing their behavior. For example, we used contests and social rewards (plaques and free publicity to outstanding church/synagogue groups) to provide minimal external incentives for participation (see Cook & Berrenberg, 1981, for more on social rewards). As another internal social incentive, we suggested that a product exchange could be an opportunity for members to network in an informal way.

The original plan was to use a combination of in-depth, ethnographic analysis, and traditional questionnaire-based data gathering. We intended to draw a random sample of 300 churches, let them operate for 3 months, and then select the least and most active churches (as defined by percentage of congregation involved, number of activities, pooled purchases, and shared leftovers, etc.); we would interview them intensively to see what they thought of the program, prospects for future activities, and their perceptions of what had worked/not worked in their group. In these interviews, we could also learn more about their levels of social integration, organizational structure, and theology to see how these are related to participation and enthusiasm for the project. We also planned to use the "data" generated by the church groups (the lists of products available for exchange; submissions to any essay and poster contests) as further information about how the groups had undertaken this project and how successful they had been. In summary, our strategy was to use persuasive information to set the groups in motion, let the groups organize themselves, and follow-up with analyses of groups products and some interviews with particularly successful and unsuccessful groups.

Each of these theoretical frameworks and research decisions reflects one or more of the issues raised at the beginning of this chapter. In the next section we provide more information about the details of the project and their philosophical underpinnings.

3.2. Guiding Philosophy and Assumptions

For the most part, development of the project was guided by a transactional world view. That is, we treated the individual, group, and their environmental behaviors as the unit of analysis, we assumed that time and temporal qualities were integral to events, we understood that our observers could become part of the group process, and we devised ways of tracking attitudes and behavior at different levels of analysis. Following Fisher's (1982) advice, we pursued a combination of efficient and formal cause, using efficient cause to initiate change but allowing formal cause to operate as groups attempted to get organized, behaviors began to change, and phenomenal experiences were transformed.

3.2.1. Unit of Analysis. The first issue or research decision concerns the unit of analysis. In accord with a holistic view, we assume that people's attitudes and behaviors are not autonomous or isolated, but rather are grounded in their social milieu of friends, families, and social organizations. This is not to say that people are mindless adherents to whatever opinions they hear or whatever behaviors they see around them. Rather, we assume that people influence and are influenced by the significant social contexts in which they are embedded. It is extremely difficult to change individuals' attitudes and behaviors if their social milieu—or their beliefs about their social milieu— remains unchanged.

A key feature of our persuasion program was an assumption about the unity of people and their social context. We aimed to embed the new attitudes and behaviors in an important reference group: residents' religious organizations (Cook & Berrenberg, 1981). As noted above, people use their home's appearance and upkeep to express individual and cultural values. We expected that our persuasion intervention would lead individuals to want to change their home's and yard's appearance, but to worry—perhaps without being aware—about what others would think of them. This pressure to change would intensify the dialectic tension between individual and communal identities. To reduce this tension, we wanted the reference group to encourage and sanction the new ways of cleaning and decorating, and/or to be a role model for members. We hoped that endorsement from the groups would make it more permissible and easier for individuals to try out alternative images of home. Indeed, we anticipated that individuals would be surprised by the level of support expressed by others in their church (people often underestimate popular support for new ideas). We acknowledged the individual/group tension in our written materials by emphasizing that the nontoxic alternatives would help maintain the homeowners' typical standards of appearance and cleanliness. In addition, by asking for a liaison who was an "opinion leader," we hoped to signal that the behavior change would be sanctioned by the group. We also hoped that people who were more independent of the group might see this as an opportunity to be different (e.g., to begin xeriscaping) but in a culturally sanctioned way.

3.2.2. Temporal Qualities. The second issue that distinguishes world views is time and temporal qualities. We made several considerations with respect to time. Some of these were simply assumptions, background for how we approached the project; others were or will be collected as relevant data. First, we assumed that time and temporal qualities would be integral parts of the processes involved in attitude and behavior change at the individual and group levels. We knew that individuals' change processes would unfold slowly, and that there might be reversals with people returning to chemicals on occasion. With respect to groups, we assumed that churches would organize their activities at different paces, depending upon the interest and motivation of the liaison, other ongoing activities (such as local and national holidays), and number of members. So we encouraged individuals and groups to work at their own pace, to pursue activities with as much depth as they liked, to involve members gradually over time, and so on. However, because we had a limited period of funding, we brought some coordination to the pace of events by introducing an external temporal framework: We gave the churches a common target date via a contest deadline (a county-wide opportunity for public recognition to highly effective churches).

Another key issue in thinking about time is whether to emphasize linear or cyclical qualities. In some ways the project will emphasize linear qualities and in other ways, it will emphasize cyclical aspects of time. We aimed to achieve long-term, internalized attitude, and behavior change (linear). But the particular target behaviors were seasonally linked, inviting a cyclical perspective. Pesticides (herbicides and insecticides) are considered to be the most toxic and overused household chemicals, so we targeted them in particular and presented the education program during the summer, when pesticides were most likely in use. Unfortunately, summer is also a time when people go out of town and are busy with other activities, making it difficult to make our project a central concern to many church groups. This is something we will probe carefully in the follow-up interviews to see whether participants can recommend a good temporal flow to future interventions.

Although the project was focused on a single season, the goal is to put in place mechanisms that will sustain the behavior over the long-term. Here again, we combine linear and cyclical qualities of time. We will ask whether individuals and groups have set up regular routines (cyclical) that support the new behavior (linear change) (e.g., do they have a box of baking soda handy for use; do they routinely pour boiling water down their drains to prevent clogs; do church groups have a system in place for continuing the product exchange). We hope to reinterview ministers and liaisons next year to see where they are with respect to long-term maintenance and follow-through. We also hope to interview noncooperating churches to see whether they changed their minds and developed product exchange groups.

A final opportunity to consider temporal qualities will be in the stories and essays submitted for a county-wide writing competition. Do people use temporal qualities when describing nature; are there patterns to the ways in which people describe recurrent (cyclical) or unique (linear) events? Do they combine the two, such as saying "My favorite time of day to be in my yard is early morning (cyclical)". One time (linear), 5 deer made their way cautiously into my yard and spread out across the lawn". Thus, in several ways, the project maintained a sensitivity to time and temporal qualities.

3.2.3. Philosophy of Science. The third issue that differentiates world views is general philosophy of science, including levels of analysis, unique or universal principles, objectivity of observers, and form of determinism. We used an eclectic mix of strategies in this project, depending upon the question. We used individual and group levels of analysis (e.g., as described above with respect to temporal qualities), we assume we will find both unique and universal principles about how individuals and groups change their attitudes and behaviors towards toxic products, and we used multiple undergraduate and graduate student observers, aware that they might differ in their styles. We also adopted Fisher's (1982) strategy of drawing on both efficient and formal causal principles.

For example, we intend to remain open to both unique and universal principles. Our primary "dependent variables" are personal essays/short stories and in-depth interviews about activities, group values and beliefs, social relations, and so on. This sort of ethnographic analysis would provide a rich set of information about group processes. It could highlight unique features of different individuals and congregations, but it could also suggest common features that underlie participation. Both unique and common principles can be used to identify ways in which other groups could improve their participation levels.

Another research decision will enable us to be receptive to unique and universal principles. Because of a high rate of refusal early on, we began trying out different strategies for inviting groups to participate (e.g., we developed a "quick start" program to show that this program could be simple). We also opened up the project to garden clubs and other settings as a way to increase our opportunities for outreach and data gathering. These changes in strategy became part of our "data" to be used for informing other cities how to initiate similar programs. Thus, rather than adhering rigidly to an original plan and attempting to be uniform in how we invited the church groups to participate, we added new, more effective strategies in response to reactions from the church members. We are learning that there is a great deal of variability in how church leaders perceive our program and we will try to develop training manuals that address this variability. In addition, by expanding the program to other kinds of groups, we may have information about effective ways of initiating toxics reduction programs in other settings.

An additional philosophical issue is the nature of observers and their observations. Are observers thought to be invisible or are they thought to influence behaviors of those being observed, and do different observers have different but equally accurate perceptions. In the current project, this is a background issue because we are using a research approach that does not allow us to learn much about it. Six undergraduate and graduate students are making the contacts, distributing the materials, and working with the groups. They are all working with different groups of unknown qualities, so there is no way to ask if the researchers differ in success rates or in their inferences about their groups—that is there is no way to address the question of whether different observers see things differently or affect participants differently. Furthermore, in selecting researchers, we selected for homogeneity. We hired people based primarily on their interpersonal skills and genuine enthusiasm for the project, yielding a uniform set of researchers. There were additional practical pressures towards homogeneity: We met regularly to discuss strategies for eliciting cooperation from the church leaders, we used similar persuasive messages, conveyed similar levels of enthusiasm, and so on. We did not do this for experimental uniformity, but rather to learn from each other and capitalize on strategies that were effective at eliciting cooperation. At the same time, we encouraged flexibility and diversity. The research assistants were of varying ages, appearances, and interpersonal styles. They were encouraged to emphasize different aspects of the project (the home safety message, the opportunity for members to network, opportunities to reduce waste and save money, environmental concerns, etc.) if it became apparent that one message was more effective than the others. Thus the nature of the project—an exploratory attempt to determine how best to effect toxics reduction—allowed for and even required flexibility in how researchers behaved, acknowledging the principle that they are part of the processes that occur in the group.

The final philosophy of science issue is our view of determinism. The project has elements of both efficient and formal causality. Efficient causality is reflected in the ideas that educational materials can change attitudes and behaviors, and that different participants react differently to the materials or even to the research assistants with whom they meet. Formal causality is reflected in the general idea that groups can organize themselves in many different ways. Formal cause is also reflected in the basic assumption behind Sansone's model, that if people have a reason to persist, they will figure out a way to make the task phenomenally interesting. Thus both with respect to individuals and with respect to groups, we are following Fisher's (1982) suggestion and pushing on the system from the outside. We expect to initiate attitude change, but then we want to see how events unfold inside individuals and inside groups. Because this is an exploratory study, we have the freedom to adopt a "minimalist" approach. We are not being controlling or insisting on uniformity in how groups organize themselves or what activities they undertake. We want to learn from the groups what good strategies for long-term attitude and behavior change might be.

4. DISCUSSION

The basic message of the present chapter is that there is no "best way" to study environment and behavior phenomena. Choice of assumptions and methodology depend on what one is trying to do. Research quality is determined by many factors: Is this an interesting question? Has it been approached in an interesting and penetrating way? Are the biases and assumptions acknowledged? Is the work carefully and systematically done? Does it make sense for this particular problem in this time and place?

The project described in the present chapter drew on interactional and transactional perspectives, although the overriding emphasis was transactional. To some extent, the fluid, ethnographic approach was chosen because it was what "worked," given our large-scale research goals. But for the most part, we adopted a transactional perspective because it gave us a more complex handle on the phenomenon of socially sanctioned behaviors (use of toxic products to fulfill societal expectations of home images). Indeed, one of the reasons we embraced the transactional world view is that it lends itself well to the kinds of large scale, naturalistic problems that interest us. It suggests that researchers watch how processes unfold rather than trying to control all aspects of the situation. It frees us to conduct in-depth analyses and to understand phenomena from the inside rather than to using settings to test pieces of larger theories.

To highlight a few of these advantages in the current project: The transactional world view forced us to think about the social context of attitude and behavior change. It forced us to think about how people appropriate the physical environment and invest it with meaning. It suggested the combining of efficient with formal cause. It led us to think through the different temporal qualities and how they might be factors in the project. It enabled us to be flexible in adapting our information outreach program to make it optimally successful. It guided our crafting of a variety of research materials and ways of gathering data. This does not mean that the transactional world view provides the most appropriate handle for every environment and behavior project, but it should certainly be considered as a possible framework.

This volume was stimulated by the question of whether there is a field of environment and behavior (Altman, 1997). Our answer is an enthusiastic "yes—but it may not be what you think". We think there is a lot of vital and vibrant work being done on environment and behavior issues, all anchored around the general question of quality of life. And although a good deal of it concerns the built environment, there is a very large and growing literature on humans and nature. Some of this work can be found in traditional environment and behavior programs, but there is even more in natural resource and environmental studies programs (e.g., see Hartig, 1993, for review). What is surprising is how rarely those who study the built environment consider human interdependence with nature. How can we study human quality of life without taking into consideration the larger natural milieu? Is civilization really independent from and outside of nature, or are humans and their cities integral with nature? How do humans benefit from nature? How do we try to insulate ourselves from nature and what are the consequences of such attempts at separation? These issues emphasize human embeddedness in nature, and are essential components of a transactional view.

Researchers in a number of disciplines are exploring topics of human/nature embeddedness, albeit not with a specifically transactional world view. There are growing literatures on the *restorative qualities of nature*, inspired in part by biological ideas about *biophilia*, or the notion that humans evolved in and therefore need nature (Wilson, 1984). Landscape architects are particularly interested in learning whether providing natural areas in medical facilities, psychiatric hospitals, and rest homes contributes to well-being and recovery. Others study human bonds to nature more abstractly, such as by comparing empty walls with nature photographs or windows to the outside. These studies find that people recover more quickly and require less pain medication with these visual reminders of nature (e.g., Ulrich, 1984). Other research asks whether spending time in nature is more restorative than engaging in other pleasant activities (e.g., Hartig, Mang, & Evans, 1991). Another setting in which to study human relation to nature is the garden, whether at a private home or in a community garden plot. Interviews indicate that a significant component of the motivation to garden is in the relaxation and pleasure people experience while working with soil and plants (Francis & Hester, 1990). This natural connection to nature is being encouraged by organizations who help people select plantings and landscape designs that encourage desirable wildlife (e.g., Nordstrom, 1991).

A related topic is *place attachment in nature*. Many of these studies involve people whose homes are at the edge of forests and natural areas. The 1998 meeting of the International Symposium for Society and Resource Management contained numerous sessions on a host of outdoor settings. Researchers explore the kinds of activities that connect people to nature, their daily or regular activities that involve nature, their use of nature for economic gain, their family history in the area, and so on. Place attachment is also studied with respect to occasional and frequent recreational users, such as backpackers and white water river runners. Researchers ask about favorite places in nature and the activities and feelings that bond people to place, bring them back repeatedly, or make it a never-to-be-forgotten experience[2].

---

[2] We would be remiss not to comment on the world views implicit in research on nature as restorative. Many studies adopt an interactional world view, asking "How does being in nature influence human affect and functioning?" A transactional world view would suggest a different way of framing the issues, such as: What are the common dimensions that define both people and nature? How do people use nature? Do different people use nature differently? How do transactions in nature unfold? Indeed, the research on place attachment in nature appears to be developing in a more transactional way. Researchers' emphasis on qualitative analysis and open-ended interviewing allows this research to have a transactional flavor even when that was not the researcher's explicit intention. For example, participants tend to describe themselves and settings as mutually defining; they describe ongoing transactions between themselves and nature; often these descriptions reflect "formal" rather than "efficient" cause; temporal qualities are integral to the experience; individuals have unique experiences, but one can also discern common themes of what connects people to place, and so on.

A casual review of research on human behavior in cities suggests that these qualities of human-nature relations have often been left out of the picture, whether we look at developments inside of the city or how the city interfaces with the surrounding open spaces in which it is embedded. Certainly, inside cities, there are studies of parks as areas of relaxation and the use of greenery such as trees and plants to enhance aesthetic qualities of built environments. But groomed parks are hardly natural; lawns and nonnative species do not provide habitat for native wildlife. New urban designs include green spaces only for people (and in some instantiations, include green spaces only for the residents of the development). Similarly, planners and developers often ignore the impact of new developments on existing natural and agricultural areas that are traditional habitats for wildlife. We allow "leapfrog" developments (but no real frogs!) which put housing in the middle of large open spaces instead of growing outward slowly, and preserving wildlife habitat and agricultural lands for as long as possible. The transactional world view invites scholars to take broader, more holistic views of phenomena, to consider the larger spaces, such as nature, in which we are embedded.

Emerging opportunities for environment-behavior researchers also argue for a more careful articulation of research assumptions. More and more funding agencies expect research proposals to adopt a multidisciplinary approach to solving environmental problems, such as protecting and restoring watersheds; reducing air pollution by increasing use of mass transit, bicycling, and walking as modes of transportation; reducing urban sprawl with new city designs; and addressing the whole issue of "sustainability" as population and consumption increase worldwide. In order for researchers to communicate effectively across disciplinary boundaries, it is essential that we be aware of our fundamental research assumptions, know how to select methodologies appropriate for those assumptions, and how to select assumptions and methodologies most appropriate for different problems.

There are no "silver bullets" for solving these complex, multiple-stakeholder problems; no single discipline or point of view has the best answers or can stand alone in providing solutions; no single world view provides the best tools for analyzing and understanding these issues. We would like to see the transactional world view take its legitimate place beside the other world views. Even more fruitful would be continued efforts to bring underlying assumptions to the fore so that they can be used specifically and effectively, depending upon the research purpose.

REFERENCES

Altman, I. (1997). Environment and behavior studies: A discipline? Not a discipline? Becoming a discipline? In S. Wapner, J. Demick, T. Yamamoto, & T. Takahashi (Eds.), *Handbook of Japan-United States environment-behavior research: Toward a transactional approach* (pp. 423–434). NY: Plenum Press.

Altman, I., & Chemers, M. M. (1980). *Culture and environment.* Monterey, CA: Brooks/Cole.

Altman, I., & Rogoff, B. (1987). World views in psychology: Trait, interactional, organismic, and transactional perspectives. In D. Stokols & I. Altman (Eds.), *Handbook of environmental psychology* (Vol. 1, pp. 7–40). New York: John Wiley & Sons.

Bormann, F. H. (1993). *Redesigning the American lawn: A search for environmental harmony.* New Haven, CT: Yale University Press.

Brown, B. B., & Werner, C. M. (1985). Social cohesiveness, territoriality, and holiday decorations: The influence of cul-de-sacs. *Environment and Behavior, 17,* 539–565.

Brown, B. B., Werner, C. M., & Altman, I. (1998). Choice points for dialecticians: A transactional/dialectical perspective on personal relationships. In B. Montgomery & L. A. Baxter (Eds.), *Dialectical approaches to studying personal relationships* (pp. 137–154). Mahwah, NJ: Lawrence Erlbaum Associates.

Cook, S. W., & Berrenberg, J. L. (1981). Approaches to encouraging conservation behavior: A review and conceptual framework. *Journal of Social Issues, 37,* 73–107.

Deci, E. L. (1992). Interest and the intrinsic motivation of behavior. In K. A. Renninger, S. Hidi, & A. Krapp (Eds.), *The role of interest in learning and development* (pp. 43–70). Hillsdale, NJ: Erlbaum.

Deci, E. L., & Ryan, R. M. (1987). The support of autonomy and the control of behavior. *Journal of Personality and Social Psychology, 53,* 1024–1037.

Fisher, L. (1982). Transactional theories but individual assessment: A frequent discrepancy in family research. *Family Process, 21,* 313–320.

Francis, M., & Hester, R. T. (Eds.). (1990). *The meaning of gardens: Idea, place, and action.* Cambridge, MA: MIT Press.

Gauvain, M., Altman, I., & Fahim, H. (1983). Homes and social change: A cross-cultural analysis. In N. R. Feimer & E. S. Geller (Eds.), *Environmental psychology: Directions and perspectives* (pp. 80–219). New York, NY: Praeger.

Ginsburg, G. P. (1980). Situated action: An emerging paradigm. In L. Wheeler (Ed.), *Review of personality and social psychology* (Vol. 1, pp. 295–325). Beverly Hills, CA: Sage.

Hartig, T. (1993). Nature experience in transactional perspective. *Landscape and Urban Planning, 25,* 17–36.

Hartig, T., Mang, M., & Evans, G. W. (1991). Restorative effects of natural environment experiences. *Environment and Behavior, 23,* 3–26.

Jenkins, V. S. (1994). *The lawn: A history of an American obsession.* Washington, D. C.: Smithsonian Institution Press.

Nordstrom, S. (1991). *Creating landscapes for wildlife.* Logan, UT: Utah State University.

Oxley, D., Haggard, L. M., Werner, C. M., & Altman, I. (1985). Transactional qualities of neighborhood social networks: A case study of "Christmas Street". *Environment and Behavior, 18,* 640–677.

Sansone, C., & Harackiewicz, J. M. (1996). "I don't feel like it": The function of interest in self-regulation. In L. Martin and A. Tesser (Eds.), *Striving and feeling: Interactions between goals and affect* (pp. 203–228). Mahwah, NJ: Erlbaum.

Sansone, C., Weir, C., Harpster, L., & Morgan, C. (1992). Once a boring task always a boring task? Interest as a self-regulatory mechanism. *Journal of Personality and Social Psychology, 63,* 379–390.

Ulrich, R. S. (1984). View through a window may influence recovery from surgery. *Science, 224,* 420–421.

Werner, C. M., & Altman, I. (1998). A Dialectical/transactional framework of social relations: Children in secondary territories. In D. Gorlitz, H. J. Harloff, G. Mey, & J. Valsiner (Eds.), *Children, cities, and psychological theories: Developing relationships* (pp. 123–154). Berlin, Germany: Walter De Gruyter & Co.

Werner, C. M., Altman, I., & Oxley, D. (1985). Temporal aspects of homes: transactional perspective. In I. Altman & C. M. Werner (Eds.), *Home environments: Vol. 8. Human behavior and environment: Advances in theory and research* (pp. 1–32). New York: Plenum.

Werner, C. M., Altman, I., Oxley, D., & Haggard, L. M. (1987). People, place, and time: A transactional analysis of neighborhoods. In W. H. Jones & D. Perlman (Eds.), *Advances in personal relationships* (Vol. 1, pp. 243–275). Greenwich, CT: JAI.

Werner, C. M., Brown, B. B., & Altman, I. (1997). Environmental psychology. In J. W. Berry, M. H. Segall, & C. Kagitcibasi (Eds.) *Handbook of cross-cultural psychology: Vol. 3. Social behavior and applications* (2nd edition, pp. 255–290). Needham Heights, MA: Allyn & Bacon.

Werner, C. M., & Makela, E. (in press). Motivations and behaviors that support recycling. *Journal of Environmental Psychology.*

Werner, C. M., Peterson-Lewis, S., & Brown, B. B. (1989). Inferences about homeowners' sociability: Impact of Christmas decorations and other cues. *Journal of Environmental Psychology, 9,* 279–296.

Wilson, E. O. (1984) *Biophilia.* Cambridge, MA: Harvard University Press.

4

NATURAL DISASTER AND RESTORATION HOUSINGS

Role of Physical and Interpersonal Environment in Making a Critical Transition to a New Environment

Masami Kobayashi and Ken Miura

Department of Global Environmental Engineering
Kyoto University
Yoshida-Honmachi, Sakyo-ku, Kyoto
606-8501, Japan

1. INTRODUCTION

Natural disaster destroys people's calm daily life, and makes survivors change their perspective about the relationship between man and environment (Raphael, 1988). When one is forcibly moved from one place to another because of a natural disaster, refugees have to face many problems in adjusting to their new surroundings. The environmental transition of relocation to restoration housing is conceptualized as consisting of two major elements which are essential for understanding environment-behavior transactions (Miura, 1995).

Natural disaster has two impacts on individuals. First, there are the physical changes that involve both physical damage and relocation and second, there are psychological reactions induced by the traumatic events of disaster. Those who have been exposed to the traumatic events in natural disaster are in grief and continue suffering from the pathological condition known psychologically as Post-traumatic Stress Disorder (e.g., Okonogi, 1979; Noda, 1992; Underwood, 1995). To cope with this condition, it is important for survivors to be able to express how they feel. Expressing their feelings presumable helps them go through the grieving process and cope with the new life which will never be exactly as it was before the disaster (Parkes, Murray, & Weiss, 1983).

Theoretical Perspectives in Environment-Behavior Research, edited by Wapner *et al.*
Kluwer Academic / Plenum Publishers, New York, 2000.

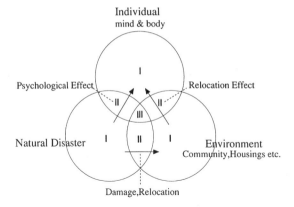

Figure 1. Environmental transition due to a natural disaster.

2. THEORETICAL ORIENTATION

Studies on human response to a natural disaster are huge, but it could be organized into three categories from environment-behavior view point: victims' short-term response to natural disaster, victims' long-term effects of natural disaster, and the recovery process from natural disaster. First are the studies on victims' behavioral and psychological short-term response to natural disaster (Beach & Lucas, 1960). These studies aim to mitigate disaster by understanding human response to a natural disaster that is called "disaster syndrome" (ex, Bolin & Klenow, 1982). Second are studies which have interest in long-term effects on survivors' mental and physical health (ex, Goldberg, 1978). These studies focus on long-term survivors' stress, and aim to make progress in understanding the ways in which people are psychologically affected by a natural disaster, and to develop more human and effective crisis management to survivors (Gleser, Green, & Winget, 1981). Third are studies which focus on a recovery process from natural disaster that involves consultation with survivors with reference to psychological theory described by Lindemann (1944) and Caplan (1964). This theory has emphasized internal processes and the importance of time for individuals to recover from their grief or traumatic experience.

Thus studies on human responses to natural disasters are shifting from studies on direct effects of natural disaster to studies on prolonged effects on the survivor; indeed the area extends its concern to understanding the manner of psychological recovery from traumatic events of a natural disaster and how to cope with the survivors' stress.

Figure 2. Shift of research viewpoints on human response to natural disaster.

Viewed in this light, it could be regarded that these three categories mainly focus on the survivors' psychological or internal processes, but no reference has been made to the relationship of physical "environmental settings." With the lack of the transactional view point the questions arose: Are the survivors only the weak who are one-sidedly affected by environmental impacts. Is time the only prescription for survivors to recover from traumatic experiences of a natural disaster? This is because the survivors' environment behavior aspects on recovery process become the next subject of scientific investigation to develop a new theory about the human being and natural disaster.

In this paper restoration housings after natural disasters are empirically explored 1) to clarify the role of the present environment in making a transition to a new environment; 2) to discuss person environment relationships by showing some cases of relocation to restoration housing projects in Asian countries; and 3) to consider the long term effects of a restoration housing project.

3. IMPORTANCE OF PARTICIPATION IN RECONSTRUCTION: SELF-HELP VERSUS MASS HOUSING METHODS IN RESTORATION PORJECTS: INDONESIA

3.1. Self-Help Housing Restoration: Indonesia, Bali

After the earthquake which occurred in Indonesia, Bali Island in 1976 housing materials were provided by the government, organized by the Bali Emergency Housing Project. Reconstruction was done by the survivors participating in the construction of their houses. As a result, the survivors rebuilt new houses where their houses were originally built. This reconstruction of new houses has made progress. Now, 17 years after the earthquake, the houses that were built according to the Bali Emergency Housing Project, in which survivors participated, merge into the landscape so well that it is quite difficult to tell the difference between the reconstructed houses and other houses. The method of self-help gave the opportunity for victims themselves to participate in coping with the disaster and thereby helping them make the transition to the new environment.

3.2. Mass-Produced Housing On a Hill Away from Sea: Indonesia, Flores Island

In contrast self-help was not applied to restoration in case of the tidal wave in Indonesia, Flores Island. The Bugis tribe was prohibited from living near the seashore by the government after the tidal wave that was caused by the earthquake with a magnitude 7.5 in 1992. The military constructed the housing units using the mass-produced Housing method and forced the Bugis tribe to settle down on the hill.

The tribe, which originally lived in houses with high floors, tried to adapt to living on the hill by extending their kitchen with high floors. However, they gradually left the mass-produced residential area, which is far from the sea, where they make a living. One year later the area on the hill turned into a ghost town. On the other hand, the people who went back to the seashore started to rebuild new houses on their own with traditional devices were successful (Maki, Miura, & Kobayashi, 1994).

3.3. Migration: Philippine Pinatubo Volcano

In the case of the Philippine Pinatubo volcano, migration was also used as a policy. The Aeta tribe, which lives in the mountains of the Philippines, is one of the tribes that lived in the Philippines from the old times; they earn their living by slash-and-burn agriculture and hunting. Since the eruption in 1991, the government banned the Aeta from living in the mountainous district and took the policy to offer them farmland and housing under the cooperation of Tindig Porac Development Foundation.

The Aeta tribe, however, did not have enough agricultural technology; they became spiritless because of the aid, and began to devote themselves to gambling. As a result, some Aeta tribe members went back to the mountains and their former life, aware of the danger (Toyoshima, Maki, & Kobayashi, 1995).

4. FUNCTION OF PRIVATE SPACE TO RECOVER FROM GRIEF

Okushiri island is a solitary island in Hokkaido which experienced a tidal wave in 1993, two years before the Great Hanshin-Awaji earthquake. The tidal wave washed away everything. The area fell into ruins by the tidal wave and fire. Many people lost all of their properties and relatives. The Aonae area is a fishing village on the island where there are 504 households and where 107 people lost their lives because of the tidal wave.

After the disaster the survivors lived in Temporary Housing for about two years. Although there were limits in the area for Temporary Housing, the survivors constructed a hand-made family Buddhist altar and mourned the deceased. For the first two years, they were limited to name tablets, family altars, and a relatively simple household shrine were mostly. When fishing resumed, they rebuilt their houses with a subsidy. People's lives improved, the survivors strengthened their approach towards physical environments.

People who lost their family were given pictures of the deceased from their friends, and placed them in the living room where they could be readily seen. They made a family altar (Fig. 3), and put a keepsake of the deceased there (Fig. 4). Through these mental dialogues with the physical environment, people tried to recover from grief. In the solitary islands, the relations among people are strong, Needless to say, this strong relationship among people supported their recovery.

But some adults, who lost a lot of relatives at once and became alone, were violently traumatized by the disaster. Thus they neither felt at home in the new house nor could they effectively arrange their residential space. Expression of mourning is evident in the private space of survivors who express mourning by placing such mementos and photographs of the deceased at a meaningful focal point (Fig. 5). Survivors mourned lost families by controlling physical settings in such a manner as to project remaining mementos on meaningful points of their private space. So far psychology has been emphasizing internal processes; it has not referred to the relationship with physical environment. The person who suffers from disaster should be understood not as a being who merely psychologically recovers from grief through internal processes or time, but

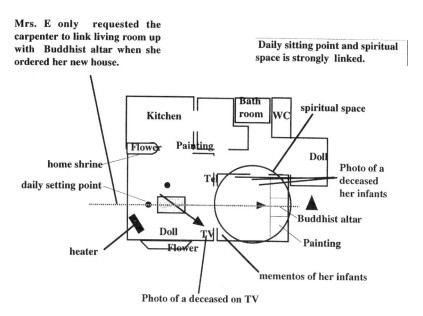

Figure 3. Rebuilt house and personalization of Mr. & Mrs. E. (young couple) who lost their two infants.

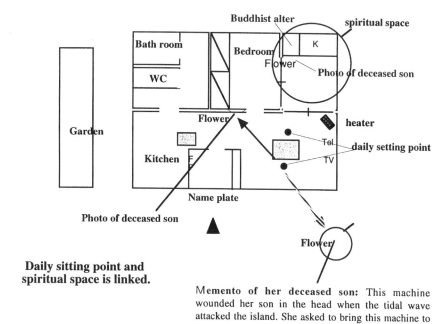

Figure 4. Rebuilt house and personalization of Mr. & Mrs. JQ (elderly couple) who lost their son.

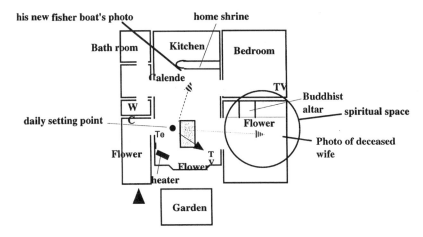

Figure 5. Rebuilt house and personalization of Mr. J (76 years old) who lost his wife.

as an independent being who psychologically recovers from grief by creating a space where he/she can express his or her feelings of grief.

5. IMPORTANCE OF CONTROLLABILITY OF PHYSICAL AND INTERPERSONAL RELATIONS IN A CRITICAL TRANSITION: GROUP-LIVING IN TEMPORARY HOUSES CONSTRUCTED FOR THE ELDERLY AND THE HANDICAPPED AFTER THE GREAT HANSHIN-AWAJI EARTHQUAKE

After the great Hanshin-Awaji earthquake all the house damaged by the earthquake were wiped away and demolished without considering a possibility of repairs; 49,681 temporary housing units have been newly constructed. Temporary Group-Living units for the elderly and handicapped were constructed among the Temporary Housing units, which were built for the survivors. These Temporary Group-Living units differ other general Temporary Houses. First of all, it is the area that differs. General temporary housings have a 6-mat room, a 4.5-mat room, a kitchen, and a bathroom.

In Temporary Group-Living units, although the housing space is small,—they only have a 6-mat room, a washroom, and a sink—, there are life support advisers who are stationed in the units for 24 hours. Moreover, there is a common living and kitchen where the tenants and the life support advisers can exchange greetings. On moving into the housing there were things that were inconvenient for the elderly and physically disabled tenants. The tenants created their own association, collected signatures, petitioned the city, and tried to improve their environment.

The elderly and the physically disabled people worked to adjust their own environment and asked volunteers to help with necessary improvements. Consequently, after a year, the bathrooms, washing machines, passageways, and many other parts of the physical environment were improved. For example, there were things hung on the walls of the passageway, which show the relations of the residents to their physical environment.

Moreover, cooperation among the elderly were seen. There were some old people who go out and shop for the elderly and weak people, who help prepare meals with each other, plant flowers, feed the fish, and so on. These kinds of movement to help make their environment better were evident. As reported in some newspapers, there are problems such as dying alone in General Temporary Houses. However, these kinds of houses which promote connections among people are favorably received. Some residents prefer Group-Living rather than the life before the disaster.

Concrete facilities often possess "hardness" that cannot readily be remodeled physically. However, since temporary houses allow one independently to control the environment, temporary houses may reinforce the feeling of one's own place but never reduce the feeling toward one's own place. That is, temporary houses showed "flexibility" in a positive sense. What we need to learn from temporary housing is that we should construct a soft residence with flexibility and individuality rather than hard buildings to which residents cannot physically relate.

6. CONCLUSION

Human beings are involved in various issues such as human relations, jobs, contact with objects, houses, and other people, in the environment. After those mutual relationships with the environment are destroyed by a disaster, the remaining personal, physical, and social relations become more important and meaningful for the survivors than before.

A series of studies concerning environmental transition of the aftermath of natural disaster made clear that remaining people-environment relations become an anchor when the survivors reorganize their life. From this people-environment viewpoint, two important points are derived for concepts of restoration housings: 1) offering private space under independent control of the survivors has significance not only

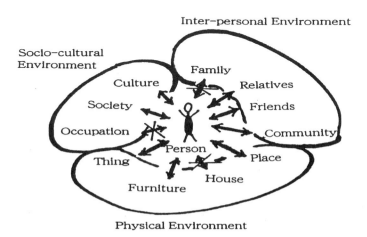

Figure 6. After person-environment relations were cut off by a natural disaster, survivors restore their new stable life gradually, and use the remaining person-environment relationships which are more important than before.

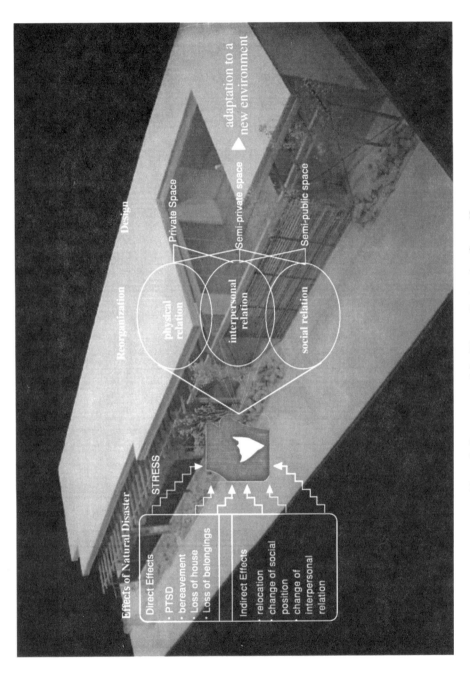

Figure 7. Key concept of the Disaster Restoration Group House.

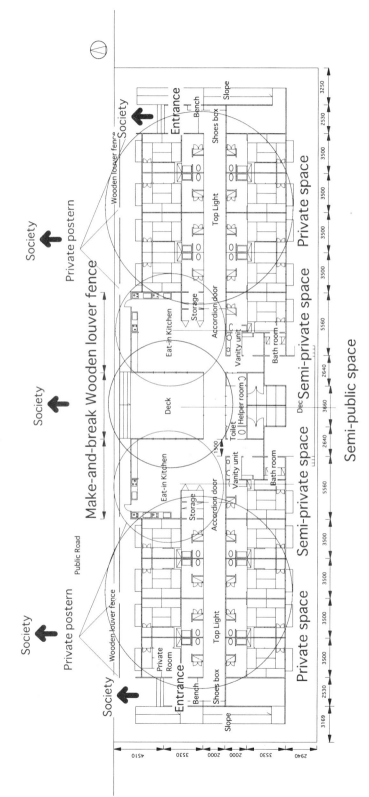

Figure 8. Plan of the Disaster Restoration Group House: The plan of private space, semi-private and public were decided based on our proposal.

to secure their privacy but to promote their psychological recovery by their being able to control the physical environment; and 2) providing some semi-private or semi-public spaces where the survivors can share space with others is also important to help survivors reorganize their interpersonal environment. These days, psychological consequences of natural disaster have been reported by the press. What we need to be concerned about in those reports is that they give a false impression that the survivors of natural disaster all need to undergo psychological treatment. Of course, if there is a severe psychological consequence in a suffer, it goes without saying that he or she should be treated medically. However, most of people have healing power of their own. It is important to create an environment where they can put their healing power into action. For the survivors to adapt to a new environment, it is necessary not only to secure their privacy in individual space but also to construct an environment designed to allow communications with others. In such an environment, the social position of survivors of a natural disaster becomes insignificant. The future form of disaster restoration housings should not only offer privacy, but also offer building functions that allow the survivors to communicate with others and stay in touch with the society. Architects and the administrators should recognize that by improving the quality of residences built after disaster the psychological healing process of the survivors can be improved.

7. RESEARCH APPLICATION

Amagasaki City, one of the cities in the disaster area of the great Hanshin-Awaji earthquake, was planning to have the elderly who require nursing and live in temporary housings move into nursing facilities for the elderly, as well as to scrap the huge numbers of temporary housing units three years after the earthquake. But vacant rooms in nursing facilities for the elderly were not found by the end of September, 1998 which was scheduled to be the deadline for the dissolution of temporary houses. As a desperate measure, the city decided to build prefab single-story houses with the limited occupancy period of 5 years as a Disaster Restoration Group House. We submitted a summary of our study on the designs of living environment for the elderly in Disaster Restoration Group House.

This proposal was adopted by Hyogo Prefecture Housing Supply Public Corporation and was completed as Disaster Reconstruction Group House in September, 1998.

REFERENCES

Beach, H., & Lucas, R. A. (1960). *Individual and group behavior in a coal mine disaster.* Disaster Study No.13, Publication 834. Washington, DC: National Academy of Science, National Research Council.

Bolin, R., & Klenow, D. (1982). Response of the elderly to disaster: An age-stratified analysis. *International Journal of Aging and Human Development, 16*(4), 283–296.

Caplan, G. (1961). *An approach to community mental health.* London: City Tavistock Publications.

Gleser, G. C., Green, B. L., & Winget, C. N. (1981). *Prolonged psychosocial effects of disaster.* New York: Academic Press.

Goldberg, D. (1978). *Manual of the general health questionnaire.* Windsor, England: N.F.E.R. Publishing Company.

Lindemann, E. (1944). Symptomatology and management of acute grief. *American Journal of Psychology, 101*, 141–148.

Maki, N., Miura, K., & Kobayashi, M. (1994). *A study on the temporary housing system after natural disasters in Indonesia* (pp. 276–292). Annuals of the Disaster Prevention Research Institute, Kyoto University.

Miura, K. (1995). *Study on relocation due to the eruption of Mt. Unzen-Fugendake: Environmental Transition in case of natural disaster.* Unpublished master's thesis, Division of Global Environmental Engineering, Graduate school of Engineering Kyoto University.

Noda, M. (1992). *Process of mourning* (in Japanese). Tokyo: Iwanami Publishing Company.

Okonogi, K. (1979). *Object loss* (in Japanese). Tokyo: Chukou Shinsho Publishing Company.

Parkes, C., Murray, S., & Weiss, R. (1983). *Recovery from bereavement.* New York: Basic Books.

Raphael, B. (1988). *When disaster strikes; How individuals and Communities Cope with Catastrophe.* New York: Basic Books.

Toyoshima, M., Maki, N., & Kobayashi, M. (1995). *Study on the restoration housing after disaster in Philippine; Restoration houses for the victims of Mt. Pinatubo Eruption* (pp. 126–135). Annuals of the Disaster Prevention Research Institute, Kyoto University.

Underwood, P. R. (1995). *Dealing with Trauma Response Syndrome* (in Japanese). Tokyo: Misuzu Shobou Publishing Company.

5

REFLECTIONS ON THE ASSUMPTIONS AND FOUNDATIONS OF WORK IN ENVIRONMENTAL PSYCHOLOGY

Leanne G. Rivlin

City University of New York Graduate School
New York, NY 10016

1. INTRODUCTION

In my teaching, a frequent admonition to students when they are reviewing literature is that they consider the authors' implicit and explicit assumptions as a way of understanding its origins, methodologies, and interpretations. This is not a simple task and is even more difficult when applied to one's own work. However, at this stage in the development of environment-behavior studies some reflection by researchers on the underpinnings of our work can be valuable both to ourselves and the field with which we identify.

2. THE UNDERLYING ASSUMPTIONS

According to dictionary definitions, an assumption, as used in scholarly work, refers to something that is taken for granted, "a fact or statement (as a proposition, axiom, postulate, or notion taken for granted" (Merriam-Webster, 1993, p. 70). Most people carry numerous unexamined assumptions about the world and their lives but for scholars and researchers these assumptions form the core of their work and should be subject to review and criticism. Examination of these assumptions strengthens the quality of scholarship and offers another criterion for the quality of research in addition to the frequently used reliability and validity of findings. But how does a researcher begin to extract the assumptive underpinnings of a lifetime of work? With difficulty and considerable concern for partially recovered memories!

Theoretical Perspectives in Environment-Behavior Research, edited by Wapner *et al.*
Kluwer Academic / Plenum Publishers, New York, 2000.

2.1. Assumptions about the Nature of the Environment

There are some clear convictions that I find in my research, reflecting assumptions about the environment that are drawn from William Ittelson's (1973) distinction between perception of objects and perception of the "large scale environment." This conceptualization of the environment has proven to be a robust one, covering my work in institutions (psychiatric hospitals for children and adults, facilities for developmentally disabled persons, schools, day care settings, shelters for homeless persons) as well as research on public spaces (parks, plazas, playgrounds, streets, and "found spaces") and two current interests, one on the impacts of residential and workplace relocation on people and a second on the factors leading to socio-spatial conflicts.

Ittelson's formulation begins with the recognition that "the environment surrounds, enfolds, engulfs" (Ittelson, 1973, p. 13), that it is not possible to be outside and observe an environment. Rather, people "explore" the environment, and research on environment experiences has to tap this exploration process. The environment also is "multimodal" providing information that is received by all the senses despite the emphasis on the visual in most environmental research. In my studies on the qualities of institutional settings the olfactory, thermal and sonic elements provided telling qualities of these places. Ittelson pointed out that there always is peripheral as well as central information. Environments offer more information than any individual can possibly process and this information is received simultaneously, and is often redundant, inadequate, ambiguous, and conflicting. Environments also involve actions, they provide, in Ittelson's terms, "the arena for action" (Ittelson, 1973, p. 14). The environment also contains "symbolic meanings and motivational messages" (p. 15) which influence the direction of action. The final characteristic is that environments have an "ambiance" or "atmosphere" which derives from their social activity, esthetic quality, and systemic quality. Although the ambiance is difficult to assess, frequently the researcher's field notes can be reviewed for details on this dimension.

Taken together these qualities offered a complex and enduring view of the environment that went beyond the individual stimuli that were the concerns of psychology when we began our work, although there have been many changes in the research perspectives of other specializations, over the years. The integration of Ittelson's conceptualization into my work has shaped the research questions that were developed as well as the research designs and methodologies that were used. Whether I studied the lives of urban squatters, the experiences of developmentally disabled persons in group homes or the use of public spaces, the lesson was that the context was multidimensional and certainly more than what would be covered in the research.

2.2. Assumptions about Historical Perspectives

In 1960 William Ittelson produced a report, *Some Factors Influencing the Design and Function of Psychiatric Facilities*, that included a section on the history of psychiatric care, relating the nature of treatment and its setting to the conceptions of the people whose behavior was perceived as "different." This document proved to be a critical influence on my future work and thinking and forms a basic, explicit assumption concerning issues in the world.

It would be impossible to ignore the heritage of the past in understanding both the role of settings in people's lives as well as the historical roots of the thinking about spatial constructs, the built environment, and place meanings. This suggests that every study, whether on partial or total institutions, housing, public space, or the environmental constructs such as privacy, crowding, place connections, place identity, and nationalism, must be considered against an historical framework. Moreover, the historical literature that is cited has to go beyond what Mills (1959) criticized as "padding" and become a substantive dimension of the work.

As an example, in research on contemporary homelessness it was essential to understand how homelessness occurred in previous times, how homeless people were perceived, if at all, and how they were treated. In a study of homelessness in the second half of the nineteenth century (Rivlin & Manzo, 1988) we found that there were different attitudes toward poverty which were directed toward the poor elderly widows, orphans, and abandoned children and unemployed men. The label of the "deserving" and "undeserving" poor (Katz, 1989) has been applied to different populations, over the years, but rests on a basic view that some people are worthy of assistance while others are not. This conception leads to what Katz (1989) described as a "language of difference" that is "both philosophic and practical" with the verbal categories assumed to be the "natural or inherent qualities of people" (p. 5). It is most informative to see who falls into these categories, over the years, and how the attributions impact the nature of the help that is offered. The historical perspective assists in clarifying the assumptions that underlie the creation of services and institutions and the selection of those who will receive help.

2.3. Assumptions about the Multi-Disciplinary Nature of Environment and Behavior Studies

The historical perspective was not the only cross-disciplinary dimension of our work despite its identity as "environmental psychology." From the onset we were stretching into a range of areas beyond psychology in attempting to conceptualize the field and its constructs, and to locate colleagues who were working on related issues. The first efforts at reaching others were reflected in the book of readings that Harold Proshansky, William Ittelson, and I compiled, published in 1970a. It covered a variety of areas that included our parent discipline, psychology, as well as sociology, anthropology, geography, architecture, planning, urban design, psychiatry, zoology, ethology, biology, resource management and wilderness studies, systems analysis, and technology. The range of disciplines and specializations in the field has increased, over the years, and is reflected in the multi-disciplinary doctoral program in environmental psychology at the City University of New York as well as other programs.

The multi-disciplinary approach has energized my own research, presenting both a challenge to master a broad knowledge base for understanding the problems under study but also a refreshing voyage into the literature that is not limited to work within a single discipline. This kind of engagement with the issues offers a challenging but more complex understanding that, in the end, benefits the work. For example, in our early efforts to clarify the nature of privacy, an issue that has reappeared in most of my studies, we found some literature and theory in the social sciences but also useful legal analyses (see, for example, Westin, 1967). In much of my recent work on homelessness and on public spaces, the policy-related implications of the findings had to be addressed, requiring an understanding of public policy. In my work on socio-spatial conflicts (Rivlin, 1995), in addition to the use of historical sources, analyses by political scientists of nations and nationalism have helped to clarify some of the molar connections to place (Smith, 1991). When this level of political analysis is examined against work on place connections in the environment and behavior field (Altman & Low, 1992), a clearer and deeper view of spatial conflicts is possible.

Open-minded approaches to sources offer useful directions for conceptualization that transcend the restrictions of work in the traditional areas of psychology. For environmental studies this orientation has been essential for understanding the complex underpinnings of the research questions that we pose. The challenge is to place the various facets of the issues, gleaned from other disciplines, into a coherent whole.

2.4. Assumption Regarding the Complexity of Environment/Behavior Issues

Historical and multi-disciplinary thinking offer a complex approach to issues, one that brings out the different levels and dimensions of a problem. This occurs through a process of consecutive analysis, an iterative process that clarifies meanings by drawing on relevant literature, looking for explanatory systems, theory, and constructs that shed light on the issue under study. Often it is necessary to reopen thinking on issues that seemed to be clear but were challenged by subsequent reading. The procedures can be facilitated by drawing on a range of professionals in various specializations, a true engagement with the intellectual community.

The procedures are similar to those recommended in qualitative analyses in which the researcher returns to the data with new perspectives that enable the formation of a credible and verifiable explanation of findings (Miles & Huberman, 1994 and Strauss & Corbin, 1990). When used in the formation of research problems, selection of methods, and interpretation of findings, this approach to research can yield useful directions for thinking that move beyond single-level approaches. These strategies are demanding but they offer a richer and multi-level understanding of environmental problems.

3. RESEARCH PROBLEMS AND METHODOLOGY

3.1. Selection of Research Topics

The institutional studies that formed the core of my initiation into environmental work were anchored in some very real issues facing various levels of government. In a period of building in the 1950's and 1960's, questions were raised concerning the physical forms of psychiatric hospitals, schools, day care centers, and facilities for developmental disabilities. Empirical data were required to place the design questions into a larger, conceptual frame and a number of government agencies supported this research. What emerged from the studies were not blueprints for construction but series of issues that required consideration in terms of the goals of each of these places. The findings also identified some critical concepts, among them concerns regarding privacy, density, personalization of the environment and the evolution of setting-specific behavior patterns.

All of the research was conducted in the field, addressing problems that were real ones in contrast to the laboratory-staged, experimental forms that characterized much of the work in psychology. The environmental work that began when I joined Proshansky and Ittelson's psychiatric hospital research initiated a sequence of field studies that addressed theoretical issues, at the same time raising questions that reflected daily-life concerns. By focusing on the interplay of the activities of daily life with the demands of the institution, evident in the over-time observations and interviews with patients and staff, it was possible to understand how the institutions attempted to control behavior and how people there resisted the constraints. In addition, by studying both total and partial institutions we were able to formulate a conception of institutions that built on Goffman's (1961) earlier work and identified the ways the physical qualities of the places were components of the institutional process (Rivlin & Wolfe, 1985).

3.2. Research Design and Methodology

The psychiatric hospital research revealed the value of research designs that addressed the time dimensions of the problem under study and the value of multiple methods in research. The time factor loomed first with the recognition that the institutions under study had 24 hour lives that extended over years. To sample a time bit that suited the convenience of the 9AM to 5PM workdays of researchers would not capture the changes that occurred over the course of a day, a week, over months and seasons. In our early hospital studies we had interviewers and observers covering a 24 hour day and our data captured the sense of being a patient or staff member in the facilities studied. They also revealed the changes in the use of space, over time, providing a connection between people's needs, the physical realities of the space and the efforts by users to maximize their freedom of choice (Proshansky, Ittelson, & Rivlin, 1970b).

The use of more than one method or triangulation (Webb, Campbell, Schwartz, & Sechrest, 1965), represents an approach to validate findings by using multiple "imperfect measures." There have been expansions in the conceptualization of triangulation offered by Miles and Huberman (1994, p. 267), who were building on Denzin's (1978) formulation. They offer a view of multiple types of triangulation, those emerging from the "data source" (persons, times, places), the "method," the "researcher," the "theory," and the "data type" (qualitative, quantitative). For the environmental work in which I have been engaged triangulation served the need to provide both theory and data on different aspects of the research question, to enable a view into the problems that could not be offered by a single voice.

The multiple methods we used generally included time-sampling and longitudinal observations of the use of the settings, the use of logs, maps, and checklists to assess their physical qualities, as well as interviews with the different users in the places. "Users" included the space managers, administrators, and staff, as well as the patients, teachers, students, and clients who were occupants of the various places studied. At different points the techniques also included the collection of archival sources, mapping techniques, the use of models of places, participatory planning to make environmental changes (with pre-post observations and interviews), and participant observations and ethnography (Rivlin, 1982).

Examining settings and situations using multiple data sets offers a way of looking at research problems from many windows through which different views of the environment can be exposed. A study, now in progress, of the relocation of a university, began with a survey sent to students, staff and faculty of all offices and departments, as well as to members of the administration, and moved on to in-depth interviews with a selected group of respondents. The interviews have clarified and expanded upon many of the issues raised in the survey responses, probing areas a questionnaire could not address. Taken together, the information offers a strong base for generating the implications of the responses, something important to a future university move that has been planned. From the vantage point of environment and behavior studies, the work has offered another perspective on a number of environmental constructs, including place attachments, place identity as well as the role of planning and support in the process of relocation.

In earlier research on open education and open school design we looked at the relationship between educational philosophy and use of classrooms. A multi-method, over-time research design was used which included systematic behavior mappings of classrooms in two schools for full days, three times in the academic year. Prior to each observation physical assessments of the furniture and equipment were made for each room so that the behavioral maps could be examined against the physical maps for each day. In addition, interviews on educational goals were undertaken with the school administrators, teachers, students, and parents. Environmental workshops were conducted with groups of students to provide needed furniture and resources for their classrooms, and using models we obtained the students' images of their own classrooms and preferred classrooms (Rivlin & Rothenberg, 1976; Rivlin & Wolfe, 1985). This two-year study, which was accompanied by an analysis of historical sources on school design and educational philosophy, made it possible to place the information gathered into a complex framework of setting and goals.

Taken together, the early work offered a comprehensive way of gathering data, an approach that has characterized my research on a range of different problems. This multiple perspective model has opened opportunities for study, including community assessments (Rivlin, 1982) and public space research (Carr, Francis, Rivlin, & Stone, 1992).

3.3. Cautions Concerning the Use of the Multiple Perspective Research Model

In-depth work on specific settings raises questions regarding the ability to generalize the research findings across geographies, cultures, and physical sites. Is research that is conducted in New York City applicable to any other area? Are the research questions unique to a large urban area and irrelevant or different from other kinds of places?

Clearly, global cities have scales of problems and physical conditions that differ from other types of communities. However, multiple perspective research of the kind described offers a view into the findings that can enable translation to other sites. The studies of homelessness in which I have been engaged offer an example of this applicability. Homelessness is a serious concern in the United States and in other countries. Although there may be more homeless persons in New York City than other areas, and the measures used to deal with them may differ in type and scale as well, there is an underlying experience of homelessness resulting from the loss of long-term housing, that impacts those affected, wherever they may be. Although the differences in available resources as well as the nature of the climate are important factors, the loss of housing is a challenge to the ability of people to function whether they are children or adults (Rivlin, 1990; Rivlin & Imbimbo, 1989).

There are multiple levels of analysis of the data on the experience of homelessness that offer a complex view beyond the first generalization of the trauma faced by people. This seems to transcend location and points to specific details that can be used to examine homelessness, across sites. For example, research on homeless children in New York City has identified their irregular attendance in school and their misplacement in classes (Imbimbo, 1996). For some children, certainly not the majority, who are able to keep going to school, it can be the most "normal" component of the child's life. The findings offer a view of the resources that are needed to buffer the impacts of homelessness on children, and school stability is one of them. This could apply to any kind of community and transcends the specific location.

Multiple perspective research also requires attention to the conceptual and methodological concern of ecological validity that Winkel (1987) has addressed. His view goes beyond internal, external, and construct validity concerns and represents "a more integrated approach to the validity question" (p. 80) that considers whether the research design and the measures that are used "yield accurate estimates of the multiple dimensionality of the phenomenon that is the focus of the study and those components of the context that may be expected to influence variation in the phenomenon" (Winkel, 1987, p. 83). This demanding but comprehensive perspective on validity offers an enlightened view that is useful to the critical assessment of the field studies in which many researchers in environmental psychology engage.

3.4. Ethical Concerns

Basic to any research are the ethical issues that underlie the work, issues of responsibility that are now specified by the statements of professional organizations and review boards of various institutions (see, for example, the American Psychological Association, 1992). These are extremely important acknowledgments of the rights of people who are involved in our research. However, I would argue that even they do not go far enough. Ethical concerns begin with the first conceptualization of the research, before entering the field. They start with the question of whether the work should be done at all, whether the study is worth the time and efforts of those participating, the persons being interviewed, observed, or evaluated and the research staff, including the principal investigator.

Environmental field research using the multiple dimension mode, whether funded or done independently, is generally an intensive investment of people's energies and time. Often the in-depth and over-time nature of the work builds connections among the persons involved, connections that are likely to end once the research is completed. There are ways of engaging the participants by offering more than payment, for example, the workshops held with teachers and parents in the schools that were our study sites, or the physical changes made in some of the psychiatric hospitals that we studied that made them more comfortable, human places in which to live (Rivlin & Wolfe, 1985). But these were small offerings for people who had opened their lives to strangers and, in the case of our work on homelessness, welcomed us into the places where they were living.

This is not a plea to stop sensitive work but rather to plan research with an early recognition that the research agreements, including those that have received institutional review board approval, generally do not represent an equity with all the persons involved. Careful reflection is necessary on the research questions and goals and consideration of the balance of the threats from inconvenience, intrusion and temporary friendship, against the end value of the findings. For those of us who send students off to do "exercises," which are small research projects, there must be concern regarding what is left behind, and whether there is any "beneficial reciprocity" (Sieber, 1982, p. 5) between researcher and those researched.

4. CONCLUSIONS

For the environmental problems that I address in my research, issues related to homelessness, public space, community, and spatial conflicts, a particular stance is needed. This emerges from the complex conceptualization of the environment and the need for a multi-disciplinary approach to developing the problems, including the use of multiple methods. This paradigm may not be essential for all research but it has been valuable in the conduct of environmental studies.

Researchers are in privileged positions in society, dealing with work that is exciting and challenging. However, the community of scholars must deal with a new level of ethical concerns that are part of their roles. This requires discussion and reflection beyond the institutional forms now in place.

Consideration of the assumptions of one's work, the challenging task of this paper, is one way towards understanding the conceptual, methodological, and analytic issues that form the core of research. Few if any journals require a section on assumptions for papers but there may be value in considering this addition. As we move toward a new millennium it is an appropriate time to take on new challenges in our work. Rethinking our conceptual and methodological foundations can stimulate a refreshing journey into the new century.

REFERENCES

Altman, I., & Low, S. M. (Eds.) (1992). *Place attachment.* New York: Plenum.

American Psychological Association. (1992). Ethical principles of psychologists and code of conduct. *American Psychologist, 47,* 1597–1611.

Carr, S., Francis, M., Rivlin, L. G., & Stone, A. M. (1992). *Public space.* New York: John Wiley.

Denzin, N. K. (1978). *Sociological methods: A source book* (2nd ed.). New York: McGraw-Hill.

Goffman, E. (1961). *Asylums: Essays on the social situations of mental patients and other inmates.* Garden City, NY: Doubleday Anchor.

Imbimbo, J. (1996). *Ties that bind: The roles of family, community, and school in the educational experiences of homeless children.* Unpublished doctoral dissertation, City University of New York.

Ittelson, W. H. (1960). *Some factors influencing the design and function of psychiatric facilities.* Brooklyn College.

Ittelson, W. H. (1973). Environment perception and contemporary perceptual theory. In W. H. Ittelson (Ed.), *Environment and cognition* (pp. 1–19). New York: Seminar Press.

Katz, M. B. (1989). *The undeserving poor: From the war on poverty to the war on welfare.* New York: Pantheon.

Merriam-Webster's Collegiate Dictionary (10th ed.). (1993). Springfield, MA: Merriam-Webster.

Miles, M. B., & Huberman, A. M. (1994). *Qualitative data analysis* (2nd ed.). Thousand Oaks, CA: Sage.

Mills, C. W. (1959). *The sociological imagination.* New York: Oxford University Press.

Proshansky, H. M., Ittelson, W. H., & Rivlin, L. G. (1970a). *Environmental psychology: Man and his physical setting.* New York: Holt, Rinehart & Winston.

Proshansky, W. H., Ittelson, W. H., & Rivlin, L. G. (1970b). Freedom of choice and behavior in a psychiatric setting. In H. M. Proshansky, W. H. Ittelson, & L. G. Rivlin (Eds.), *Environmental psychology: Man and his physical setting* (pp. 173–183). New York: Holt, Rinehart & Winston.

Rivlin, L. G. (1982). Group membership and place meanings in an urban neighborhood. *Journal of Social Issues, 38*, 75–93.

Rivlin, L. G. (1990). Home and homelessness in the lives of children. *Child and Youth Services*, 14(1), 5–17.

Rivlin, L. G. (1995). *Perspectives on place attachment and the politics of ethnic and religious conflicts.* Presented at the meeting of the Environmental Design Research Association. Boston, MA.

Rivlin, L. G., & Imbimbo, J. E. (1989). Self-help efforts in a squatter community: Implications for addressing contemporary homelessness. *American Journal of Community Psychology, 17*(6), 705–728.

Rivlin, L. G., & Manzo, L. C. (1988). Homeless children in New York City: A view from the 19th century. *Children's Environments Quarterly, 5*(1), 26–33.

Rivlin, L. G., & Rothenberg, M. (1976). The use of space in open classrooms. In H. M. Proshansky, W. H. Ittelson, & L. G. Rivlin (Ed.), *Environmental psychology: People and their physical settings* (pp. 479–489). New York: Holt, Rinehart & Winston.

Rivlin, L. G., & Wolfe, M. (1985). *Institutional settings in children's lives.* New York: John Wiley.

Sieber, J. E. (Ed.). (1982). *The ethics of social research: Fieldwork, regulation, and publication.* New York: Springer-Verlag.

Smith, A. D. (1991). *National identity.* Reno: University of Nevada Press.

Strauss, A. L., & Corbin, J. (1990). *Basics of qualitative research: Grounded theory procedures and techniques.* Newbury Park, CA: Sage.

Webb, E. J., Campbell, D. T., Schwartz, R. D., & Sechrest, L. (1965). *Unobtrusive measures.* Chicago: Rand McNally.

Westin, A. (1967). *Privacy and freedom.* New York: Atheneum.

Winkel, G. H. (1987). Implications of environmental context for validity assessments. In D. Stokols, & I. Altman (Eds.). *Handbook of environmental psychology*: Vol. 1. (pp. 71–97). New York: John Wiley.

6

ASSUMPTIONS, METHODS, AND RESEARCH PROBLEMS OF ECOLOGICAL PSYCHOLOGY

Robert B. Bechtel

Department of Psychology
University of Arizona
Tucson, Arizona 85721

1. INTRODUCTION

The basic assumption of ecological psychology is that the person is *not* the basic unit of human behavior, instead, the most basic unit is the Behavior Setting. This is a very radical statement and contrary to almost all of psychology which sees the person as the most fundamental unit of study and the source of all behavior.

The "method" of ecological psychology has often been misunderstood. First, there still exists a confusion between the first method developed, the *behavior specimen record* and the second method, the *behavior setting survey*. The behavior specimen record was first developed at the Midwest Psychological Field Station in the late forties and early fifties and was made public in the book *One Boy's Day* (Barker & Wright, 1951). Behavior specimens were simply observations of daily behavior made in every day language by observers who followed children around with clipboards for taking notes. About seventeen of these daily records were made and stored in the old bank vault of the Midwest Psychological Field Station at Oskaloosa, Kansas.

It was these records that led to the questioning of the cornerstone of psychological research, the experiment. Clifford Fawl, a student of Roger Barker (Fawl, 1978), compared the behavior of children in the experiment designed to demonstrate Freudian regression (Barker, Dembo, & Lewin, 1941) and found no equivalent in the daily behavior of children. Thus, the behavior created by the experiment, regression, could not be demonstrated to exist in the common, daily observable behavior of children. This raised the question of whether experiments as a genre create behavior which cannot be found in daily life.

Of course, this is the question of *ecological validity*. The requirement for ecological validity is that any phenomenon created in the laboratory must be demonstrated to exist outside the laboratory in the behavioral repertoire of the species in

Theoretical Perspectives in Environment-Behavior Research, edited by Wapner *et al.*
Kluwer Academic / Plenum Publishers, New York, 2000.

question. Thus, for any experiment to be valid, it must be demonstrated that the independent variable can be found outside the laboratory, and operating as it did in the laboratory.

What is often not understood or accepted is that the method of the behavior setting survey underlines the findings of the behavioral specimen record. Since the behavior specimen method was dropped and taken up by very few, the behavior setting survey is currently seen as the sole method of ecological psychology.

2. ASSUMPTIONS

2.1. Inseparability of Behavior from Environment

In a behavior setting survey, the basic assumption is that the prime unit of human behavior is the *behavior setting*. Barker's original definition is, "A behavior setting has been defined as a standing pattern of behavior and a part of the milieu which are synomorphic and in which the milieu is circumjacent to the behavior." (Barker & Wright, 1955, p. 45). In other words, this standing pattern of behavior is a part of the stream of behavior at the same time it is surrounded by it. You can go to a place and see it at specific times. But always remember that the time, place and behavior are all tied together and inseparable. Barker & Wright were the first to emphasize the environmental dimension to behavior. Behavior is *always* tied to a specific place. In fact, Barker said that it is more important to go to the *place* where behavior occurs than it is to look inside the head. Here is real heresy! Psychology as a profession has had great difficulty accepting it. To this day nearly all psychologists believe the origin of all behavior is in the head. Skinner, of course would be an exception. Skinner (1972, p. 185) said something very similar to Barker: " People are extraordinarily different in different places, and possibly just because of the places." He believed that behavior was "shaped" by the environment. Yet among his followers very few accepted this radical posture.

Barker and his followers stood virtually alone in believing that environment and behavior are inseparable until the transactional and organismic theories evolved. The organismic and transactional views are presented by Wapner and Demick (pp. 7–19, this volume). The fundamental difference is the broader view that these two views take compared to the setting specificity of ecological psychology. Ecological psychology would not deny the larger context, but it would insist that all influences are mediated within and through the behavior setting boundaries.

An interesting fact about ecological psychology is that the ordinary person in any culture is able to recognize and describe behavior settings. They are the repeated events of every day life, the meetings, school classes, stores, offices, and games. This enables the observer to use any participant of behavior settings as an informant. How many people attended your meeting? How long did it last? Who were the chief actors? These are the kinds of questions that can be asked to collect the data for a behavior setting survey. And this survey is nothing more than a census of all behavior settings in a community for a period of one year. After a year, the settings repeat themselves.

By collecting records of behavior settings, it is possible to account for virtually *all* of the waking behavior of a town's residents. And a remarkable result of these data is that one can predict the behavior of any one person for the next year with remarkable accuracy, exceeding 90%! There is no other method that can claim such accuracy. What does it mean?

Essentially, it means that people tend to behave the same from year to year in the environment. There are changes, of course, but these tend to be insignificant in the larger pattern. Human behavior is a very regular, predictable phenomenon. Ah, you say, but it doesn't seem that way. Yes, from reading newspapers and some psychological reports one would often get the impression that human behavior is all exception and no rule. That is because newspapers define news as the unusual and many psychologists are just like them, they, too, want to study the unusual. But if one goes to the places where people live and watches without interfering, behavior is seen to be a regular, predictable pattern.

This raises more questions about the experiment. If one accepts the fact that all of human behavior occurs in behavior settings, then it follows that a psychological experiment is just another behavior setting. The consequences of this posture are profound. In order for an experiment to be valid it must be generalizable to other behavior settings.

If we go back to the Clifford Fawl study (1978), we can now see that the reason why the Barker et al. experiment did not have validity is because it created a behavior setting like no other. The more profound question to ask is: Are *all* psychological experiments unique behavior settings?

A way to approach answering this question is to consider the most fundamental kinds of experiments, those done in perception. Experiments done in depth perception, for example, create conditions that could be generalized easily to other settings. Or could they? Gibson, (1979) for instance, would argue that depth perception experiments done in the reductionistic tradition are too static to be valid in the real world. This is because in the real world people are always moving and it is the motion that enables them to perceive depth rather than the static conditions found in the reductionistic experiments.

More telling are the experiments done in social psychology. How can subjects, who know this is an experiment, act "naturally" in an experimental condition?

2.2. Behavior Settings and Larger Units

At first it was assumed that communities naturally separated themselves into behavior settings in order to get the business of living done. Then it was discovered that segments of communities acted in a relatively independent fashion and could be analyzed on their own. For example, schools (Barker & Gump, 1964) were found to be susceptible to size variations which had an influence on the behavior of persons within the settings of the schools. For example, students tended to have more activities and exhibit better attendance at events in smaller schools as opposed to larger ones. This was also found for other entities like churches, work, and small towns.

Behavior settings are discriminated, one from the other, by use of the K-21 scale. This scale has seven additive dimensions and whenever two entities are being measured if the score reaches twenty one they are considered separate.

Communities and organizations are analyzed according to how inclusive or exclusive they are in their various settings. For example, the *penetration level* of occupants is measured by the degree to which each occupant engages in leadership. Single leaders are given a score of six, shared leaders five and so on to onlookers who get only a one. Leadership can then be analyzed within or across settings in an organization by looking at the average penetration scores of any group like children, elderly, or blacks. Thus, Binding (1969) was able to demonstrate that across the everyday settings in a community women have more leadership roles than men. This is a startling discovery, given the gender bias of our culture.

2.3. Context

All behavior must take place within some behavior setting. There is no behavior "outside" a behavior setting. The context *defines* and *prescribes* the behavior independent of individual differences. This is not to say that there are no individual differences. But it is to say that they become redirected by the setting program. Individuals who disrupt or behave contrary to the setting are made to conform or excluded. Variations are so rare as to be statistically insignificant.

2.4. Development

The development of children and adults is determined by the context of the settings they inhabit. It has been discovered that certain macro features of settings influence development markedly. Such features as size have a definite bearing on the quality of behavior and its effect on the behavior of the persons in any context. Smaller settings tend to include people more in their events and make more demands on the inhabitants. Larger settings tend to exclude and make less and different demands.

2.5. Functioning

Ecological psychology assumes all persons function at more or less the same level. Exceptions throughout a population are not statistically significant.

3. MODES OF ANALYSIS

There are over ninety ways behavior settings can be measured and analyzed. The various scales developed all quantify different aspects of behavior within settings. For example *richness* is an index of variety of behavior and variety of different kinds of persons times penetration level. The measure seeks to quantify the mixture of behavioral variety, ethnic and age variety, and leadership opportunities as a resource. Thus a very rich behavior setting is one that has lots of different kinds of behavior going on, lots of different kinds of people doing it and many of these in leadership roles. City markets have lots of richness.

4. PROBLEM FORMULATION AND METHODOLOGY

The most basic problem of ecological psychology is how to study human behavior *in situ* and without influencing it. If this is done carefully enough, the secrets of human behavior will come to light. One does not have to impose theories or methods on the behavior itself. Barker was very careful to try to eliminate the psychologist as *operator* and sought instead to act as a *transducer* of the data. Data are not taken in samples but *in toto*. All the behavior in a town or school is recorded for one year. Inferences can be made from these data. Generalization to the population is not an extrapolation. The data are from the whole population.

Validity and reliability are contained within the data collection process itself. Since observation by the researcher is matched with observations or reports from participants, unreliable data are eliminated. And, since virtually all independent sources of data are used to count behavior settings, the independent agreement of the various sources guarantees validity. For example, an initial behavior setting list may be made from a high school year book. This list is then checked against bulletin board announcements, statements of informants, and rosters that teachers kept. Thus, all the independent sources of data are, in effect, used up in doing the behavior setting survey.

5. PARADIGMATIC RESEARCH PROBLEMS

5.1. Communities and Organizations

Because ecological psychology emphasizes the community context of behavior and its larger time perspective, it is difficult, time consuming and expensive. A typical behavior setting survey will take one year just to collect, without considering community preparation or analysis of data. Two years could easily be consumed.

These reasons make larger units like cities almost impossible to use for a behavior setting survey. While parts of larger towns have been surveyed, no complete behavior setting survey has ever been done of a community larger than a thousand or so persons.

The most useful research has been in comparisons of institutions like schools and churches within small vs. large communities. Results of these studies have produced the *undermanning* or *underpersoning* theory. This theory stipulates that in smaller organizations the pressures to perform are greater and thus have a favorable effect on the persons within them. By contrast, larger organizations have less pressure to perform and have deleterious effects on personnel. The benefits in smaller organizations go contrary to popular beliefs that large size is superior.

5.1.1. Person-unit size interaction. Smaller organizations tend to make the persons in them feel more important, more necessary, and more obligated to do a job well. Fewer people in a setting result in more immediate feedback on any person's performance. Smaller organizations avoid the common problem of isolation in today's large organizations.

5.1.2. Environmental demand. Behavior arises out of the interaction between human need and environmental demand. A human is hungry and must either pick fruit or go hunting in a pre industrial society or go to the food store in a more industrial setting. The search requires finding or foraging behavior. The environment gives this behavior its shape.

5.1.3. Environmental-social organization. In order to accomplish the large tasks of survival, humans must sort themselves into groups with special behaviors. Some must grow food, others must sell it, and so on to produce a complex society. Each group is either a behavior setting or group of settings. These settings, in turn become organized into communities and the communities into large entities like counties, states, etc. All of these settings continue beyond the life of any of the individuals in them and virtually regardless of the differences within each individual.

6. RELATIONS BETWEEN EXPERIENCE AND ACTION

People willingly join settings, in fact, are born into them, learn the proper pattern of behavior, and graduate to other settings. A person's life is defined by the series of settings entered. There is more choice about entering a setting then there is about how one will behave after entering. One enters a setting because one wants to do the behaviors in it. While the behavior may be experienced as originating from self, the demand of the setting is seen as the cause.

7. CONCLUSIONS

Ecological psychology has produced a body of literature that challenges conventional psychology. At the same time it has produced findings about institutions that are useful to the betterment of all institutions. Shadish (1984) points out that ecological psychology has not been widely accepted because the data are contrary to unsupported conventional wisdom.

Nevertheless, the data stand. The evolution of the transactional and organismic views tend to incorporate much of the assumptions and methods of ecological psychology and to keep many of the assumptions of conventional psychology also. It will be interesting to see what the future brings from such a mix.

REFERENCES

Barker, R., Dembo, T., & Lewin, K. (1941). *Frustration and regression: A study of young children* (University of Iowa Studies in Child Welfare, No. 18). Ames: University of Iowa.

Barker, R., & Gump, P. (1964). *Big school, small school*. Stanford, CA: Stanford University Press.

Barker, R., & Wright, H. (1951). *One boy's day*. New York: Row Peterson.

Barker, R., & Wright, H. (1955). *Midwest and its children*. Evanston, IL: Row Peterson.

Binding, F. (1969). *Behavior setting leadership in an American community*. Unpublished dissertation. Lawrence, KS: University of Kansas.

Fawl, C. (1978). Disturbances children experience in their natural habitat. In R. Barker & Associates (Eds.), *Habitats, environments and human behavior*. San Francisco, CA: Jossey-Bass.

Gibson, J. (1979). *The ecological approach to visual perception*. Boston, MA: Houghton-Mifflin.

Shadish, W. (1984). Policy research: Lessons from the implementation of deinstitutionalization. *American Psychologist, 39*, 725–738.

Skinner, B. (1972). *Beyond freedom & dignity*. New York, NY: Alfred A. Knopf.

7

SOCIAL-PSYCHOLOGICAL APPROACHES IN ENVIRONMENT-BEHAVIOR STUDIES

Identity Theories and the Discursive Approach

Marino Bonaiuto and Mirilia Bonnes

Università degli Studi di Roma "La Sapienza"
Dipartimento di Psicologia dei Processi di Sviluppo e Socializzazione
00185 Rome (Italy)

1. MOVING FROM AN INDIVIDUALISTIC TO A SOCIAL PSYCHOLOGICAL APPROACH

Environment-behavior studies need to strengthen theoretical formulations and to clarify methodological assumptions. This is a valid option in general, but a particularly important one when psychological approaches are adopted. Actually, psychology is increasingly a very large umbrella under which dramatically different theoretical assumptions and methodological practices co-exist. And the same is true, perhaps on a smaller scale, for that part of psychology specifically devoted to environment-behavior research, namely environmental psychology. With respect to this, environmental psychology, studying people-environment relations or transactions, aims on one side to establish itself as an applied psychological discipline; and on the other, it aims to understand psychological processes "in the real world" acquiring internal relevance for psychology in general.

Traditionally, environmental psychology originated as an experimental-perceptual approach that characterized specific environment-behavior research. Typically, this approach has attempted to explain and study environment-behavior regularities with reference to both theoretical constructs and methodological devices focused on the single individual. In some ways, this can already be considered a significant development with respect to some traditional psychological perspectives that isolate individual psychological processes from any sort of environment. Still, at the theoretical level this approach to people-environment research emphasized the explanation of environment-behavior regularities in terms of intra-individual

Theoretical Perspectives in Environment-Behavior Research, edited by Wapner *et al.*
Kluwer Academic / Plenum Publishers, New York, 2000.

constructs, preferably cognitive or psychophysiological ones (say, individual perceptual processes, schemata theory, etc.). At the methodological level, experimental settings or surveys were preferred and were adopted in such a way as to guarantee observations and data consistent with individualistic assumptions. Therefore single individual responses, whether behavioral or cognitive, were recorded and linked to the relevant environmental stimulus condition. The classical design comprised environmental conditions as independent variables and individual response(s) as dependent variable(s), within the context of linear causal models (rather than within a systemic logic, as advocated in Bronfenbrenner (1979). This epistemological circularity holds true for traditional psychology in general (Harré, 1989), not only for the psychology traditionally adopted in environment-behavior research.

More recently, environmental psychology has also been characterized by social-psychological approaches. Elsewhere we emphasized the importance of moving from an individualistic perspective to a psycho-social one, giving importance to the interface between the individual and the collective and social-cultural processes (Bonnes & Secchiaroli, 1995). In a social psychological perspective the individual should not be considered as an isolated entity, but as a part embedded in a social network or system, that is, an entity materially and symbolically connected to other individual and social entities, to be studied and understood with respect to the social processes such as those of communications, groups, institutions, etc. This general view has been increasingly advocated in cognitive and social psychology in recent years (e.g., Doise, 1982; Harré, 1980; Harré & Gillett, 1994; Moscovici, 1984; Still & Costall, 1991). The point here is not to deny any kind of cognitive or psychological reality but, rather, to connect individual psychological constructs and processes to socially shared processes and activities and to study their interfaces.

This has implications for environment-behavior studies, both in terms of which constructs to refer to in order to conceptualize and explain psychological regularities and in terms of which methodologies to use in order to operationalize constructs and to gather and analyze data. More specifically, looking at the psychological theories which have been adopted by environment-behavior research, it should be stressed that social-psychological approaches have often been relatively overlooked as compared with perceptual or cognitive approaches. Indeed, even if theoretical pleas in favour of more socially, contextually, and culturally enlightened approaches have been generously produced (e.g., Altman & Rogoff, 1987; Canter, Correira Jesuino, Soczka, & Stephenson, 1988; Graumann & Kruse, 1990; Saegert & Winkel, 1990; Stokols, 1987; Wapner, 1987), some of the main current approaches of social psychology are not employed or are under-utilized in environment-behavior research. It is assumed here that the adoption of these social-psychological approaches could help environment-behavior research to consider psychological processes in relation to social ones, that is, to offer a richer and more complex picture than that of the single individual reacting to a given environment on the basis of intra-individual cognitive processes.

In the present chapter, we would like briefly to illustrate two social-psychological approaches which can be used to study people-environment transactions: identity theories, on one hand, and discursive approaches, on the other. Both are major contemporary approaches in social psychology, especially in Europe. The reference to discursive approaches can help to show how people's experiences of environmental features are socially constructed. Further, the use of identity theories, especially social identity theories, can help to show how people's evaluations of the environment may be linked to intrinsically social dimensions and factors.

2. IDENTITY THEORIES

Generally speaking, identity defines who or what an individual is. Some aspects of identity pertain to personal and unique conceptions or characteristics of the person and constitute what is commonly called personal identity. But identity also involves membership in social groups or categories; these aspects of identity are called social identity. According to identity theories, we derive much of the sense of who we are and much of our self-esteem from our personal and unique aspects as well as from our group memberships; however, the emphasis on one element or the other may vary among different theories. Of specific social-psychological interest is the idea that an individual identity can be defined according to one's own group membership(s). Such an idea has a long tradition in social psychology (e.g., Mead, 1934) and was formally elaborated by Tajfel (1978, 1981) and Tajfel and Turner (1979, 1986) as "Social Identity Theory" (SIT). This general idea was subsequently developed by Turner and others (Turner, 1985; Turner, Hogg, Oakes, Reicher, & Wetherell, 1987) as "Self-Categorization Theory" (SCT) in order to specify cognitive processes underlying social identity and group processes. Generally, this theory was meant to explain the processes underlying group behavior and intergroup relations, as well as related phenomena such as those of social stereotypes and prejudices. Obviously, there is neither time and space to go into details of one the most diffuse social-psychological theories, nor to support it with empirical data. Rather we will provide its main ideas and assumptions and will then summarize some research showing the relevance of people's social identity in evaluating the features of their environment.

2.1. Social Identity Theory (SIT) and Self-Categorization Theory (SCT)

Social Identity Theory (SIT) revolves around two basic tenets: "categorization process" and "social comparison". Regarding categorization, our social identity depends on our belonging to one social group or category (gender, ethnicity, nationality, religion, social class, etc.) rather than another. As to social comparison, the theory assumes that people tend to see themselves in positive rather than negative terms, in order to gain and preserve self-esteem through social comparison. These two aspects together (psychological processes of categorization and self-enhancement) constitute the motivational core of the theory: since part of our self-image is defined by our group memberships, the preference towards positive self-evaluation will be extended to the group to which we belong, because it contributes towards defining our identity. In other words, there will be a tendency to make biased intergroup comparisons, with a preference to see our group (the "ingroup") in a positive light with respect to another group (the "outgroup"). The main hypothesis of the theory is derived from these assumptions: group members will pursue various forms of positive distinctiveness for their ingroup with respect to the outgroup in order to develop or maintain a satisfactory identity. If this is impossible, they may seek alternative group memberships, offering greater chances for positive self-evaluation. Such ingroup positive distinctiveness can be achieved either by discriminating the outgroup or by positively reinterpreting the ingroup's negative features. Self-Categorization Theory (SCT) particularly stresses the importance of the contextual salience of intergroup self-categorization in order to have depersonalization of self-perception (i.e., group phenomena, such as social stereotyping, ethnocentrism, but also group cohesion, cooperation, and altruism).

Intuitively, this social-psychological theory may be useful for studying environmental perceptions and actions in socially conflictual contexts. In general, several perspectives and authors both in psychology (e.g., from James, 1890, and Mead, 1934; to Proshansky, Fabian, & Kaminoff, 1983, and Neisser, 1988) and in other disciplines (philosophy, history, geography, anthropology, sociology) have argued that the environment is relevant for people's identity. Some authors (e.g., Proshansky et al., 1983; Proshansky & Fabian, 1986) explicitly wrote about place identity in order to refer to environmental elements contributing towards defining people's identity. However, most of these contributions focused on personal rather than social identity. On the contrary, SIT and SCT, which have been major social psychology approaches in the last three decades, have rarely been applied to environment-behavior research. This may seen quite paradoxical, given the increasingly social conflictual nature of many environmental issues, often involving group conflicts triggered by identity and/or interest threats (at the local, national, and international level). In general, it could be hypothesized that social identity principles and strategies of coping with identity threats used in relation to places and environments appear similar to those operating in the case of social identification with a social category or group. For example, environmental evaluations of the place where a person or group lives could be more positive than evaluations of an outgroup's place, showing that the struggle for a positive social identity, which offers positive self-esteem through self-enhancement, could also be achieved through a positive ingroup-place distinctiveness.

2.2. SIT Research

Here we would like to summarize some studies using constructs and assumptions generally derived from social identity theories. The first study aimed to describe the evolution of the role played by social identity in the context of place identity development emerging from inhabitant-city relations, and the second, to predict inhabitants' perceptions of beach pollution on the basis of their own social and place identity.

2.2.1. Place and Social Identity: Inhabitant-city Relation. In a study addressing the issue of identity/urban environment relationships (Bonnes, Giuliani, & Bonaiuto, 1995), we specifically focused on the temporal development of place identity, trying to discover different ways different individuals develop their own place identity when they move to a new large urban area (such as the city of Rome). Therefore, personal and social identity were assumed to be non-static constructs, which could vary qualitatively and quantitatively with time. The research had a descriptive aim: to identify different kinds of place identity developed by 385 Italian non-native Roman residents and to study their associations with a wide pool of sociodemographic and residential variables.

In this case, place identity was conceived and measured by six yes/no items covering three different domains of place experience, each considered both in a present and a future perspective: *physical rootedness*, in Tuan's (1980) terms (present: "I feel at home in Rome"; future: "I would find it hard to leave Rome"); *personal place identification* (present: "Rome has become part of me"; future: "Rome will always be alien to me for some reason"); and *social place identification* (present: "I feel completely Roman"; future: "I will never feel Roman").

Among other things, the results deriving from cluster analysis revealed five groups.

A "full alienation" group, characterized by lack of identification at all three levels, i.e., physical, personal, social (very short- or medium-term residents, working males).

A "partial alienation" group, characterized by partial lack of physical and personal level of identification and total lack of social level of identification (young, medium-term residents).

A "prevalent rootedness" group, characterized by full physical rootedness and only present level of identification, with no present social identification and weak future personal and social identification (long-term residents).

A "personal identification" group, characterized by full rootedness and personal levels, with social level of identification not in the present, only in the future perspective (they are of medium age, high education and medium- or long-term residents).

A "full identification" group, characterized by full identification at all levels, i.e., physical, personal, and social, both in the present and as future perspectives (old, educated, and long-term residents).

On the whole, these results illustrate different kinds—which could be interpreted as different levels—of inhabitants' identification with the city they moved to. The five identified groups and their sociodemographic and residential characteristics might also suggest a possible, but not necessary, chronological development of place identity, moving from an earlier stage of no identification at all or of partial rootedness or personal identification, to a first stage of full rootedness, to a second stage of full identification of the person with the city (Rome), to a third stage of full social identification with citizenship (Roman).

2.2.2. Place and Social Identity: Perception of Beach Pollution. In a second study, making specific reference to SIT and SCT, as well as to Place Identity Theory (PIT), Bonaiuto, Breakwell, & Cano (1996) carried out a study to show that people's evaluation of a feature of their environment is affected by relevant aspects of their own social identity, in a context in which these environmental features are at the core of a relevant social conflict. Specifically, the research focused on an internationally disputed environmental issue: the degree of beach pollution. In the European Union (EU) this issue is regulated by central (European) standards and norms which are "imposed" at the local level in each nation. In this context, conflict between the nation and the international body about specific local places permits for a salience of specific aspects of an individual's social identity, namely, national identity and local identity. Therefore, if the national and local environments people belong to constitute relevant parts of their social identity (at various levels of self-categorization, such as nationalism and localism), they should be treated according to the same social-cognitive processes which work for other traditional stimuli relevant for their social identity, namely, biases in the perception and evaluation of environmental features which are relevant for one's own self-concept and social identity, according to a general differentiation in search of positive ingroup distinctiveness. For example, in order to strive for a positive social identity people might deny negative characteristics of their own local and national environment, especially if these were initially fostered by a powerful and disliked outgroup institution.

British beach pollution, as defined by the EU regulations, was selected as a salient environmental issue because it is internationally disputed and it has manifestations both at the national and town level. Authors expected that a negative attribution like "pollution", attached to the place where people live (town or nation) by an external/international institution, would have an impact on the individual's environmental evaluations according to the strength of local identity and/or nationalism, respectively. Since neither the environmental situation nor the local/national identity of the person can change in a short time, the authors hypothesized that the strategy people would adopt to cope with the threatening definition of their own environment-identity would be to perceive environmental conditions in more positive terms (Breakwell, 1986); that is, to minimize the degree of negative environmental impact and therefore its negative implications. The two main hypotheses were: *a*) the stronger people's local identity, the less they will perceive their own local town beach to be polluted; and *b*) the greater their nationalism, the less they will perceive their own national beaches to be polluted.

Data were gathered according to standard methodologies. Paper-and-pencil scales measured degree of nationalism and local identity as well as a number of sociodemographics and environmental knowledge and attitude variables, all considered as predictors of two main criteria: perception of local beach pollution and perception of national beach pollution. A path model tested on a sample of 347 adolescents from six different British seaside resorts confirmed the two main hypotheses: the two levels of self-identification with town and nation have the significant effect of moderating perception of pollution. On one hand, the stronger the local identity of residents, the less polluted they perceived the beaches of their own towns to be (with nationalism also exerting the same effect). On the other hand, the stronger the national identity of residents, the less polluted they perceived the beaches of their own nation to be. Interestingly, traditional social-demographic and environmental concern variables did not prove to be significant predictors of environmental perception, once social and place identity constructs were brought into the picture, that is, when specific case studies of environmental change are considered by taking into account the specific social-political context, the relevant matching dimensions of inhabitants social identities become key-factors. Results also show that multiple social identities referring to different geographical levels can co-exist in the same person at the same time, as a multi-level place identity in which each level partly "works" at the proper geographical scale (town or nation).

On the whole, the results offered empirical evidence about the importance of self-identification processes for people's environmental representations. These results showed the environment people live in and belong to can be considered part of their self-concept because it is treated according to similar principles operating for other aspects of their social identity. When national (or local) social identity is salient in the context, nationalism (or local identity) affects environmental perceptions and evaluations which become ingroup stereotypical and normative. If people can treat place characteristics as if they were group characteristics, the authors argue that "the struggle for a positive social identity, which offers positive self-esteem through self-enhancement, can therefore also be achieved through, what we might call, positive 'in-place' distinctiveness" (Bonaiuto et al., 1996, p. 172). The specific denial option adopted by the most identified subjects (a positive bias in the perception of pollution in their own places) is also similar to one of the strategies adopted to cope with common identity threats, as outlined in the Identity Process Theory (Breakwell, 1986).

2.2.3. Implications of the Two SIT Studies. On the whole, we believe the two above-mentioned studies show that the use of a social-psychological approach in the environment-behavior research area may have two advantages. On one hand, as in the second study, this approach permits application and extension of the validity of already known general principles to the specific case of environment-behavior research (thus, generalizing an existing social-psychological theory to the new realm of people-environment relationships). On one hand, as in the first study, it suggests new lines of development for the theoretical approach adopted, that is, specific environment-behavior phenomena foster the internal development of a social-psychological theory.

The second study represents an example of using a specific social-psychological theory to predict and explain a specific aspect of the environment-people relationship. It shows how environment-behavior phenomena can be subsumed and integrated within a general theoretical framework, more specifically, how the perception and evaluation of the features (pollution) of one's own environment can be explained according to the general social-psychological principles offered by SIT (positive ingroup or inplace distinctiveness). In this way, environment-behavior research can be connected to more general explanatory social-psychological systems, therefore benefiting from their theoretical and methodological resources, as well as linking environment-behavior phenomena to a wide range of other social-psychological processes.

The first study shows how a specific environment-behavior phenomenon can offer stimuli and suggestions for the development of social-psychological theories. In particular, it shows how the content of place identity can be different for different groups of inhabitants, and it suggests that these differences may be related to social factors or to inhabitants' residential past experiences, implying both possible lines of development. But it also shows a possible general development trend, where rootedness is the simpler level and social identification the more complex one. Therefore, this kind of research can envisage the need to include these kinds of phenomena in a more general identity theory, as well as to stress the integration between place and social identity theories.

On the whole, both studies show the importance of specific social-psychological factors and processes, namely social identity, in mediating people-environment relationships; they also show how environmental features acquire importance for people's identity. More generally, they illustrate the importance of studying social-psychological processes without detaching them from specific relevant contexts.

3. DISCURSIVE APPROACH

Interest in and study of "discourse", defined as a social practice aimed at construing meanings and realities through the production of oral or written "texts" (as well as visual or non-verbal "texts"), is common to several contemporary social sciences. In the last ten years there has been a growing interest in communication and language practices in social psychology. More specifically, in European, and especially British social psychology, there has been a growing interest in the study of discourses and texts, and in their implications for psychology. There is no time and space here to describe the reasons for and the context of the emergence of this approach. We can just briefly mention that the general theoretical and methodological dissatisfaction with traditional experimental approaches and the first social constructionist criticism from authors such as Kenneth Gergen (e.g., 1985) and Rom Harré (e.g., Harré & Secord, 1972; Harré,

1979) during the seventies and eighties, prepared the field and offered the basis for the new proposal that emerged in the eighties and the nineties. In the U.K., authors such as Michael Billig (1987/1996) or Derek Edwards and Jonathan Potter (1992) advocated a "rhetorical approach" or a "discursive psychology", to be adopted in social and cognitive psychology. This approach not only drew upon social constructionist criticism in psychology, but also upon non-psychological approaches.

For example (see Potter, 1996), philosophers of language such as Wittgenstein or Austin are referred to because of their emphasis on the uses of language, on its role as social action and practice, and on its being both context dependent and context creator. Ethnomethodology and conversation analysis are referred to because of their emphasis on the importance of folk means to define and produce the meaning of their own world and interactions. And particularly conversation analysis (starting from the pioneering works of Sacks, Schegloff, & Jefferson, 1974) is referred to because of its specific empirical approach and methodology developed to study verbatim transcripts of interaction as the main tool for analyzing participants' own categories and structures in managing their everyday social interactions. Semiotics, post-structuralist and postmodernist authors (e.g., Barthes, Derrida, Foucault; e.g., Potter, 1996, chapter 3) are referred to because of their emphasis on the role of discourse in the creation of subjective and objective reality and in power management, rather than as a neutral way to represent a given reality. Sociology of scientific knowledge and rhetoric are referred to because of their emphasis on the strategies of "fact constructions" and on the pervasiveness of controversies between facts and interpretations.

Some authors stressed the importance of discourse practices also for environmental issues (Bonnes & Secchiaroli, 1995; Graumann & Kruse, 1990). For example, some authors showed the relevance of verbal language and communication processes in affecting people-environment relationships, especially at the level of environmental perception and evaluation (e.g., Eiser, Reicher, & Podpadec, 1993; Macnaghten, 1993; Rydin & Myerson, 1989) as well as their relevance in mediating the communication of environmental issues in mass media (e.g., Bell, 1994).

More generally, anthropological contributions stressed the culture-specific nature of environmental constructions, as evidenced by both the selection of specific environmental issues and the way they are described and represented in each society (e.g., Douglas, 1966; Douglas & Wildawsky, 1982). In psychology, attention has also been drawn to the importance of social-cultural factors for environment-behavior research (e.g., Bonnes, 1998; Graumann & Kruse, 1990). From a psychological point of view, Graumann and Kruse (1990) noted that "what the individual consumer of media news (in Western democracies) receives is what is presented as facts as well as a choice of pro and con evaluations of these facts by spokespersons of the different social and political forces of a given society. [. . .] this input of information tends to be more important than direct individual experience" (p. 217). Therefore people experience socially mediated and constructed information and reality. According to these authors, our experience of many environmental changes has two fundamental psychological characteristics: "gradualness" and "sensory amodality". First, environmental processes, especially global changes, have a long temporal dimension along which they manifest (implying gradualness of manifestation). Second, often we have no direct sensory experience of the physical or chemical processes constituting environmental change and we can only experience the consequences (sensory amodality). These two aspects of environmental changes imply that we mostly refer to mediated information about environmental issues, particularly by the mass media which offer a *"pattern of mediated environmental experience"* (Graumann & Kruse, 1990, p. 220).

According to this, some of our recent research lines have been directed towards studying the way different mass media sources deal with environmental issues, in order to show which environmental phenomena and changes are *selected* and given attention, and to show how they are *framed* in terms of causes, consequences, and remedies at different temporal, spatial, and social levels. The aim is to look at *which* environmental changes are selected and *how* they are framed, according to different agents/agencies with different perspectives (whether political, economical, personal, etc., at the level of individuals, collectivities, groups, institutions, etc.). If the concern becomes the production of environmental representations or versions, and both their cognitive effects and social implications, the focus of the analysis can be turned towards the discursive strategies through which these representational practices are concretely realized.

3.1. Research Study: Social Actors and Discursive Strategies

Within a broader international research project funded by the European Union a few years ago (Levy-Leboyer, Bonnes, Chase, Ferreira-Marques, & Pawlik, 1996), we carried out both a content analysis and a lexical analysis of the environmental news published for a period of one year by different Italian newspaper headings representing different political leanings. The main aim was to identify systematic associations between selection or framing of environmental issues and newspaper headings, with the general hypothesis that agencies with different political leanings will construe different environmental representations (see Bonnes, Bonaiuto, Metastasio, Aiello, & Sensales, 1997; Bonnes, Bonaiuto, Metastasio, Sensales, & Aiello, 1996; Metastasio, Bonaiuto, Sensales, Aiello, & Bonnes, in press).

From November 1993 to November 1994, once a week according to a day schedule, we sampled environment related articles appearing in three Italian newspapers. At least during the sample period, they could be considered to represent different political areas: *il Giornale* (center-right), *la Repubblica* (center-left), il *Corriere della Sera* (center-governmental).

First a content analysis was carried out for each article according to several dimensions pertaining to the kind of environmental issue focused on and the kind of framing in terms of responsibility and effects. The associations among these and the three different headings were also analyzed. A second analysis was carried out on the lexicon used in each article's headline and subtitles, and associations with the three different headings.

Results showed that each heading is characterized by specific kinds of content and lexicon selected to report on environmental events. For example, the center-governmental newspaper (il *Corriere della Sera*) focused mainly on destructive natural environmental issues or on decay of historical-artistic heritage, both framed in terms of natural causes; the lexicon was not specifically characterized. The center-right newspaper (*il Giornale*) focused mainly on health risk framed in terms of natural or scientific referents, or on bio-physical pollution framed in terms of human and industrial causes; the lexicon was characterized by economic and nationalistic terms or by natural terms. The center-left newspaper (*la Repubblica*) focused mainly on risk and decay of artistic-architectonic heritage, framed in terms of human responsibility and of political agencies for causes and remedies; the lexicon was characterized by terms referring to legislative-administrative aspects, and by terms referring to local-regional aspects, generally referring to human-political intervention. Other results also showed that the same kind of environmental event (i.e., changes in the built environment) can be framed according to different kinds of responsibility, that is, political causes in the center-left newspaper or natural causes in the center-governmental newspaper.

On the whole, content and lexicon analyses converged in showing that different social-political agencies can select and/or frame differently the environmental events on which they choose to focus. In so doing, each agency contributes to construing a specific representation of a certain environmental event and/or a specific version of possible responsibility. The systematic associations among different kinds of environmental information and newspapers belonging to different political areas suggest that the representations of environmental issues are socially constructed according to broader social-political-economical aims.

4. CONCLUSION

Social identity theory and a discursive approach share the tendency to give importance to social-cultural factors in studying psychological processes. Therefore their application to the people-environment relationship giving relevance to the social and cultural dimension of the way people perceive, evaluate, and behave with respect to the environment(s) they live in or relate to. More specifically, social identity theory assumes that environment-behavior transactions can be affected by people belonging to specific social groups and categories and, consequently, the classical social-psychological dynamics in terms of ingroup-outgroup relations can affect the way people perceive, evaluate, and behave with respect to their own environment, their group's environment and others' (groups) environment. The discursive approach assumes that the same environmental event, and therefore its causal explanation, are intrinsically a social-cultural product, at least partly defined and re-defined, constructed and re-constructed through the use of linguistic resources. Moreover, such a discursive construction of the environment and its characteristics can be quite different according to the source of this construction, showing that environmental discursive constructions are rhetorically oriented and serve the interests of people and groups which are part of a wider argumentative context (which is intrinsically cultural, social, political, and economical).

Both approaches offer an example of how environment-behavior research can be enriched by applying specific social-psychological approaches. In particular, they show that people-environment transactions work according to social-psychological processes. On one hand, people's experience of the environment tends to be socially constructed; consequently, it can be different according to the specific mediation agency considered by the individual. On the other hand, the way people relate to their own environment can respond to social-psychological principles, showing that environmental features are not separate from people's social and personal identity; rather, the way people consider and relate to their own environment is similar to the way they treat other significant features of their social worlds.

REFERENCES

Altman, I., & Rogoff, B. (1987). World views in psychology: Trait, interactional, organismic, and transactional perspectives. In D. Stokols, & I. Altman (Eds.), *Handbook of environmental psychology, Vol. 1* (pp. 7–40). New York: Wiley.

Bell, A. (1994). Climate of opinion: public and media discourse on the global environment. *Discourse and Society, 5*, 33–64.

Billig, M. (1987/1996). *Arguing and thinking. A rhetorical approach to Social Psychology*. London: Sage.

Bonaiuto, M., Breakwell, G. M., & Cano, I. (1996). Identity Processes and Environmental Threat: the Effects of Nationalism and Local Identity upon Perception of Beach Pollution. *Journal of Community and Applied Social Psychology, 6*, 157–175.

Bonnes, M. (1998). The ecological-global shift, environmental sustainability, and the 'shifting balances'. In J. Teklenburg, J. van Andel, J. Smeets, & A. Seidel (Eds.), *Shifting Balances. Changing Roles in Policy, Research, and Design. Proceedings of the 15th Bi-Annual Conference of the International Association for People-Environment Studies (IAPS)* (pp. 165–174). Eindhoven: EIRASS European Institute of Retailing and Services Studies.

Bonnes, M., Bonaiuto, M., Metastasio, M., Aiello, A., & Sensales, G. (1997). *Environmental discourse and ecological responsibility in media communication in Italy*. In R. Garcia-Mira, C. Arce, & J. M. Sabucedo (Eds.), *Responsabilidad Ecològica y Gestiòn de los Recursos Ambientales* (pp. 99–135). A Coruna: Diputacion Provincial de A Coruna.

Bonnes, M., Bonaiuto, M., Metastasio, M., Sensales, G., & Aiello, A. (1996, July). *The social construction of environmental issues: an analysis of Italian newspaper discourse*. Paper presented at the 11th General Meeting of the European Association of Experimental Psychology, Social Psychology in Europe, Gmunden, Austria.

Bonnes, M., Giuliani, M. V., & Bonaiuto, M. (1995, July). *Place Identity development in the city of Rome*. Paper presented at the 4th European Congress of Psychology, Athens.

Bonnes, M., & Secchiaroli, G. (1995). *Environmental psychology. A psycho-social introduction*. London: Sage.

Breakwell, G. M. (1986). *Coping with threatened identities*. London: Methuen.

Bronfenbrenner, U. (1979). *The ecology of human development*. Cambridge: Harvard University Press.

Canter, D., Correira Jesuino, J., Soczka, L., & Stephenson, G. M. (Eds.). (1988). *Environmental social psychology*. Dordrecht: NATO ASI Series.

Doise, W. (1982). *L'explication en psychologie sociale*. Paris: Presses Universitaires de France.

Douglas, M. (1966). *Purity and danger. An analysis of the concepts of pollution and taboo*. London: Routledge and Kegan Paul.

Douglas, M., & Wildawsky, A. (1982). *Risk and culture. An essay on the selection of technological and environmental dangers*. Berkeley, CA: University of California Press.

Edwards, D., & Potter, J. (1992). *Discursive psychology*. London: Sage.

Eiser, J. R. , Reicher, S., & Podpadec, T. J. (1993). What's the beach like? Context effects in judgements of environmental quality. *Journal of Environmental Psychology, 13*, 343–352.

Gergen, K. (1985). The social constructionist movement in modern psychology. *American Psychologist, 40*, 266–275.

Graumann, C. F., & Kruse, L. (1990). The Environment: Social Construction and Psychological Problems. In H. T. Himmelweit, & G. Gaskell (Eds.), *Societal psychology*. London: Sage.

Harré, R. (1979). *Social Being: A Theory for Social Psychology*. Oxford: Blackwell.

Harré, R. (1980). Man as rhetorician. In A. J. Chapman, & D. M. Jones (Eds.), *Models of man*. Leicester: The British Psychological Society.

Harré, R. (1989). Metaphysics and methodology: Some prescriptions for social psychological research. *European Journal of Social Psychology, 19*, 439–453.

Harré, R., & Gillett, G. (1994). *The discursive mind*. London: Sage.

Harré, R., & Secord, P. F. (1972). *The explanation of social behaviour*. Oxford: Blackwell.

James, W. (1890). *The principles of psychology*. New York: Holt.

Levy-Leboyer, C., Bonnes, M., Chase, J., Ferreira-Marques, J., & Pawlik, K. (1996). Determinants of pro-environmental behaviour: a five countries comparison. *European Psychologist, 1*, 123–129.

Macnaghten, P. (1993). Discourses of nature: argumentation and power. In E. Burman, & I. Parker (Eds.), *Discourse analytic research. Repertoires and readings of texts in action* (pp. 52–72). London: Routledge.

Mead, G. H. (1934). *Mind, self, and society*. Chicago: University of Chicago Press.

Metastasio, R., Bonaiuto, M., Sensales, G., Aiello, A., & Bonnes, M. (in press). La comunicazione di eventi ambientali nella stampa quotidiana: esame di tre principali testate italiane (Environmental issues communication in newspapers: examination of three main Italian headings). *Rassegna di Psicologia*.

Moscovici, S. (Ed.). (1984). *Psychologie sociale*. Paris: Presses Universitaires de France.

Neisser, U. (1988). Five kinds of self knowledge. *Philosophical Psychology, 1*, 35–59.

Potter, J. (1996). *Representing reality*. London: Sage.

Proshansky, H. M., & Fabian, A. K. (1986). The development of place identity in the child. In C. S Weinstein, & T. G. David (Eds.), *Spaces for children* (pp. 21–40). New York: Plenum.

Proshansky, H. M., Fabian, A. K., & Kaminoff, R. (1983). Place identity: physical world socialisation of the self. *Journal of Environmental Psychology*, *3*, 57–83.

Rydin, Y., & Myerson, G. (1989). Explaining and interpreting ideological effects: a rhetorical approach to green belts. *Environment and Planning D: Society and Space*, *7*, 463–479.

Sacks, H., Schegloff, E. A., & Jefferson, G., (1974). A simplest systematics for the organization of turn-taking for conversation. *Language, 50*, 696–735.

Saegert, S., & Winkel, G. (1990). Environmental Psychology. *Annual Review of Psychology*, *41*, 441–477.

Still, A., & Costall, A. (1991). *Against cognitivism. Alternative foundations for cognitive psychology.* Hemel Hempstead: Harvester Wheatsheaf.

Stokols, D. (1987). Conceptual strategies in environmental psychology. In D. Stokols, & I. Altman (Eds.), Handbook of environmental psychology, Volume 1 (pp. 41–70). New York: Wiley.

Tajfel, H. (Eds.). (1978). *Differentiation between social groups.* London: Academic Press.

Tajfel, H. (1981). *Human groups and social categories.* Cambridge: Cambridge University Press.

Tajfel, H., & Turner, J. C. (1979). An integrative theory of intergroup conflict. In W. G. Austin, & S. Worchel (Eds.), *The social psychology of intergroup relations.* Monterey, CA: Brooks/Cole.

Tajfel, H., & Turner, J. C. (1986). The social identity theory of intergroup behaviour. In S. Worchel, & W. G. Austin (Eds.), *Psychology of intergroup relations* (2nd ed.). Chicago: Nelson-Hall.

Tuan, Y. F. (1980). Rootedness versus Sense of Place. *Landscape, 24*, 3–8.

Turner, J. C. (1985). Social categorization and the self-concept: A social cognitive theory of group behaviour. In E. J. Lawler (Ed.), *Advances in group processes: Theory and research, Vol. 2.* Greenwich, CT: JAI Press.

Turner, J. C., Hogg, M. A., Oakes, P. J., Reicher, S. D., & Wetherell, M. S. (1987). *Rediscovering the social group: A self-categorization theory.* Oxford: Basil Blackwell.

Wapner, S. (1987). A holistic, developmental, system-oriented environmental psychology: Some beginnings. In D. Stokols, & I. Altman (Eds.), *Handbook of environmental psychology, Vol. 1* (pp. 1433–1465). New York: Wiley.

8

PERSONS, CONTEXTS, AND PERSONAL PROJECTS

Assumptive Themes of a Methodological Transactionalism

Brian R. Little

Department of Psychology
Carleton University
Ottawa, ON K1S5B6, Canada

1. INTRODUCTION

Our research program on personality and social ecology is based explicitly on a set of assumptive propositions about the nature of persons, the nature of contexts, and the transactional features of persons in context (Little, 1976, 1983, 1989; Little & Ryan, 1979). Our perspective shares many of the assumptions of other transactional or social ecological approaches (e.g. Altman & Rogoff, 1987; Bronfenbrenner, 1979; Endler, 1983; Moos, 1973; Wapner, 1987) but those commonalities will not be the primary concern of this chapter. Rather, I wish to emphasize the most distinctive feature of our approach: its emphasis on ways of *measuring* person-context relations This approach assumes the need for isomorphism between conceptual units of analysis and their measurement operations. This *methodological transactionalism* is the superordinate assumptive theme in our work and undergirds each of the other core themes to be discussed in this chapter.

Central to our perspective is the assumption that personal projects, extended sets of personally salient action, serve as "carrier" units for the study of the transactional processes of persons acting in context. A personal project is neither exclusively a person unit, nor a contextual unit; it is a "person-in-context" unit of analysis (Little, 1987; Wapner, 1981).

By way of overview, personal projects analysis (PPA) as a methodology comprises several assessment modules. Individuals first generate a listing of their own personal projects. They then appraise each project on a set of dimensions (typically between fifteen and twenty) chosen for their theoretical and applied importance (e.g., how

Theoretical Perspectives in Environment-Behavior Research, edited by Wapner *et al.*
Kluwer Academic / Plenum Publishers, New York, 2000.

enjoyable, stressful, and how much under their control each project is). Optional procedures examine the impact of projects on each other (as well as those of other people), and the hierarchical position of each project within the larger system.

Introductions to the methodological components of personal projects analysis have appeared elsewhere (Little, 1989, 1993, in press a, in press b) as have detailed treatments of the social ecological model (Little & Ryan, 1979). My goal in this chapter will be to *interrelate* these two aspects of our work by showing that, common to both the social ecological theoretical perspective and the methodology of personal projects analysis, are four major assumptive themes: constructivism, contextualism, conativism and consiliency. Consistent with the goals of this volume, I will show how these four core assumptions are linked with twelve measurement criteria and how these linkages have generated a novel set of theoretical, research and policy implications for environment-behavior research and for the study of human well-being.

2. ASSUMPTIONS

2.1. Constructivism: Reflexive, Personally Salient, and Evocative Assessment

Constructivism is the assumption that the individual's personal construal of self, context, and daily transactions is of fundamental importance, and that a comprehensive examination of person-environment transactions must include the systematic exploration of the individual's personal viewpoint. Although this assumption is consistent with phenomenological and narrative perspectives in environmental psychology (e.g., Sarbin, 1983; Seamon, 1987) the root source of our particular social ecological perspective is Kelly's (1955) personal construct theory (Little, 1972, 1983). Kelly emphasized the critical importance of understanding the distinctive "personal constructs" through which individuals view themselves and their contexts. Personal constructs are bi-polar templates or conceptual "goggles" through which individuals view their worlds. One of the distinctive contributions of Kelly was his development of a methodology, repertory grid technique, through which one can obtain actual samples of the personal constructs used by individuals (Kelly, 1955).

Kellians have long argued that there is an intimate link between the tacit assumptions undergirding a theory and the stances we take toward our research subjects. Kellians regard their research subjects as co-scientists who are actively trying to make sense of their world by erecting and testing hypotheses about it. This assumption, formally known as *reflexivity*, is also part of the assumptive structure of our research program on personality and social ecology. It proposes that the analytic units through which we explain the actions of our research subjects should be the same as those through which we explain our own conduct as scientists. Both constructs and projects are common elements in the pursuits of professional and "lay" scientists alike and are, in that sense, reflexive units of analysis.

One consequence of the reflexivity assumption is the adoption of what personal construct theorists call a credulous approach to assessment. This assumes that individuals are privileged sources of information about their transactional experiences. We thus approach our undergraduate student subjects as inquisitive co-investigators rather than as sleepy data points in baseball caps. If we are puzzled about how these subjects think about a particular situation, context or transition, the credulous approach enjoins us, in Allport's (1937) classic phrase to "ask them" (to which Kelly, 1955 adds "They might just tell us").

Our own adoption of the constructivist assumption and its corollary of credulousness is seen in the elicitation phase of personal projects analysis (PPA) (Little, 1983). In PPA our concern is with determining what individuals are currently doing in their lives. Such "doings" are transactional units of analysis affording us an image both of the person and of the context within which the action occurs. In lay terms we are asking "What are people up to?". In the project elicitation procedure, respondents generate a list of their personal projects which may be at the inception or planning stages, in full flight, or nearing completion. The listings we have received and stored over the years have been informative and often intriguing. They have ranged from quotidian routines (e.g., "pick up the newspaper", "put out the cat") to the prepossessing commitments of our lives (e.g., "Help my Mom cope with Alzheimers"; "figure out whether I still want my relationship with Suzanne to work"). The essential characteristic of this methodological reflection of a constructivist theme is that the unit of analysis with which subjects work is *personally salient* to them. The contrast to such a methodological approach would be the use of assessment techniques in which individuals are probed via questionnaire items that reflect the *investigator's* professionally salient constructs, rather than the respondents' personally salient ones. As we shall see, it is possible to conjoin both of these legitimate concerns in a transactional methodology.

An important consequence of the constructivist assumption, both in the study of personal constructs and of personal projects, is that it provides a distinctive solution to a problem that I believe has received insufficient attention. We might call this the "winnowing problem"—the need to deal with the potentially unmanageable volume of information that is generated if one takes a constructivist approach seriously. Consider, for example, the sheer volume of cognitions and acts in which individuals engage in their daily lives. To attempt a complete elicitation and cataloguing of these would be impossible. The question thus arises as to how to winnow appropriately for the purposes of understanding important aspects of human conduct. One approach to this is to allow the theoretical or applied interests of the researcher to dictate and delimit the types of cognitions or acts that are going to be studied. This is a time honored and appropriate approach, and the social ecological perspective that we have been developing can be adapted for such theoretical or applied research purposes. However, we believe that there is a more fundamental, propaedeutic task that needs to be carried out, at least in concert with, if not before, theory testing research. We need to know the content, frequency, and nature of the personally salient actions in which individuals are engaged, without respondents being specifically primed to elicit information in any particular domain. Our position is that until such research is carried out we will remain ignorant of the actual distribution of projects and activities in daily life and therefore have a truncated or distorted base upon which to construct an empirical science. Environmental psychologists will recognize this as essentially the same argument employed by Roger Barker in admonishing researchers to catalog exhaustively the activities going on in behavior settings, before proceeding to the logically subsequent task of discovering and explaining regularities by causal models (Barker, 1968).

The final implication of basing our methodology on a set of constructivist assumptions is that it becomes possible to make the assessment experience itself something that is *personally evocative* for the participant in our research. One of the heartening aspects of adopting methodological tools such as personal projects analysis is that participants frequently report that the assessment experience is evocative and provides opportunities for personal reflection (see, for example, Omodei & Wearing, 1990). Relatedly, I have recently suggested that psychologists be encouraged to explore the considerable potential of multimedia technology in their assessment methodologies, so that the coming together of scientist and subject or professional and patient can actually be an engaging aesthetic experience. (Little, in press c).

2.2. Contextualism: Ecologically Representative, Temporal, and Social Indicator Assessment

The environmental counterpart to our assumption of constructivism is contextualism—the assumption that human conduct is explicable only if the contexts within which it is embedded are systematically explored. These contexts range in scale from the microecology of action to the macroecology and historical epochs within which much of our behavior can be understood. The methodological implication of this is that our assessment units must be *ecologically representative*. By this we mean that they should bring into focus those aspects of their context that shape, facilitate and frustrate the daily lives of those being assessed. A number of methodological implications flow from this assumption. It entails the elicitation of information about the personal contexts of daily lives as well as of the objective environments, a distinction captured decades ago in Murray's distinction between alpha and beta press (Murray, 1938). These personal contexts may comprise both social and physical environmental features. We solicit information relevant to our respondents' personal contexts in several different ways. First, the content of the projects frequently is explicitly concerned with context relevant material (e.g. "Prepare the garden for fall", "clean up after the ice-storm"). Second, we use appraisal dimensions that are contextually relevant (e.g. "To what extent does your work environment frustrate or facilitate each of your projects?" (Phillips, Little, & Goodine, 1996). Third, we have used "open columns" in which respondents tell us "With Whom" and "Where" their projects are being undertaken (Little, 1983; Little, Pychyl, & Gordon, 1986).

Another aspect of our contextualist assumption is that our units of analysis need to be sensitive to the *temporal* aspects of behavior. Personal projects are *extended* sets of personally salient conduct and their dynamic nature means that we can examine the stages through which they progress as they move from the nascent thought of a course of action to reflection on its success or anguish over its failure. This feature of our methodology contrasts most clearly again with trait units, which are conceived of as relatively static features of individuals (e.g. Costa & McCrae, 1994). The psychometric implications of these differences are notable. High test-retest reliabilities of the indices derived from the assessment of dynamic projects could, in rapidly changing systems, actually be evidence of measurement *invalidity*. As it turns out, many of the features of project systems such as their overall stressfulness, meaningfulness, etc. have surprisingly high levels of stability. But this is simply an observed empirical fact, not a psychometric requirement (Little, in press b).

A third aspect of the contextualist assumptive theme is that the assessment of persons in context can supply important information about the social ecologies inhabited by people as well as information about the individuals themselves. In traditional personality research, for example, the primary goal of assessment is the valid ascription of psychological predicates to people (e.g., Joachim has a high internal locus of control; Kimiko is stressed). Quite literally, after assessment is completed and an individual's normative score is calculated, the items on the test can be discarded. In contrast with this "wastebasket" approach to the measured units of analysis, our own approach, derived from the contextualist assumption theme, is that the same unit of analysis used for personal ascriptions can also be saved and stored as potential *social indicators*. For example we have stored in our SEAbank (Social Ecological Assessment Data Bank) literally thousands of personal projects and their appraisals on approximately twenty dimensions (e.g., stressfulness, control, enjoyment, efficacy) as perceived by the project pursuer. We also keep important demographic information such as age, gender, and place of residence. We are thus able to ask social ecological questions such as the following: what are the most stressful personal projects engaged in by teenagers in urban centers in Ontario? How do these compare with those of the elderly, or those in other countries? In what kind of work environments are individuals likely to experience a sense of efficacy in their project pursuit. In which do they seem to be constantly frustrated? In short, the contextualist assumption guides us to use the context not simply as a way of refining our understanding of individuals, but as a substantive, policy relevant domain of substantial importance in its own right.

2.3. Conativity: Systemic, Middle-Level, and Modular Measurement

A third assumptive theme of our research program is that *conative* processes (trying, striving, seeking after, etc.) offer a particularly effective vantage point from which to view the dynamics of persons in context. This assumption contrasts with purely cognitive perspectives, as well as with those that adopt more behavioral or affective lenses through which to view person-environment transactions. As units of analysis personal projects are explicitly conative: they are volitional undertakings, the meaningful pursuits to which individuals are committed.

Three features of our methodological approach reflect the assumptive theme of conativity.

First, one of the most challenging aspects of daily living arises from the fact that we are typically managing not one, but a whole set of projects. Were these merely plans or cognitive expectations this would pose no particular difficulties, other than tradeoffs in one's attentional economy. But the conative nature of projects, the fact that they are volitional pursuits and commitments means that they form personal action systems with the potential for temporal, social, and ethical conflicts. This assumption requires that our methodology be designed to facilitate *systemic measurement*. We achieve this by having individuals directly rate the degree of impact of projects upon each other, using a cross-impact matrix. This allows us to calculate, within the single case, the degree of coherence or conflict among the personal projects. This matrix can also be expanded to look at similar relationship between the project systems of two or more individuals.

Second, personal projects are *middle-level* units of analysis (Little, 1987). By this we mean that personal projects fall in an intermediate zone between specific behavioral acts and higher order values and aspirations. One of the important features of using a middle level unit of analysis is that it allows us to access both higher and lower constructs in the action hierarchy by the use of various probes (see Little, 1983 for details on "laddering" procedures). Taken in conjunction with the systemic measurement criterion, personal projects methodology allows us to examine, within the single case, the hierarchical nature of project systems. For example, one project may be linked with all other projects in a given project system, such that if it runs into difficulties or loses meaning for the individual, the rest of the project system is very much at risk. We call these *core personal projects* and we hypothesize that they are central to the overall sense of coherence experienced by individuals. Much of our current research is examining how human well-being is influenced by the successful and balanced pursuit of these core personal projects (Little, in press; McGregor & Little, 1998).

Third, in some assessment methodologies the measurement units, such as items on trait inventories, are fixed features that cannot be changed with impunity. Despite the protestations of the occasional obsessive-compulsive respondent, items on inventories cannot be changed, modified, shaded, or rewritten. To do so would vitiate the whole measurement philosophy that requires complete standardization in the measurement instrument. Our approach to conative assessment is based on a very different assumption: that the measurement system, much like a motherboard in computers, should be *modular*. This means that the various elements in the methodology are flexible, can be removed, modified or added to as required by the particular research question being explored. This feature of our methodology reflects the fact that personal project systems are not closed systems but open systems and more specifically, *personal* systems (Little, 1972, 1976). This means that while there may be relatively stable features of a person's project system, such systems are frequently in flux, in transition, or on occasion, in utter chaos. A well articulated, hierarchically integrated, coherent personal project system, however much desired, is unlikely to be experienced over long periods by most people. Exceptions might be found, however, with megalomaniacs, hermits or psychopaths, for whom the single minded pursuit of projects is enabled by their lack of awareness or concern for the projects of others. But in the muddling through of middle class lives, most of us have project systems that are vulnerable to the macro-level exigencies of economic downturns, the meso-level pressure of moving house, yet again, or the micro-level annoyances of incontinent pets or malfunctioning modems. And even if the surrounding context is relatively benign and stable, the pursuit of our projects can be brought to a dead stop by the caustic voice of self-recrimination or the inexplicable ache of incipient depression.

The modular nature of personal projects methodology can be illustrated by the use of "*ad hoc*" dimensions which are added to the standard PPA matrix. Typically, as alluded to earlier, respondents rate their personal projects on a set of approximately twenty appraisal dimensions that reflect theoretically important constructs such as how much enjoyment, control, and sense of efficacy people have with respect to their projects. But researchers have frequently created appraisal dimensions designed to capture some distinctive aspect of the social ecology of their respondents. We have asked Indo-Chinese refugees about the extent to which their personal projects require the extensive use of English, and senior managers in the public and private sectors about the impact of their organization's climate on their projects. Yetim (1993) has created dimensions examining the impact of economic factors on project pursuit, and Omodei and Wearing (1990) showed that well-being was closely linked with the extent to which personal projects were rated as satisfying the basic needs espoused in the motivational psychology of Murray (1938). Neil Chambers, in our Social Ecology Laboratory, has recently completed a richly documented and annotated Compendium of literally hundreds of special dimensions that have been used in PPA methodology over the past two decades (Chambers, 1997).

It should also be mentioned that not only can project appraisal dimensions be added or deleted from PPA methods, but projects themselves can be "added" to a person's matrix. For example, in study of eating disorders, all subjects were "supplied" with the project of "controlling my weight". The appraisal ratings on this project alone were sufficient to significantly discriminate between eating disorder patients and a "weight-pre-occupied" control group, as well as between anorexic and bulimic patients (Goodine & Little, 1986).

2.4. Consiliency: Conjoint, Integrative, and Directly Applicable Measurement

The final assumptive theme undergirding our research program is that integrative and transdisciplinary analytic units and methods are required in order to address the complexities of human conduct. Wilson (1998) has recently invoked the term "consiliency" (literally, a "jumping together") to characterize the vital need for linkage between the humanities and the life sciences. I believe the term is also appropriate for depicting a set of methodological issues in environment-behavior research.

The first methodological issue relates to the need for *conjoint* individual and normative levels of measurement. There has been a long standing debate between researchers over the status and significance of single case studies in behavioral research, in which those favoring idiographic studies are pitted against those concerned with normative or nomothetic analysis.

Since the outset of our research program we have emphasized the need for conjoint measurement at both levels of analysis, and the statistical properties of PPA afford us this opportunity. For example, it is possible to examine the relationship between the stressfulness and sense of control over one's personal projects at the individual level by correlating ratings on these two dimensions across the personal projects for a single person. But it is also possible to take the average stress and control ratings (across projects) for each individual and run these correlations at the group level. Depending upon the goals of the assessment, for example whether the main focus is for clinical or for institutional purposes, either or both of these levels of analysis can be adopted. The ability to carry out such conjoint analysis enables us to address some important psychometric issues which have been of concern to statisticians and methodologists. For example, Simpson's paradox in mathematical statistics holds that the relationships between a set of variables measured at one level of analysis are not mathematically constrained to hold at other levels of analysis. To assume that there is isomorphism between levels can give rise to a number of inferential fallacies, the most famous of which is the "ecological fallacy" in which it is assumed that normatively measured relationships will hold for individual cases. But the fallacy can also work in the reverse direction. Richards (1990) has demonstrated a counterpart fallacy in the environmental psychology literature which he refers to as the "individual difference fallacy". This fallacy assumes that the reliability of assessments of environmental climate, which are measured at the level of the individual, are isomorphic with the reliability of *aggregated* climate scores. In our personal projects methodology the issue of isomorphism between relationships at the two levels is approached the same way as we do the issue of reliability: it is a matter of open empirical inquiry rather than a psychometric constraint.

Indeed, Travis Gee has recently completed an intensive examination of the possible operation of Simpson's paradox in a comprehensive analysis of individual level versus normative level measurement of personal project systems (Gee, 1998). Using multidimensional scaling analysis, he found clear and striking evidence that the underlying nature of the personal project space for normative level analysis is strongly iso-morphic with that obtained at the individual level, suggesting that there may be mutual informative transfer of research findings between case studies and the more traditional normative inquiry in projects analysis.

A second consiliency goal in our research is the requirement that our methods be *integrative*. By this we mean that assessment of persons in context not be restricted to one domain, for example, the cognitive or behavioral or affective domains, but allows each of these aspects of human action to be incorporated into the assessment proce-dure. For example, personal projects are clearly cognitive phenomena in the sense that they represent plans or goals that can be examined through the lens of classical cog-nitive theory. But they have equal claim to providing information about the affective experiences and behavioral undertakings of individuals. Thus we are increasingly con-cerned with asking individuals about the extent to which their ongoing projects are a source of passion for them (e.g., Phillips, Little, & Goodine, 1996; Pychyl, 1995), and we are beginning to explore the social or behavioral impact of personal projects in terms of the extent to which they might create "social capital" for their ecosystems. The central point of consiliency, then, is that we are able to access cognitive, affective, and behavioral aspects of the lives of individuals not by the use of *separate* methods from each sector, but through a single consilient assessment methodology that is explicitly designed to afford us an integrative view of people in action.

Finally, consiliency also characterizes the applied aspects of our social ecological perspective. Once again, the critical role of the unit of analysis arises. Many of the ana-lytic units currently influential in both personality and environmental psychology have only an oblique relevancy to attempts to change or ameliorate the conditions of people's lives. For example, neither traits nor macro-level economic factors influencing human well-being can be easily changed. Personal projects, on the other hand, allow for *direct applicability* in clinical, counseling, organizational change or community development interventions; in essence they afford tractability for change attempts. The implications of whether a unit has such interventional utility are considerable. For example, a recent review of the genetic base of the disposition to be happy concludes "it may be that trying to be happier is as futile as trying to be taller and therefore is counterproductive" (Lykken & Tellegen, 1996). But it can also be shown that appraisals of personal projects serve as excellent proxy indicators for overall life satisfaction (Omodei & Wearing, 1990) and our own research has shown that increases in the sense of meaning, structure, efficacy, and support, and lowering the level of stress of personal project systems has a salutary effect on individuals (Little, 1989). By dealing with person-context transactional units that provide at least some tractability for change we may not be able to make people happy, but at least we may help them be happier.

3. CONCLUSIONS

Environmental-behavior research is only as secure as the tacit pre-suppositions undergirding it. I have presented the case that the methodological tools adopted in the field is a crucial factor in the viability of the research enterprise. Our social ecological perspective emphasizes that an assessment methodology needs to be constructivist, contextualist, conativist and consilient. These high C's are the assumptive themes that have generated twelve specific criteria for methodological development and they will continue to animate our explorations of persons-in-context.

REFERENCES

Allport, G. W. (1937). *Personality: A psychological interpretation*. New York: Holt.

Altman, I., & Rogoff, B. (1987). World views in psychology: Trait, interactional, organismic, and transactional perspectives. In D. Stokols, & I. Altman (Eds.). *Handbook of environmental psychology* (pp. 7–40). New York: Wiley.

Barker, R. G. (1968). *Ecological psychology: Concepts and methods for studying the environment of human behavior.* Stanford, CA: Stanford University Press.

Bronfenbrenner, U. (1979). *The ecology of human development.* Cambridge, MA: Harvard University Press.

Cantor, N. (1990). From thought to behavior. "Having" and "doing" in the study of in personalty and cognition. *American Psychologist, 45,* 735–750.

Chambers, N. (1997) Personal Projects Analysis: The Maturation of a Multi-dimensional Methodology. Unpublished manuscript, Carleton University, Ottawa, Canada.

Costa, P. T., & McCrae, R. R. (1994). Set like plaster? Evidence for the stability of adult personality. In T. Heatherton, & J. L. Weinberger (Eds.), *Can personality change?* (pp. 21–40). Washington, DC: American Psychological Association.

Emmons, R. A. (1986). Personal strivings: An approach to personality and subjective well-being. *Journal of Personality and Social Psychology, 51,* 1058–1068.

Endler, N. S. (1983). Interactionism: A personality model, but not yet a theory. In M. M. Page (Ed.), *Personality: Current theory and research* (pp. 155–200). 1982 Nebraska Symposium on Motivation. Lincoln, NB: University of Nebraska Press.

Gee, T. (1998). *Individual and joint-level properties of personal project matrices: An exploration of the nature of project spaces.* Unpublished Ph.D. Dissertation. Department of Psychology, Carleton University, Ottawa, ON.

Goodine, L. A., & Little, B. R. (April, 1986). *Personal projects analysis: A contextual approach to measurement in anorexia and bulimia.* International Association for Eating Disorders, New York, NY.

Kelly, G. A. (1955). *The psychology of personal constructs.* New York: Norton.

Little, B. R. (1972). Psychological man as scientist, humanist, and specialist. *Journal of Experimental Research in Personality, 6,* 95–118.

Little, B. R. (1976). Specialization and the varieties of environmental experience: Empirical studies within the personality paradigm. In S. Wapner, S. B. Cohen, & B. Kaplan (Eds.), *Experiencing the environment* (pp. 81–116). New York: Plenum.

Little, B. R. (1983). Personal projects: A rationale and method for investigation. *Environment and Behaviour, 15*(3), 273–309.

Little, B. R. (1987c). Personality and the environment. In D. Stokols, & I. Altman (Eds.), *Handbook of environmental psychology.* New York: Wiley.

Little, B. R. (1989). Personal projects analysis: Trivial pursuits, magnificent obsessions, and the search for coherence. In D. Buss, & N. Cantor (Eds.). *Personality psychology: Recent trends and emerging directions* (pp. 15–31). New York: Springer-Verlag.

Little, B. R. (1993). Personal projects and the distributed self: Aspects of a conative psychology. In J. Suls (Ed), *Psychological perspectives on the self.* (Vol. 4, pp. 157–181). Hillsdale, NJ: Erlbaum.

Little, B. R. (1998). Personal project pursuit: Dimensions and dynamics of personal meaning. In P. T. P. Wong, & P. S. Fry (Eds.), *The human quest for meaning: Theory, research, and clinical application.* Mahwah, NJ: Erlbaum.

Little, B. R. (in press a). Free traits and personal contexts: Expanding a social ecological model of well-being. In W. B. Walsh, K. H. Craik, & R. Price (Eds.). *New directions in person-environment psychology* (2nd Ed.). Mahwah, NJ: Erlbaum.

Little, B. R. (in press b). Personality and motivation. In L. Pervin, & O. John (Eds.), *Handbook of personality theory and research* (2nd edition). New York: Guilford.

Little, B. R. (in press c). Personality psychology: Havings, beings, and doings in context. In S. Davis, & J. Halonen (Eds.), *The different faces of psychology in the twenty first century.* Washington, DC: American Psychological Association.

Little, B. R., Pychyl, T., & Gordon, C. (May, 1986). *What's a place like this doing in a project like me?* Eastern Psychological Association: New York, NY.

Little, B. R., & Ryan, T. J. (1979). A social ecological model of development. In K. Ishwaran (Ed.), *Childhood and adolescence in Canada* (pp. 273–301). Toronto: McGraw-Hill Ryerson.

Lykken, D., & Tellegen, A. (1996). Happiness is a stochastic phenomenon. *Psychological Science, 7,* 186–189.

McGregor I., & Little, B. R. (1998). Personal projects, happiness and meaning: On doing well and being your-self. *Journal of Personality and Social Psychology*, ••.

Moos, R. H. (1973). Conceptualizations of human environments. *American Psychologist, 28*, 652–665.

Murray, H. A. (1938). *Explorations in personality*. New York: Oxford University Press.

Omodei, M. M., & Wearing, A. J. (1990). Need satisfaction and involvement in personal projects: Toward an integrative model of subjective well-being. *Journal of Personality and Social Psychology, 59*(4), 762–769.

Phillips, S. D., Little, B. R., & Goodine, L. A. (1996). *Organizational climate and personal projects: Gender differences in the public service*. Ottawa, ON: Canadian Centre for Management Development.

Pychyl, T. A. (1995). *Personal projects and the lives of doctoral students*. Unpublished Ph.D. Dissertation, Department of Psychology, Carleton University, Ottawa, ON.

Richards, J. M. (1990). Units of analysis and the individual difference fallacy in environmental assessment. *Environment and Behavior, 22*, 307–319.

Sarbin, T. R. (1983). Place identity as a component of self: an addendum. *Journal of Environmental Psychology, 3*, 337–342.

Seamon, D. (1987). Phenomenology and the environment. *Journal of Environmental Psychology, 7*, 367–377.

Wapner, S. (1981). Transactions of person-in-environments: Some critical transitions. *Journal of Environmental Psychology, 1*, 223–239.

Wapner, S. (1987). A holistic, developmental, systems-oriented environmental psychology: Some beginnings. In D. Stokols, & I. Altman (Eds.). *Handbook of environmental psychology* (pp. 1433–1465). NY: Wiley.

Wilson, E. O. (1998). *Consilience: The unity of knowledge*. New York: Knopf.

Yetim, U. (1993). Life satisfaction: a study based on the organization of personal projects. *Social Indicators Research, 29*, 277–289.

9

WOMEN AND THE ENVIRONMENT

Questioned and Unquestioned Assumptions

Arza Churchman

Faculty of Architecture and Town Planning
Technion-Israel Institute of Technology
Haifa Israel

1. INTRODUCTION

The purpose of this paper is to analyze questioned and unquestioned assumptions about the topic of women and the environment specifically, but within the larger context of the questioned and unquestioned assumptions about science, research, and practice from a feminist perspective. Thus, before I present my justification for discussing the topic of women and the environment, I will first relate to the broader and more theoretical issues.

1.1. General Science Issues

We are living at a time when many underlying assumptions of science in general, and of psychology in particular, are being questioned. For example:

- the assumption that scientific theories are built up from a foundation of secure, unquestionable, objective, and theory- neutral observations (Phillips, 1987). The argument is that all facts of science are theory laden, and that all theories in the behavioral sciences make assumptions about people that are most often not explicit. The scientific method itself is seen as a speculative guess about how one gains credible knowledge, because method is also a type of theory (Slife & Williams, 1995).[1]

---

[1] They examine and question the assumptions of different theories-determinism, reductionism, empiricism, structuralism, post modernism, etc.

Theoretical Perspectives in Environment-Behavior Research, edited by Wapner *et al.*
Kluwer Academic / Plenum Publishers, New York, 2000.

- There is much talk of paradigm shifts, and increased visibility of research designs that are interactive, contextualized, and humanly compelling, because they invite joint participation in the exploration of research issues (Lather, 1987).
- Assumptions about human learning ability, seen as a process that can be isolated from its volitional and emotional components, or as a relatively passive process of exposure and acquisition (Salomon, 1996).
- There is the argument (Gergen, 1994) that psychological theory has undergone a major transformation from a behavioral to a cognitive base, but that there has been no transformation at the level of metatheory or methodology.

1.2. Feminist[2] Theory and Research

One of the major challenges to the assumptions of science has come from the work of feminist theorists and researchers. Feminists have argued that history has been male-defined, predicated on the assumption that the male experience is universal (for example, 'universal suffrage' was meant for men only), that it fairly represents all of humanity, and that it constitutes an adequate basis for generalizing about all human beings. James (1997) calls for the rejection of false universalism, and of the tyranny of averages.

Worell (1996) argues that 25 years ago, psychology as a discipline reflected a male model of reality. Researchers and populations studied were disproportionately male, topics studied related to stereotyped male concerns, such as aggression and achievement, and results from male samples were typically generalized to women too. (The same criticism can be made of medicine, and probably other fields).

Rakover (1992) argues, in return, that the research on sex differences that is criticized for being influenced by male stereotypes and prejudices which distorted the choice of research questions, the hypotheses, the research methodology, data gathering, and interpretation- was simply bad science, and that this can be corrected without changing the whole manner in which research is conceptualized and conducted.

The guiding themes of feminist research are, according to Worell:

1. To challenge traditional scientific inquiry- make values explicit because science is never value free; to identify and monitor bias in research procedures. Sandercock and Forsyth (1990) also challenge the assumption in planning that the intentions and outcomes of planning are gender-neutral, are experienced in the same way by women and men, and affected in the same way by them.
2. To focus on the experiences and lives of women.
3. To consider asymmetrical power arrangements- It views subordinate status via power, not deficiency; sees differences among women as mediated by power. Spain (1992) declares that throughout history and across cultures, architectural and geographic spatial arrangements have reinforced status differences between women and men. Segregation reduces women's access to knowledge and thereby reinforces their lower status.
4. To recognize gender as an essential category of analysis, and consider gender

---

[2] I particularly like this definition of feminism: Feminism is the radical notion that women are people.

as socially constructed. I will shortly discuss more fully what the concept of gender means and implies.

5. To attend to language and the power to 'name'- asserts that language frames thought; restructures language to be unbiased and inclusive. Here too I will expand somewhat a little later on.
6. To promote social activism and societal change.

A feminist theoretical perspective tries to explain why people are gendered in particular ways, what that means for their ability to live full and productive lives, and what practical strategies are needed to overcome oppressive roles and relationships (Sandercock & Forsyth, 1990).

1.3. Parallels between Environment-Behavior Studies and Feminist Research Approaches

There are many interesting parallels between the field of Environment-Behavior Studies and feminist research approaches. I will briefly mention a few of them:

1. A social constructionist perspective- in which humans are seen as actively engaged in their perceptions of the world, and thus 'construct' their view of the world. Beall (1993) says that culture provides a set of lenses through which people can observe and understand their environment, and that gender is such a socially constructed category. Ahrentzen (1992) accepts the approach that an individual's subjective knowledge of the world is derived not only from her own lived experience, but also from the shared realm of ideas known as culture, that are often institutionalized in the political and economic system.

   Schneekloth (1994) criticizes our use of the term users, as implying consumption, passivity, and homogeneity, rather than describing active, producing and varied individuals in groups.
2. The importance of freedom of choice in the lives of individuals.
3. Contextualism, as elaborated by Stokols (1987). McDowell (1993b), from a feminist position, writes that we must get away from generalizations about women as an undifferentiated category, towards more particular understandings of the historically specific processes that produce the particular range of gender relations in a range of places. Women's lives are characterized by interdependence to such an extent that one is forced to consider their responsibilities to and for others as significant and critical elements in their environmental transactions. Women are clearly embedded in and inseparable from their physical, social, and temporal contexts, thereby reflecting the complex, holistic, and transactional quality of women's lives (Churchman & Altman, 1994).
4. The transactional approach (Altman & Rogoff, 1987)- which is the study of changing relations among environmental aspects of holistic entities, and is characterized by the following principles:

   - the whole is composed of inseparable aspects that simultaneously and conjointly define the whole
   - incorporate temporal processes
   - start with the event as the fundamental unit of study
   - understand the perspectives of the participants in the event. Feminists, in

parallel, call for situated and embodied knowledges that acknowledge the partial vision of everyone's seeing and knowing (Schneekloth, 1994).
– understand the observer as an aspect of the event
– study process and change
– accept relativity of indicators and measures of psychological functioning
– emphasize methodological eclecticism

These are all principles that exist in feminist research, even if not stated exactly in these words. What the feminist research adds is the political-ideological level, which is basically absent from the Environment-Behavior field. The way we think about context could be extended to include this, but on the whole, it isn't at the moment.

Riger (1998) argues that a focus on the immediate social situation may overlook the larger social system, ignoring economic, political or historical forces that shape women's and men's behavior. She argues that the ahistorical nature of social psychology relates to behaviors that are the product of contemporaneous conditions as if they were universal, timeless principles of human behavior.

Despres (1991) found that most studies on the meaning of home overlook the impact of structural, societal and formal forces on the individual's perceptions, judgments, behaviors, and experiences of and about the home.

Schneekloth (1994) tried to link the two fields by arguing that the basis for the link exists in the historical fact that Environment Behavior Studies started as a utopian project. However she sees us as having been captured by the scientific trap, and thus we did not do this utopian work with what she terms as the critical self-reflection that acknowledges our locations within the culture, that criticizes our ways of working and that attempts to be passionate and inclusive. If we discuss women as a special group, this implies that men are the norm.

We have argued that the distinction between private and public spheres that feminists have identified as problematic and as a political and value-laden act, is untenable within a transactional approach, because individuals and groups function in contexts that are embedded in and inseparable from all larger contexts relevant to their lives (Churchman & Altman, 1994). Once we understand that women cannot be detached from their social, cultural, political, and physical environmental contexts, then it becomes clear that changes that take place only within the individual are not sufficient. Change must be eventually introduced throughout all levels of systems, including the physical environment (Churchman & Altman, 1994).

2. WHAT IS GENDER?

2.1. Gender Defined

Gender is defined as a complex set of principles that organize male-female relations in a particular culture or social group, and a marker of hierarchy that, in concert with others, determines relations of power (Maracek, 1995). Gender is identified by differential power, and associated with prescribed roles, and with implicit and explicit meanings that cultures provide the necessary conditions for learning and maintaining. Beliefs about gender are learned early, subscribed to widely, and difficult to change because they are reinforced by social consensus, by structural arrangements that support and demand them, and by the operation of self-fulfilling prophecies (Lott,

1997). The human attributes differentially associated with gender systematically corre-
spond to those generally regarded as prerequisites for occupying positions of power
and status in society (Berscheid, 1993).

Gender is the earliest, most central, and most organizing component of everyone's self concept (Geis, 1993). Gender is not an external overlay for the individual. It
is an essential, defining aspect of the person, with far-reaching implications for that
person's transactions with the environment (Churchman & Altman, 1994).

Gender roles are culturally defined sets of behavior differentiated by gender.
Gender role ideology is the introduction of value judgments about the roles men and
women should occupy, or traits they should differentially posses (Gibbons, Hamby, &
Dennis, 1997). Feminists have long argued that this ideology is harmful to women, and
now men are beginning to realize that this ideology can be detrimental to them also.
Levant (1996) reports that men's studies scholars have begun to examine masculinity
not as a normative referent, but as a complex and problematic construct.

The argument is that women and men, or girls and boys, may at present act differently and may exhibit different abilities, but they do so because they are differently
situated, and because they are significantly affected by different attitudes, rules, and
expectancies. However, this does not mean that they are incapable of acting otherwise,
or that intrinsic differences in abilities necessarily determine men and women's place
in society (Riger, 1998; Saegert & Hart, 1978). While one might argue that this cannot
be "proven" in a traditional sense; the counter argument is that the other position can
similarly not be "proven".

Feminists question the autonomous masculine identity, with its emphasis on independence, self government, self reliance, assertiveness, selfishness, and separateness, as
the universal model of adulthood and humanness. Chodorow argues that these are
based on male children's experiences of separation from the mother, and that other
models, such as caretaking and incorporation could be a different kind of autonomy
(Yanay & Freedman, 1995).

2.2. Gender as a Variable

Can we assume that gender is always *the* relevant variable for women? No. All
of us have multiple identities and reference groups and common interests of various
kinds. Forester (1992) argues that privileging any one facet of identity, including gender,
can have its problems. Of course we could argue back that the saliency of gender in
women's lives in not necessarily a matter of choice. Yeandle (1996) says that both those
experiences and characteristics which women share, and those which differentiate them
require analysis and explanation.

Indeed women and men share many attributes, qualities, and characteristics. These
include basic biological and physiological characteristics, basic psychological processes,
such as perception, cognition, and development, and within a given context and to a
certain extent, a common culture, and a common physical environment. Furthermore,
at any given time in any given situation, various factors may, by choice or by force of
circumstance, link women to men who share a given feature more than to women who
do not.

The factors that can distinguish between women are: socio-economic status,
culture, age, stage in the life cycle, family status, family size, health, talent, personality,
values, ideology, danger, etc. Particular women may feel that they have more in common
with men with whom they share one or more of these characteristics, than with women

of another status, for example life style, or health, or ideology. These are personal, subjective, and sometimes situational choices.

Related to this, there is criticism from within and without the feminist field, of the hegemony of a particular kind of woman in feminism and an ignoring of variables such as race, class, and so forth. Hurtado (1997), for example, criticizes that fact that within psychology, gender has been studied independent of other group memberships. James (1997) argues that categories like sex, race, and class are too reductive to describe the complexity of social identities. Successive UN conferences on women and population have revealed as many differences as similarities among women in 1st and 3rd world countries; such that a readily shared agenda does not emerge automatically when women get together (Scott, 1996).

One counter argument is that within any class, women are less advantaged than men (Lengermann & Niebrugge-Brantley, 1990). The feminist movement (and others) argues that the personal is political, that so long as problems are seen as personal and private, and not shared by a group, they are invisible and thus not solved; we must shift the boundaries of public and private (Schneekloth, 1994). Problems must be seen as visible societal problems in order to be dealt with (Churchman & Ginsberg, 1991).

The fact is, however, that in spite of this complexity, there are factors that are common to all women These are: 1. Those biological systems that are gender-differentiated. 2. The ability to become pregnant and bear children. 3. The gender-defined norms of behavior. 4. Social definitions of the role of women as linked to home and children. 4. Social attitudes towards women as a group- which are often expressed as discrimination. 5. The possibilities for violence against them.

With regard to point 4, one very basic difference between men and women in the stage in the life cycle where there are school-age children, is in the degree of their responsibility for children and the home. Regardless of whether the woman is or is not gainfully employed, she is the one who assumes this responsibility almost completely by herself, and this is generally accepted as obvious and 'natural'. All of the recent studies in the Western world have found that, even though women are sharing more and more in the responsibilities for supporting the family, there has been no real change in their sole responsibility for the care of the home and the children. She may put in less time in housework than previously, though there is even some debate about this, and he may put in a little more time. However, the gap is still huge, and the responsibility is still hers (Churchman, 1993).

A number of writers argue, that one of the most significant impediments to equality for women in the public world, is the failure of men to assume equal responsibility in the private world of the family (Lott, 1997; Silverstein, 1996). It is interesting and telling, that early medical studies assumed that men's risk factors for coronary heart disease only related to their work lives; non-work and family life were seen as irrelevant. Whereas for women, family events were seen as critical factors (Barnett, 1997). Phares (1996) points out that by not including fathers in research, as I too have been guilty of doing, we are communicating an implicit message about the parent who is most integral to the child's functioning. Why are there so few studies on the impact of active fathering on child development, and why has the effect of paternal employment and absence on children been ignored (Silverstein, 1996). Harding (1986) states that until the emotional labor and the intellectual and manual labor of housework and child care are perceived as desirable human activities for all men, the intellectual and manual labor of science and public life will not be perceived as potentially desirable activities for all women.

2.3. Problems of Language

Earlier in this paper, I alluded to the recognition of the importance of language within feminist research. Let us take for example the symbolic aspect of the name changes that our field has undergone. Initially it was called man-environment studies, on the common assumption that the term man includes everyone. However, we now recognize that language is a particular set of symbols that defines and constructs reality (Schneider & Hacker, 1973), and that it may be taken as a diagnostic of our hidden feelings about things (Lakoff, 1975). We now call the field Person-Environment Studies or Environment-Behavior Studies, in order to be sure that everyone feels included. We no longer talk of chairmen.

Unfortunately, there are some languages in which it is much more difficult to be so inclusive, or gender- neutral. Unlike English, where it is possible to talk in ways that do not define a particular gender, it is impossible in other languages- for example in French and Hebrew. The value-laden issues here are very problematic- witness the present debate in France about whether to use a feminine form for the word minister when referring to a woman who is in that position. I have been fighting the use of the term father-house in Hebrew to refer to a household, or the word for husband that also means owner.

Recently, I read that the English word family comes from the Latin and there the word meant female servant (Lengermann & Niebrugge-Brantley, 1990). I asked some colleagues for the root of the word family in their language. So far I have discovered that in Italian and Spanish the root is also the Latin one, and in Spanish the word familia means an ensemble of servants living at home. The Arabic word has related words that refer to the people that *he* has to support. The Hebrew word for family comes from the same root as the word for maidservant. In Dutch the word familie refers to relatives, not core family. The word for the core family is used for persons you like or fancy. The Japanese word is made up of two Chinese characters, the first that of a dwelling, and the second of relatives, so that it refers to kinship under one roof.

The politics of naming arises also with regard to the question of whether one should talk of women's studies or gender studies. In IAPS-the International Association for Person-Environment Studies two years ago we discussed the question of whether to call our network women and the environment or gender and the environment. We finally decided to call it women and the environment, although the decision is not a clear-cut one. I think the point of calling it women and the environment, as I did this paper, is to explicitly state that we are dealing with women and their lives and situations. Gender suggests that one should also talk about men and their lives and situations. That is certainly a legitimate task, but not one that I wish to talk about, or that there is research that enables one to talk about it. Lees (1998) in a review of a book on gender and technology comments that the editors of the book have not allowed a mutation of women's studies into gender studies to obscure the necessary exploration of that which is particular to women's lives.

2.4. Assumptions: Justifiable and Unjustifiable

In this section, I wish to summarize some of the assumptions related to the topic of women and the environment, that in my view are either justifiable or unjustifiable.

The very basic assumptions that we will hopefully all agree to are:

1. That people are different and have different needs; that one cannot specify, identify, or posit one model of the person-environment relationship.
2. That people are active and not passive.
3. That aggregates can be identified who have some needs and characteristics in common.
4. That the physical environment has implications for people's lives.
5. That the purpose of the environment is to afford opportunities for achieving one's own definition of quality of life. The term "quality of life" is used to denote the subjective judgment by an individual as to the degree to which her or his needs in the various domains of life are met. These domains include the degree of self actualization, health, family life, social relations, dwelling place, work situation, services, income level, security, environmental quality, social justice and equality, etc. (Churchman, 1993).
6. That just about everything is context dependent in one way or another (Rosnow & Rosenthal, 1989).

Planning cannot relate specifically to the particular needs and expectations of every individual. However, the opposite policy, too often adopted, of considering the population to consist of an 'average' person or of a very few types, is equally untenable. One cannot talk of The quality of life. One must talk of Qualities of lives, differentiating between groupings of individuals who have characteristics in common that have significance for their environmental and settlement needs.

On the other hand the problematic assumptions that exist are:

1. that science and research are value free, or can be value free.
2. that the results of research can be considered universal, or what is defined as an imposed etic- a phenomenon, concept or construct assumed to be universal without evidence (Gibbons, Hamby, & Dennis, 1997). An example would be when studies on cognition or perception are done with American college students, and the results taken as describing a universal, human phenomenon.
3. the notion of the universal, or the generic human, or the unspecified subject as applying to women as well as to men.
4. that users are a unitary, undifferentiated group- Environment-Behavior Studies theoretically knows this, but in practice it has more often than not ignored all sorts of groups, including women. The point is that the decision on which groups to differentiate is subjective and value-laden.
5. that various phenomena or concepts mean the same thing in different people's lives- leisure, work, home, family, etc.
6. that there is a single public good, a single public interest.
7. that we can talk about the 'nature' of women and men.
8. that the functional asymmetry in marital roles is attributable to the biological fact that women bear children, and that this is universal and inevitable (Parsons, in Barnett, 1997).
9. that there is a simple, uni-directional, causal relationship between the environment and people's behavior, attitudes and feelings.
10. that everyone is like us: has a car, or a computer; that everyone will work at home in the future, and so forth.
11. that there is a typical household and it consists of an employed father, a homemaker mother, and children younger than 18. (In the US in 1986, these were 10% of the households).

12. that social change already includes change in responsibility for home and children.

3. WHY WOMEN AND THE ENVIRONMENT?

Now, I am ready to answer the question—What is the justification for discussing the topic of women and the environment, aside from the legitimate justification that it is a topic that interests me? People have argued with me that I should talk about all the disadvantaged groups (children, elderly, handicapped, poor, minorities), who in fact make up the vast majority of the populations of every country, rather than focus on women alone. And indeed many of the applications and implications for planning that are suggested for women would improve the lives of all of these other groups also. Thus the question is a strategic one, in terms of which kind of discourse will resonate for more people and will accomplish the end of improving the quality of life of women.

However, there are a number of reasons that I see as justifications for focusing on women:

1. If we don't start with an analysis of each group separately, we run the risk of again missing specifics and differences- in the way that we did when we talked about an average person or a typical family and so forth. There may also be contradictions between the needs of women and these other groups, that we should be aware of. When the family was taken as the focus, the needs of women or children were not necessarily considered per se, but rather the adults' notion of what those needs were. We must also be sensitive to the implications of 'solutions' for one group on the lives of others. For example, those who talk about a future in which children will not go to school, but rather work at a computer at home, not only ignore the social and emotional development of children, but ignore the question of who will be at home with the children, and what happens to families with many children who do not have one, let alone many, computers and who do not have room in their homes for all of these children to work.

   In fact, the concern for the needs of women has paralleled and strengthened concern for children, the aged, etc.
2. If we don't understand the underlying problems, we probably will not be able to find the most appropriate solution. A good example of this is the phenomenon of children's outdoor play, where when the problem is identified as children playing in dangerous areas such as streets, the solution proposed is to chase them into playgrounds. If we understand that playgrounds do not satisfy their play needs, we understand that we have to look for other solutions, such as perhaps making streets that are safe for them to play in, or providing alternative areas that are as attractive as the streets.

   Until now, only some of the activities of women have been considered, and the changes that have occurred in women's daily lives have been ignored, partly because everyday life has been a relatively neglected aspect in the literature on the environment (Michelson, 1994).
3. Women are the ones who have raised the issue; we are in the strongest position to do this, and I do not see any reason why we should hide our interest or our concern. For too long women have been asked to subordinate their needs to those of others, or for the greater good. This was true of the civil rights move-

ment in the United States, of all the national liberation movements around the world, of workers' movements and so forth. We have learned that there is no point to that, because these movements never get to the question of women's rights by themselves.

4. The questions that have been raised in the context of women's issues are very basic and encompassing ones, that challenge very basic assumptions, and require rethinking and questioning of epistemological and methodological theories, and of all sorts of concepts.

Without at the moment going into the basic epistemological and methodological questions, the challenge represented by feminist research on women and the environment requires changing the way we look at homes, neighborhoods, public space, density, urban design, transportation planning, the work environment, urban, regional and national planning, social policy planning, and the planning process itself. We will need to redefine the meaning of family, household, productive labor, the Gross National Product and unemployment, the statistics we gather and use, and zoning laws. On an even more basic level we will need to change the language we use, our basic value systems, the way in which we socialize children, and the educational system, including redefining and restructuring the social context of the practice and discipline of architecture (Ahrentzen, 1996).

We are questioning the very basic nature of the categories we use to describe the world, and our work in it. The challenge is to the underlying power structure, to seek to understand the nature of the systems and values that create certain types of environments and not others, for certain types of people and not others, and how and why this happens (Wolfe, 1990). These are very threatening questions to 'the system', whether political, social or professional, and to individuals, and for this reason, they have encountered a great deal of resistance and objections.

In the context of women and the environment, the basic feminist argument (whether explicit or implicit) is that:

1. the environment does not fit the needs of women (and lots of other groups).
2. that the environment is most suited to the needs of men.
3. that this is so because men create it and define it (Matrix, 1984; Weisman, 1992). Today some of this is intended and conscious. Some is a continuation of the past situation due in part to what Ahrentzen (1996) calls the invisibility of privilege. The difference in orientation to questions of design and planning is illustrated in the work of the Matrix group, and in Jane Jacobs' representation of the city, which clearly emerges from a woman's experience of the domestic scale of everyday life (Rosenberg, 1994). Weisman argues that space, like language, is socially constructed, and like the syntax of language, the spatial arrangements of buildings and communities reflect and reinforce the nature of gender, class, and race relations in society. Space is socially constructed, and in its turn, once bounded and shaped, influences social relations (McDowell, 1983).
4. that men are able to create and define the environment because society is patriarchal- they have the power both on macro and micro environmental and social levels, and they are also the vast majority of professionals and decision makers. Even in Scandinavia, women are only 30% of the national representatives (Horelli & Vepsa, 1994), whereas the percentage a few years ago, was 9% in Israel and 5% in France, Australia, and the United States. Architecture

works, together with other aspects of social and economic relations, to put people in their 'place', and to describe symbolically and spatially what that place is (Boys, 1984). Men create the built environment on the implicit assumption that their views are unproblematically normal, accurate, and obvious.

A feminist analysis of architecture shows the ways in which the physical arrangement of the built environment can reinforce women's differential access to resources; the way in which the built environment simultaneously legitimizes and naturalizes that inequality (puts people in their 'place'); and the way in which designers consistently construct their own socialized experience as the norm (Boys, 1984). We have systematic information that 1 and 3 are correct- that the environment is not suited to women's needs- and I will give examples of this, and that men are more likely to be in the position to make decisions about the environment. However, we do not know that 2 is correct- that the environment is most suited to men's needs. This assumption falls into the same trap as any assumption that defines a group in a general and all inclusive manner. Just as all women are not the same, so all men are not the same. The environment may be suited to certain men- we don't even have data on that, but most probably not to all of them. I once argued that decisions about the environment are made by white, middle or upper middle class, middle aged, healthy, non-minority males, that at least in Israel represent about 5% of the population, and clearly not all of the males either.

Let me expand on some questions and issues. For example, the question is being raised as to the definition of work and production (McDowell, 1993a), and the tension between fundamental aspects of social life- production and nurturance (Milroy, 1991). The present definition assumes that women don't work at home. Friedmann (1992) argues that the care of home, children, and community should be counted as part of production and not as consumption. Social reproduction is defined as maintenance of physical and mental health, meals, personal services, education, and maintenance of living conditions, and having and caring for children, emotional sustenance, and life sustaining activities. All of this consumes more time than a full time wage-paying job, and much of it is invisible to those who are not part of it (Stoecker, 1989; Lengermann & Niebrugge-Brantley, 1990).

It is work that if women did not do as unpaid labor, would have to be provided by someone else (Milroy, 1991), either privately or publicly, and then it would be considered part of the GNP. Friedmann (1992) states that if all household work done for free were paid employment, we would have to add 33%–42% to the National Income. The vision of the Scandinavian women expands the concept of work, so that paid and unpaid work are seen as equal (Horelli & Vepsa, 1994). Research has also shown that the notion that paid work at home would solve 'women's' child care problems has been shown to be true only for a small group of women in particular circumstances (Christensen, 1988).

Furthermore, women must do most of the 'patching' work that is essential to bring together a wide range of urban facilities that often are separated in space but need to be temporally adjacent (McDowell, 1993a; Michelson, 1994). Kaufman recognized in 1974 that because of the mutual dependencies between the housing, service, and employment systems and the transportation system that links them, no one system can be planned separately without considering its relationship with the others. And yet this is what still happens all the time.

Freeman (1980) argues that we need to change the assumption that one's primary responsibility should and can be to one's job. This assumption works for men who have a spouse whose primary responsibility is the maintenance of home and family obligations, but not for women, who do not usually have such a spouse. For women to be on an economic par with men, it is necessary to guarantee their right to equal labor force participation, and to equal benefits from that participation.

What about our definition of a family and our understanding of its meaning? It is argued that the family is not necessarily an active agent with unified interests, but a locus of struggle, a location where production and redistribution take place (Hartman, 1981). Ahrentzen (1992) points out that there is a false assumption that the home is a refuge and a haven for women. It is not necessarily so, either in terms of its physical characteristics, or in terms of the violence that may occur there, or in terms of the fact that it represents a second shift of work for all employed women. Hayden (1984) says that rethinking home life involves rethinking the spatial, technological, cultural, social, and economic dimensions of sheltering, nurturing, and feeding society.

4. WHAT ARE NOT ASSUMPTIONS, BUT FACTS SUPPORTED BY RESEARCH?

Very basically and significantly, we have research findings that show that there differences in the everyday life of men and women. Women are involved in a balancing of roles, a merging of role-associated interests and orientations, and through this in a weaving together of social institutions. They are in situations where they take on the tasks of monitoring, coordinating, facilitating, and moderating the wishes, actions and demands of others (Lengermann & Niebrugge-Brantley, 1990).

Women's daily lives are different from those of men. They begin the day with different tasks. If they are employed outside the home, they leave at a later hour and return earlier, they are more likely to travel to their workplace by public transportation, they work at different kinds of jobs and receive less pay for this work, they have to begin their second role of care for the home and family before they leave home and when they return home from work. Their use of public areas and services is very often together with children, making this a more difficult experience. Their use of public space is affected negatively by a number of factors: by their own concern for their security, by social norms that dictate where and when they can be outside and under what circumstances, by their poverty, and by discrimination.

I would like at this point not to get into the issue of how those differences developed or were created. However, Scott (1996) argues that in merely describing the differences one establishes the social distinction as social fact; whereas by analyzing how they were produced, one disrupts their fixity as enduring facts and recasts them as the effects of contingent and contested processes of change. Darke (1996) writes that planning professionals tend to accept prevailing social assumptions, but if these are then built in to the urban form, their effects may persist after social patterns have changed.

Therefore, I will try to somehow straddle these two approaches, because at one level, from the point of view of the implications for women and the environment, these questions could be seen as unimportant. The position I am taking now is that the fact is that the daily lives of women and men are different, and the question is how the environment can be congruent with those differences so that each person can achieve as positive a quality of life as possible.

If we want to change the situation, to do away with those differences, and to adopt a social change, utopian approach (Schneekloth, 1994: Horelli & Vepsa, 1994) then we are in the area of values, and the question becomes whose values are relevant and important, and as professionals do we have the right to impose our values on others. Furthermore, as psychologists we know how difficult it is to change peoples' behavior, let alone their values. In the field of the environment, it may be possible to achieve change on the micro scale of architectural features. However, we will see that the more serious issues are on the macro planning scale, and there the change required is a much more societal, political, and cultural one. Most of these differences are rooted in the fact discussed previously that the distribution of responsibilities for the home and family are extremely unequal, and that women bear the brunt of them.

Moser (following Molyneux) distinguishes between practical and strategic needs, and raises the question of how one decides which to address: The practical. immediate basic survival needs of women for amelioration of their situation, or the political, structural, economic, and cultural forces that have created those situations and led to the systematic disempowerment of women that is encoded in social institutions. A possible resolution to this dilemma is to assume that, since the practical problems of everyday life are socially and environmentally generated, they also need to be addressed, in anticipation of the time when they are no longer firmly linked to gender but apply equally to women and men.

For example:

1. Public Transportation: Women are more dependent on public transportation than are men, and are much more likely to use public transportation with children. Women travel at different times of day, and are more likely to need to make a number of stops on the way (Pickup, 1984). Despite these facts, most public transportation systems do not accommodate these needs.

2. Public space: Women use public space in particular ways, because of cultural norms about their use of such space, and because of problems of harassment and danger (Franck & Paxson, 1989). Women's mobility is far more restricted than that of men: they have less time, they are with children, they use public transportation; and they fear crime (Franck & Paxson, 1989; Wekerle & Whitzman, 1995).

3. Economic status: Women are more likely than men to be poor- they are the vast majority of single-heads of households, if they are old, they are more likely to not have pensions or to have very low pensions.

4. Paid employment: Women with children have particular employment characteristics that are a result of an interaction between their responsibilities and the non-congruent environment in which they live. The increase in the percent of women in this group has been very rapid. In Israel, for example, it has gone from 27% in 1970 to 63% in 1990 and 71 % in 1996. Women in this group tend to work relatively close to home, in part time work, and thus to limit their choice of employment to types that allow this, and these may not be consonant with their abilities or with their financial needs.

5. Zoning laws: One of the issues that women face is that of finding child care opportunities. Zoning laws that limit residential areas solely to residences of particular kinds, seriously affect the possibility of women's paid employment, both because child care opportunities do not exist in the near environment, and neither do work opportunities. Ritzdorf (1994) illustrates how zoning and

land-use regulation in the United States are based on the cultural assumption that the physical and geographical separation between private (home) and public (workplace, culture, politics) is the 'correct' lifestyle, and how laws and policies regulate and enforce this separation at the community level. These have been shaped by the implicit meaning of the word family previously described, and this traditional family ethic has permeated all public policy making in the United States. She argues that women don't object to this, because the spatial segregation of home and work is seen as a representation of the differences between the classes (Ritzdorf, 1992).

Although it seems clear that the working class poor and immigrants never had such a separation between home and work; this social idea of separate spheres continues to permeate residential development, public policy, and scholarship (Ahrentzen, 1992). The dichotomies of home/work, female/male, suburb/city misrepresent actual activities (e.g. that no work occurs at home, and that home activities are independent of other services and facilities), but also make those activities more difficult for both genders (Franck & Paxson, 1989).

5. PLANNING AND POLICY RECOMMENDATIONS

I will end this paper by very briefly presenting some planning and policy recommendations that we developed for national and urban planning in Israel, based on the analysis that I have just presented (Churchman, 1996; Churchman et al., 1996). These recommendations are specific to the Israeli context, but I think that you will find that many of them are appropriate to other contexts as well.

5.1. Planning Recommendations

1. Density: Give preference to settlements and neighborhoods with average higher densities. Ensure that they include infrastructure systems, services of all kinds (including public transportation and public open space), and opportunities for employment, recreation, and culture at the quantitative and qualitative levels required by the particular population.
2. Zoning: Encourage mixed land use areas that carefully combine residences with environmentally appropriate opportunities for employment, and various and varied services for children and adults.
3. Public Transportation: Give national and budgetary preference to public transportation. Invest resources sufficient to make a significant improvement in both inter-settlement and intra-settlement public transportation systems, including both underground and aboveground rail systems. Pay attention to the differential needs of large and small settlements and of metropolitan and peripheral areas of the country.
4. Roads: Develop transportation models and solutions that conceive of roads as activity locales in their own right, that can attract economic initiatives, services, and urban functions that can ease the routine of life for women in the city.
5. Individual Security: a) Implement neighborhood or community policing that could prevent crime, aid citizens in crime prevention, and be a key factor in improving the security of women in the urban environment. b) Include con-

sideration of the security concerns of women in the planning and design of the various areas of the settlement.

6. Environmental quality-Transform issues of environmental quality from a constraint and a hindrance to an inherent and indispensable part of the planning and development process.

7. Housing -Ensure sufficient and appropriate housing facilities for women who have difficulty finding housing in the private market.

8. Employment-Create appropriate employment opportunities for women in each region, including professional training courses and financial incentives for encouraging small scale businesses owned by women.

9. Rural Services-Bring services to women who live in small scale, remote settlements by means of mobile facilities (especially in education, health, banking, and cultural activities).

5.2. Policy Recommendations

None of these planning recommendations will come into being unless there are significant changes in the governmental decision-making process in general and the planning decision-making process in particular.

A. Representation in Local Government and Planning Bodies: Women are grossly under-represented in local government, where most locally-important planning and development initiatives are either initiated, or initially approved. To mitigate this problem, it should be required that women be represented on all official planning committees, and in addition, it is recommended that local governments be encouraged to set up Women's Advisory Councils to advise the official elected committees on how ongoing policies—from urban planning, through financing, to social services—are perceived by women and are likely to affect many women's lives.

B. Public Participation and Dispute Management in the Planning Decision-making Process: First, it is recommended that public participation procedures be strengthened, and public authorities be required by law or administrative guidelines to consult interest groups before making decisions—and in particular to consult representatives of women's interests and perspectives before making decisions. Second, it is recommended that alternative dispute resolution methods (ADR) be adopted as regular procedures by those public agencies that regularly deal with allocative decisions and with inter-group competition over land, financial, or good-environment resources.

C. Gather Gender-Sensitive Data: A good information base is a necessary condition for good decision-making in all areas of public policy. The task of creating gender awareness among decision-makers would become much more feasible if there were better statistical information that shows how women consume resources and services differently from men, or how they are differentially impacted by public policies. Examples of public-policy areas would be: housing (how do female-headed households benefit from publicly-assisted mortgages, or public housing, and how do they compare to men in indicators of housing quality); transportation (noted above); health (consumption of health services by location and distance from home); employment (distance from home by employment type); environmental quality (information on populations that live close to low-quality or high-hazard environmental areas); public safety (complaints to police stations by gender and place or residence, and not only on sex-related crimes as commonly reported today); and the various household-income and employment-income indicators.

D. Introduce a Gender Impact Statement as a Decision-making and Plan-Review Tool: In order to mitigate the tendency of public authorities to be blind to the impact of many decisions on women's interests and lives, a new tool is necessary, that would raise the consciousness of decision makers to the impact of their proposed or operating policies on women. We recommend using the analogy of the Environmental Impact Statement already instituted in Israel and in many other countries, as a model for a new tool—the Gender Impact Statement. Staff agencies should develop a comprehensive Gender Impact Statement to review the cross-impacts of policies. Thus, the Ministry of Interior and the National Planning Board and the District Planning Commissions would do well to develop a GenIS tool to assess statutory land-use plans submitted to them for approval by local planning commissions. Such a tool would include a set of indicators for the interrelationships between land-use, transportation, housing, public service, access to amenities, and environmental quality.

6. CONCLUSION

All of the recommendations that we have suggested for improving the environment for women are commensurate with 'good' planning in general terms. None of them are detrimental to other groups; on the contrary they are as positive for other population groups as for women. They are also clearly supportive of the principle of sustainability, whose goal is to balance concerns for the economy, the physical environment and the social/cultural integrity of the society, and the attempt to ensure that future generations have the opportunity to do the same. The recommendations for relatively high housing densities, for mixed land-use policies, for public transportation, for open spaces, and for walkable distances to services all lead to savings of land and energy resources, and less air pollution, thus all contributing to sustainability (Churchman et al., 1996).

Thus, it would appear that the strategy I have adopted in this paper, of focusing on the interests of women, is able to point to the ways in which we can achieve the goal of creating physical-social environments that are congruent with the needs and aspirations of many groups.

REFERENCES

Ahrentzen, S. (1992). Home as a workplace in the lives of women. In I. Altman, & S. Low (Eds.), *Place attachment* (pp. 113–138). New York: Plenum Press.

Ahrentzen, S. (1996). The F word in architecture: feminist analyses in/of/for architecture. In T. Dutton, & L. Mann (Eds.), *Reconstructing architecture: Critical discourses and social practices I* (pp. 71–118). Minneapolis: University of Minnesota Press.

Altman, I., & Rogoff, B. (1987). World views in psychology: Trait, interactional, organismic, and transactional perspectives. In D. Stokols, & I. Altman (Eds.), *Handbook of environmental psychology* (pp. 7–40). New York: John Wiley.

Barnett, R. (1997). How paradigms shape the stories we tell: Paradigm shifts in gender and health. *Journal of Social Issues, 53*(2), 351–368.

Beall, A. (1993). A social constructionist view of gender. In A. Beall, & R. Sternberg (Eds.), *The psychology of gender* (pp. 127–147). New York: Guildford Press.

Berscheid, E. (1993). Foreword. In A. Beall, & R. Sternberg (Eds.), *The psychology of gender* (pp. vii–xvii). New York: Guildford Press.

Boys, J. (1984). Is there a feminist analysis of architecture? *Built Environment, 10*(1), 25–34.

Christensen, K. (1988). *Women and home-based work: The unspoken contract.* New York: Henry Holt.

Churchman, A. (1993). A differentiated perspective on urban quality of life: Women, children and the elderly. In M. Bonnes (Ed.), *Perception and evaluation of urban environment quality* (pp. 165–178). Proceedings of the International Symposium. Rome.

Churchman, A. (1994). *Fitting spatial planning to the needs of various groups in the population of Israel.* Haifa: Technion, Center for Urban and Regional Studies. (Hebrew)

Churchman, A. (1996). Modeling the person-environment relationship through a gender lens: Implications for a Master Plan in Israel. *European Spatial Research and Policy, 3*(1), 87–93.

Churchman, A., Alterman, R., Azmon, Y., Davidovici-Marton, R., & Fenster, T. (1996). *Habitat II shadow report.* Jerusalem: The Israel Women's Network.

Churchman, A., & Altman, I. (1994). Women and the environment: A perspective on research, design and policy. In I. Altman, & A. Churchman (Eds.), *Women and the environment* (pp. 1–15). New York: Plenum Press.

Churchman, A., & Ginsberg, Y. (1991). Dimensions of social housing policy: An introduction. *Journal of Architectural and Planning Research, 8*(4), 271–275.

Darke, J. (1996). The man-shaped city, In C. Booth, J. Darke, & S. Yeandle (Eds.), *Changing places: Women's lives in the city* (pp. 2–13). London: Paul Chapman.

Despres, C. (1991). The meaning of home: Literature review and directions for future research and theoretical development. *Journal of Architectural and Planning Research, 8*(2), 96–115.

Forester, J. (1992). Raising the question: Notes on planning theory and feminist theory. *Planning Theory. 7–8,* 50–54.

Franck, K., & Paxson, L. (1989). Women and urban public space: Research, design, and policy issues. In I. Altman, & E. Zube (Eds.), *Public places and spaces* (pp. 121–146). New York: Plenum Press.

Freeman, J. (1980) Women and urban policy. *Signs: Journal of Women in Culture and Society, 5*(3) supple S4–S21.

Friedmann, J. (1992). *Empowerment: The politics of alternative development.* Cambridge, England: Blackwell.

Geis, F. (1993). Self-fulfilling prophecies: A social psychological view of gender. In A. Beall, & R. Sternberg (Eds.), *The psychology of gender* (pp. 9–54). New York: Guildford Press.

Gergen, K. (1994). *Realities and relationships. Soundings in social construction.* Cambridge: Harvard University Press.

Gibbons, J., Hamby, B., & Dennis, W. (1997). Researching gender-role ideologies internationally and cross culturally. *Psychology of Women Quarterly, 21*(1), 151–170.

Harding, S. (1986). *The science question in feminism.* Ithaca: Cornell University Press.

Hartman, H. (1981). The family as the locus of gender, class, and political struggle: The example of housework. *Signs: Journal of Women in Culture and Society, 6*(3), 366–394.

Hayden, D. (1984). *Redesigning the American dream.* New York: WW Norton.

Horelli, L., & Vepsa, K. (1994). In search of supportive structures of everyday life. In I. Altman, & A. Churchman (Eds.), *Women and the environment* (pp. 201–226). New York: Plenum Press.

Hurtado, A. (1997). Understanding multiple group identities: Inserting women into cultural transformations, *Journal of Social Issues, 53*(2), 299–328.

James, J. (1997). What are the sexual issues involved in focusing on difference in the study of gender? *Journal of Social Issues, 53*(2), 213–232.

Kaufman, J. (1974). An approach to planning for women. In K. Hapgood, & J. Getzels (Eds.), *Planning women and change.* American Society of Planning Officials.

Lakoff, R. (1975). *Language and woman's place.* New York: Harper and Row.

Lather, P. (1987). Research as praxis, *Evaluation Studies Research Annual, Vol. 12* (pp. 437–457). CA: Sage.

Lees, G. (1998). Book review of Processed lives: Gender and technology. *Journal of Design History, 1 1*(l), 103–105.

Lengermann, P., & Niebragge-Brantley, J. (1990). Contemporary feminist theory. In G. Ritzer (Ed.), *Sociological theory* (pp. 400–443). New York: McGraw Hill.

Levant, R. (1996). The psychology of men. *Professional Psychology: Research and Practice, 27*(3), 259–265.

Lott, B. (1997). The personal and social correlates of a gender difference ideology. *Journal of Social Issues, 53*(2), 279–298.

McDowell, L. (1983). Towards an understanding of the gender division of urban space. *Environment and Planning D: Society and Space, 1,* 59–72.

McDowell, L. (1993a). Space, place, and gender relations: Part I, Feminist empiricism and the geography of social relations. *Progress in Human Geography, 17*(2), 157–179.

McDowell, L. (1993b). Space, place, and gender relations: Part II, Identity, difference, feminist geometries and geographies. *Progress in Human Geography, 17*(3), 306–309.

Maracek, J. (1995). Gender, politics and psychology's ways of knowing. *American Psychologist, 50*(3), 162–163.

Matrix (1984). *Making space. Women and the man made environment.* London: Pluto Press.

Michelson, W. (1994). Everyday life in contextual perspective. In I. Altman, & A. Churchman (Eds.), *Women and the environment* (pp. 17–42). New York: Plenum Press.

Milroy, B. (1991) Taking stock of planning, space, and gender, *Journal of Planning Literature, 6*(1), 3–15.

Moser, C. (1987). Women, human settlements and housing: A conceptual framework for analysis and policy-making. In C. Moser, & L. Peake (Eds.), *Women, human settlements, and housing,* London: Tavistock.

Phares, V. (1996). Conducting nonsexist research, prevention and treatment with fathers and mothers. *Psychology of Women Quarterly, 20*(1), 55–77.

Phillips, D. (1987). On what scientists know, and how they know it. *Evaluation Studies Review Annual, Vol. 12* (pp. 377–399). CA: Sage.

Pickup, L. (1984). Women's gender-role and its influence on travel behavior. *Built Environment, 10*(1), 61–68.

Rackover, S. (1992). Is there substance to the feminist criticism of science? *Psychologia Israel Journal of Psychology, 3,* 5–12. (Hebrew).

Riger, S. (1998). From snapshots to videotape: New directions in research on gender differences. *Journal of Social Issues, 53*(2), 395–408.

Ritzdorf, M. (1992). Feminist thoughts on the theory and practice of planning. *Planning Theory, 7–8,* 13–19.

Ritzdorf, M. (1994). A feminist analysis of gender and residential zoning in the United States. In I. Altman, & A. Churchman (Eds.), *Women and the environment* (pp. 255–279). New York: Plenum Press.

Rosenberg, E. (1994). Public and private: Rereading Jane Jacobs. *Landscape Journal, 13*(2), 139–144.

Rosnow, R., & Rosenthal, R. (1989). Statistical procedures and the justification of knowledge in psychological science. *American Psychologist, 44,* 1276–1284.

Saegert, S., & Hart, R. (1978). The development of environmental competence in girls and boys. In M. Salter (Ed.), *Play: Anthropological perspectives* (pp. 157–175). Cornwall, N.Y.: Leisure Press.

Salomon, G. (1996). Crucial issues in the field of educational, instructional, and school psychology. *IAAP Newsletter, 8*(2), 4–9.

Sandercock. L., & Forsyth, A. (1990). *Gender: A new agenda for planning theory.* Berkeley: University of California at Berkeley, Institute of Urban and Regional Development, Working Paper No. 521.

Schneekloth, L. (1994). Partial utopian visions. Feminist reflections on the field. In I. Altman, & A. Churchman (Eds.), *Women and the environment* (pp. 281–306). New York: Plenum Press.

Schneider, J., & Hacker, S. (1973). Sex role imagery and use of the generic "man" in introductory texts: A case in the sociology of sociology. *The American Sociologist, 8,* 1218.

Scott, J. (1996). Introduction. In J. Scott (Ed.), *Feminism and history.* Oxford: Oxford University Press.

Silverstein, L. (1996). Fathering is a feminist issue. *Psychology of Women Quarterly, 20*(1), 3–37.

Slife, B., & Williams, R. (1995). *What's behind the research?* CA: Sage.

Spain, D. (1992). *Gendered spaces.* Chapel Hill: University of North Carolina Press.

Stoecker, R. (1989). *Who takes out the garbage? Social reproduction as a neglected dimension of social movement theory.* Paper presented at the American Sociological Association Annual Meeting.

Stokols, D. (1987). Conceptual strategies of environmental psychology. In D. Stokols, & I. Altman (Eds.), *Handbook of environmental psychology* (pp. 41–70). New York: John Wiley,

Weisman, L. (1992). *Discrimination by design. A feminist critique of the man-made environment.* Urbana: University of Illinois Press.

Wekerle, G., & Whitzman, C. (1995). *Safe cities. Guidelines for planning, design, and management.* New York: Van Nostrand Reinhold.

Wolfe, M. (1990). *Whose culture? Whose space? Whose history: Learning from lesbian bars.* Keynote Address at IAPS 11, Ankara, Turkey.

Worell, J. (1996). Opening doors and feminist research. *Psychology of Women Quarterly, 20,* 469–485.

Yanay, N., & Friedman, A. (1995). Gender and identity: A psychological identity or cultural discourse? *Psychologia, 5*(1), 7–15. (Hebrew)

Yeandle, S. (1996). Women, feminisms, and methods, In C. Booth, J. Darke, & S. Yeandle (Eds.), *Changing places: Women's lives in the city* (pp. 2–13). London: Paul Chapman.

10

SCIENCE, EXPLANATORY THEORY, AND ENVIRONMENT-BEHAVIOR STUDIES

Amos Rapoport

Department of Architecture
University of Wisconsin-Milwaukee
P.O. Box 413, Milwaukee
WI 53201

1. (A PERSONAL) INTRODUCTION

This metatheoretical paper is in two major parts, the first dealing with scientific theory in general, the other with theory in Environment-Behavior Studies (EBS) based largely on a preliminary synthesis of my own work.

Part 1 is a brief outline summary of a major ongoing project begun in 1982 (a book—*Theory in Environmental Design*). The major premise is that EBS needs to be seen as the science of Environment-Behavior Relations (EBR). Moreover, it needs to be seen as a "pure" (or basic) science, with applied research and research application, while clearly important, useful, and relevant being distinct. These latter can be seen as a form of experiment and in EBS and the social "sciences" they are often the only available way to do experiments.

Although application, as a way of changing the world requires knowledge from basic research, since one cannot apply what does not exist, my emphasis on basic research is based in an interest in *understanding* the world, not changing it (Rapoport, 1990a). It is also the case that, in the presence of knowledge and theory from basic research, agreement, and convergence follow even in applied scientific fields (unlike in the design professions and even EBS). Thus Agnew (1998) makes the point that since bioengineering overlaps virtually every field of basic biological research, one might expect to find major disagreements about the importance of applied research areas among its practitioners. However, there was agreement on three areas as the most important—computational science (including bioinformatics), imaging and tissue engineering.

Given my emphasis on EBS as basic science, the ongoing project considers the nature of science and its most typical and valuable product—explanatory theory—

Theoretical Perspectives in Environment-Behavior Research, edited by Wapner *et al.*
Kluwer Academic / Plenum Publishers, New York, 2000.

because it follows that EBS must have such theory. When I began this project I expected to find some clear answers in a few weeks and began to read philosophy of science. One soon discovers, however, that philosophy is better at asking questions than at giving answers. Moreover, it should be noted that much of what had been philosophy (e.g., "natural philosophy") is now science. At the extreme it has even been suggested that this process will eventually largely replace philosophy by science (e.g., Himsworth, 1986; cf. Rubinstein et al., 1984; Nye, 1993). In any case, ontology is now clearly part of science, ever more empirical, and attempts to "naturalize" epistemology on the basis of empirical and computational studies of human cognitive abilities bids to make that part of science also (e.g., Kornblith, 1985; Shrager & Langley, 1990; Giere, 1992; Prieditis, 1988; Papineau, 1993; Hahlweg & Hooker, 1989; P. S. Churchland, 1988; P. M. Churchland, 1989, 1990) (among many others). Philosophical analysis will, of course, remain critical—asking good questions, challenging and clarifying concepts, sharpening logic.

In any case, from philosophy of science I moved to the history of science, biographies, and autobiographies of scientists, sociology and anthropology of science, journalistic accounts of science projects and then scientific work in many disciplines as reported in scientific journals (Rapoport, 1990a). The latter was in order to understand the role and use of theory in "science in use" as opposed to the reconstructed version in the philosophy of science, by analogy to Kaplan's (1964) distinction between "logic in use" and "reconstructed logic." At the same time I began to try and synthesize these bodies of literature.

Having gone through this process I discovered that I had recapitulated the development of what have variously been called 'the science of science,' and 'metascience.' This is now a large, rapidly growing and developing field with an extensive, fascinating, and expanding literature. In any case I have by now read and taken copious notes on about 2000 items and this process continues with continuous additions.

In my major study the argument will be developed in detail and in depth and all this material reviewed, synthesized, and used. This paper should be seen as a sketch that summarized some of the argument in the context of an edited book so that it can be compared and contrasted with other approaches. I will not reargue, in any detail, the need for theory and reasons for this need, nor what theory is, how it is constructed, how evaluated (so as to choose among theories), how theory is tested, etc.—important as these questions are in a full development. I will, therefore, be relatively brief and cite relatively few references, whether from metascience or my own work.[1] The paper may thus read like a series of assertions but it is, in fact, based on much material. This is important, particularly in Part 1, because it is striking how little of that work seems to be known or used, even in the one significant attempt actually to deal with theory (Lang, 1987). To give just one example, in my review of that book (Rapoport, 1988a) I pointed out that the sole reference to the important topic of objectivity of science was to Kuhn (1965), in spite of the large literature on this topic since then.

The examples used in the large literature mentioned above are either philosophical or from the natural sciences. Luckily the latter are no longer confined to physics as was the case for a long time. There are philosophical and other studies of chemistry, geology oceanography, applied science and technology, and other sciences, most impor-

---

[1] In a sense, this paper follows my (1973, 1975a, 1990a, 1997a) and begins where my (1990a, 1997a) leave off.

tantly of biology, paralleling the explosive growth of the field itself. In fact, the philosophy of biology and discussions of theory and explanation in biology have proliferated. This has emphasized the importance of domain definition and also differences between biology and physics, for example, in the course of theory development and change, the notion of emergent phenomena and the acceptability (at least by some) of functional (although not teleological) explanations which are never acceptable in physics. In EBS, however, not only are functional explanations (including *latent functions*, such as meaning) essential, but so are teleological explanations. This is because the difficult task of altering or modifying the environment is always undertaken for a purpose. This makes explanation in EBS easier. EBS has another explanatory advantage. This is that it occupies the mesoscale, that at which humans evolved, so that understanding is, in effect "guaranteed" (and epistemology is more easily naturalized). One might possibly even identify subdomains where both explanation and epistemology might be easier or more difficult. One also avoids the problems in physics of the very large and fast (relativistic phenomena) and the very small (quantum phenomena). At the scale of EBS these problems do not arise since humans evolved to understand it and perception is largely veridical (e.g., Gibson, 1950, 1968, 1979). This, I would argue, also applies to explanation and understanding, so that it might be easier and simpler in EBS than in the natural sciences rather than more difficult, as is often claimed. EBS and the social "sciences" seem "difficult" because not only do people seem to glory in obfuscation but the problems have never been *discovered* (hence, the importance of problem *identification* rather than definition—because it is easy to set oneself irrelevant (or trivial) problems (e.g., Rapoport, 1989b, 1995a, b)).

The discussion immediately above begins to refer to EBS. So far there have not been attempts to use examples from EBS and relate them to metascience although there have been some attempts to develop philosophy of social science and particularly archaeology (cf. Rapoport, 1990b, Chs. 4 and 5; there is also significant work since then). In Part 2 of this paper I will, therefore, use a very brief summary sketch of my work over the past 30 years (my "research program" (cf. Rapoport, 1990a)). This will begin to suggest what the implications of Part 1 are for the process of constructing an explanatory theory of EBR. A fuller development will have to await the larger study, and the emphasis here will also be metatheoretical, emphasizing the use of fewest assumptions; ontology will be taken to be an empirical matter and issues of epistemology will largely be ignored. As far as EBS itself is concerned, the emphasis will be on the unification among concepts, models, etc., which were developed independently and for different purposes. I will also suggest that this growing framework can not only be modified as new problems are identified and solved but is also capable of relating to, and taking advantage of, advances in other fields and benefit from intertheory and interfield support.

2. PART 1

2.1. Science and Explanatory Theory

This paper, as a sketch of my approach more generally is metatheoretical partly because there is danger in becoming too specific too early. I have, therefore, previously argued (Rapoport, 1990b, Ch. 5) that the "new archaeology," impressive and useful as its achievements have been, mistakenly adopted a *specific* position based on a particular view of explanation (Hempel's D-N model) which is no longer accepted. This point has since also been made by Salmon (1998, pp. 333–368). In this connection one needs not only to keep up with the development of new disciplines (and their relevance for the domain) but also with changes and advances in the various relevant disciplines (in this case the philosophy of science). One must use the best available knowledge/state of the art. In this case this has meant keeping up with changes that have clarified (and continue to clarify) the nature of explanation, theory and, hence, science (of which they are part).

My basic premise is that although there are other ways of *interacting* with the world (e.g., affective, mystical, religious, aesthetic, economic, ideological, meliorative, conative (action oriented), and possibly others) there is only one workable and, hence, valid way of interacting with the world *cognitively*, i.e. in terms of *knowing* and explaining and hence understanding it and how it works—and this is science. In fact, science can even address some of the other ways of interacting with the world. It can either explain and understand them (e.g., which types of attributes of environments elicit affective, mystical or aesthetic reactions; what are the sources of these, etc.); or their effects can be traced (e.g., how various ideologies or religions affect the way environments are shaped, what is the relative economic value of various environmental attributes, etc.).

This is why I have been calling for a shift away from an art metaphor of design to a science metaphor. It should be admitted that what has been called the "demarcation criterion" that distinguishes between science and non-science (or pseudo-science) is not as simple as it once seemed. One issue, as for many other concepts (such as "vernacular," "tradition," "ambience," etc.), is that it is not a question of a single (monothetic) criter*ion* (or ideal type) but rather of multiple criter*ia* or attributes, possibly constituting a polythetic set. (e.g., for vernacular see Rapoport (1990c), for spontaneous settlements see Rapoport (1988b), for tradition Rapoport (1989a) for ambience Rapoport (1992a, pp. 276–280). I have previously listed some preliminary attributes (Rapoport, 1990b, pp. 62–65) but clearly others can, and will be, added.

Moreover, one usually knows science when one comes across it, as shown by statements such as "this is more an art than a science." There are clearly certain cues. One concerns major efforts to be as objective and neutral as possible and to avoid excessive identification with the subject(s) of study (which is also important in ethnography). Understanding requires distance and detachment. This follows from the cognitive orientation, the desire to explain and understand phenomena rather than to change them (e.g., meliorative, conative, or ideological orientations). That distinction is often revealed by the use or avoidance of emotive, loaded terms whether supposedly negative ("sprawl") or positive ("home") which, in addition to creating other problems, prejudge the issue (see also fn. 13). This also means that I regard the term "normative theory" as an oxymoron (but cf. Lynch, 1981).

It also seems clear that among any other attributes of science one of the most important is the presence of explanatory theory (e.g., Nash, 1963 who makes it *the* demarcation criterion). This is why, as mentioned in the introduction, I began this project by trying to understand the nature of *explanatory* theory. There is another important reason; at several recent conference presentations I discovered that one can be completely misunderstood because one has referred to "theory" and there are different, widely divergent and sometimes almost contradictory ways in which this term, if unqualified and unrestricted, is used. This only confuses the issue. There is also a problem with the almost contradictory meaning of the term in common "folk" usage and the scientific view of it. In the former, it typically means something vague or speculative as opposed to facts. Its position as the most important product of science is well captured by the well-known quotation from the English astrophysicist Sir Arthur Eddington to the effect that "it is also a good rule not to put too much confidence in experimental results until they have been confirmed by theory."[2]

Even in philosophy of science study of theory there is much apparent (and even real) disagreement, which some scholars have described as "scandalous." There is also a considerable degree of *implicit* agreement which I hope to bring out in the book version of this paper, and more briefly below.[3]

There is even disagreement about something as basic as whether "theory" is a generic concept (with specifics at the level of components, concepts, linkages, mechanisms, models, etc.) or whether theory is domain specific. Similarly, there has been much debate about the nature of "explanation."[4] Surprisingly, explanation has only rather recently been taken to the main function of science. Earlier, description and instrumental value (simplification) were taken as the purpose of science (as in positivism).

The *locus classicus* of the discussion of explanation is taken to be Hempel and Oppenheim's 1948 paper which had been ignored for a decade, so that the emphasis on explanation really dates from the mid 1960's (Hempel, 1965; cf. Kitcher & Salmon, 1989; Salmon, 1990, 1998). The discussion of this topic has since then generated a large and complex literature and, although disagreement persists, some convergence can be detected. Kitcher and Salmon (1989) and Salmon (1990) proposed two principal views—*explanation as unification* (proposed by Kitcher and, in my view, including both intertheory and interfield, i.e. "lateral" connections (Rapoport, 1990b, esp. Ch. 3)), and *explanation as causal* emphasizing mechanisms (supported by Salmon). At the time, it seemed to me that they were complementary rather than conflicting. Since then Salmon (1998), who has worked on this topic for close to four decades, supports the view that these two positions are, in fact, complementary.

At any rate, given the variety of views, positions, definitions on all these (and other) topics in the philosophy of science one not only needs to refer to a broader body of work (as already discussed). One also needs to, and is able to, make a choice based on one's reading of the situation. One might be wrong but, after all one must start somewhere and then to modify, change, and improve that choice as one goes up what one might think of as "the spiral of development" (Rapoport, 1990b, Fig. 3.10, p. 96). In effect, given my view that, in general, philosophy is better at asking questions than giving answers, one essentially gives an answer (based on the totality of the literature read) to a series of question: What is science? What is theory? What is explanation? And so on.

Without reviewing the voluminous literature I will, therefore, take the view that "theory" is generic, although different domains will have different ontologies, concepts, constraints, mechanisms, and models linking theory to empirical data (which will also be domain-specific). Given the lack of a universally agreed upon meaning of the term

---

[2] This neat reversal of the usual view has been cited very frequently. This quote is from R. L. Weber (Ed.) (1982) *More Random Walks in Science*, London, British Institute of Physics.

[3] In this latter discussion, I will, for the moment, ignore the very real difference between what one could call the accepted view and the newer semantic view of theories (e.g., Suppe, 1989).

[4] This is a good example of the value of philosophical analysis. One tends to use the term without really thinking through what it means or implies until the question is posed.

theory, already mentioned, it appears to me that at least some of the definitions, metaphors, logical forms, etc., that have been proposed, discussed, and used will prove to be complementary and one begins to detect some, often implicit, agreement and overlaps at a general, metatheoretical level. Moreover, as already discussed, multiple attributes will most likely be needed.

Some definitions see theories as sets of interrelated high-level principles or concepts that can provide an explanatory framework for a broad range of phenomena in a domain. Other complementary positions view theories as networks of concepts eventually anchored to empirical reality (e.g., Rapoport, 1990b, Fig. 3.2, p. 69; 1997a, Fig. 28.1, p. 412). Theory has also been described as a system of ideas that explain something and which are based on principles more general than, and at least partly independent of the facts, phenomena, etc., to be explained, and as any set on constructs that is derived verifiably to represent and make intelligible (i.e. explain) specified classes of phenomena. A very useful metaphor has described theories as maps (and maps as theories!); many interesting and useful analogies follow which cannot be developed here.[5]

With regard to the equally extensive literature on explanation already discussed, I take the view that the most useful current position is that discussed by Salmon (1998). First, that explanation is not a logical argument (as the "received view" based on Hempel's D-N and I-S models maintains); second, that the two major ways of explaining—theoretical unification (cf. also Barnes, 1992; Thagard, 1989) and the identification of mechanisms (e.g., Morris, 1983; Frankel, 1980; Menard, 1986) and causes are complementary and can be combined.[6] Furthermore, I take the view that the purpose of explanation is *understanding*—one wants to known why things are as they are or why the world (or the part of it in the domain of concern) is the way it is and how it works.

I would also add that unification (or synthesis) is both internal, within a discipline or domain, and external (related to other and possibly new disciplines and their theories). In other words, that unification involves linear and lateral connections (Rapoport, 1990b, Ch. 3) so that interfield and intertheory support is extremely important in science (e.g., Bechtel, 1986; Darden and Maull, 1977; Wilson, 1998). At the same time the identification of causes and mechanisms is essential—one cannot do without them (e.g., Salmon, 1998; Humphreys, 1989; Morris, 1983 among many others).

Two recent examples (among numerous others) from "working science" may be useful. Regarding mechanisms, it was recently pointed out that aspirin, the oldest "miracle drug" (100 years old) worked, but that no one knew why or how, i.e. there was no explanation for its success. Now the mechanisms of its action have been discovered and, as a result, it becomes possible to design improved versions of aspirin (Pennisi, 1998a; Kalgutkar et al., 1998) i.e. *to apply research*. The second example shows the role of both unification and mechanisms in understanding (Pennisi, 1998b). It begins by pointing out that biologists currently often explore isolated corners of the molecular structure of cells. [This is a result of the fact that, as Jacob (1982) points out, asking small questions leads to large answers, whereas asking excessively big questions often leads nowhere]. It continues by describing how every so often the trails that they have

---

[5] Recently, there has been development of a "semantic" view of theories (e.g., Suppe, 1989). What effect this will have on the above brief discussion is not yet clear and requires more work.

[6] This is not to imply that there are not a number of other interesting and potentially enlightening views about explanation. One example is "inference to the best explanation." Although it seems to belong to a different domain, it might be worth seeing whether it and others can be incorporated into a larger synthesis.

been following converge unexpectedly and "a unified picture emerges . . . a confluence of evidence (p. 1438) [an example of consilience (Whewell, 1840; Wilson, 1998)]. This unified picture leads to a better understanding not just of embryonic development *in species ranging from fruitflies to humans*, but also of cancer (p. 1438; my emphasis). This unification is the result of a 16 year long process of research and putting together of bits of evidence from many sources. The process of unification shows every sign of continuing (p. 1441) and also leads to the development of causal explanations. In this process the "skeleton" is first worked out and then the details are filled in (p. 1441). What this confluence, or "surprising convergences" (p. 1441) have discovered is, in fact, *mechanisms* whereby a particular molecular pathway damages a particular tumor suppressor gene which, in turn activates an oncogene (cf. He et al., 1998) [both these mechanisms of cancer formation themselves had previously had to be discovered through other research]. Also, this type of understanding then immediately leads to insights enabling the *design of drugs* (as in the case of aspirin above and many other examples). In other words, it leads to clinical applications—the application of research which in EBS is design (e.g., Rapoport, 1989b, 1995a, b; see also Part 2).

Another example of this process is provided by Sellers et al. (1997) where one sees, once again, how both models and theories are built through integration and synthesis. One begins by building a large "general circulation model" (of climate) based on available data. One then uses new data, often through the development of submodels *which can be developed separately* (my emphasis) and then plugged in. Through this process of unification the original model is modified, refined, and improved over time.

Note that this need for unification is reflected in recent institutional developments in science. For example, for some time there have been departments of *integrative biology*, and the Zoological Society of America changed its name to The Society for Integrative and Comparative Biology. There are also centers of *integrative neuroscience*. That such integration ultimately demands theory is shown by the existence of a *Theory Center* (of physics) at the University of Rochester (cf. The *Theoretical Division* at the Los Alamos National Laboratory). Recently, in Japan, RIKEN established a new, and apparently unique, *Institute of Theoretical Neuroscience*. The founder, Professor Masao Ito, argues that theory is essential and that without it the field is at an impasse (Kaiser, 1997). A symposium on "synthesis in ecology" (Nov, 1996 in Santa Barbara, CA.) decided actually to meet to work at synthesis, projects of integration and so on. Synthesis is defined as the bringing together of *existing information* (my emphasis) in order to discover patterns, mechanisms and interactions that lead to new concepts and models. (Taubes, 1997, p. 311) The meeting emphasized that one can do first-class research without collecting primary data when one already has much (as I have been arguing regarding EBS (e.g., Rapoport, 1990a)). It is also suggested that this process will lead to the ability to *apply* ecology which is taken to be a *problem solving discipline* thus closely paralleling my view of design and its relation to explanatory theory. The above examples (and I have cited very few) show that there exist models for EBS.

Note that arguments have been made by a number of researchers that meta-analysis might prove useful in connection with synthesis and unification. However, this relatively new method has also been criticized (e.g., Bailar, 1997). This suggests a topic for research to determine the value of meta-analysis; possibly a *careful* use of it may be called for.

An additional reason for suggesting that science is the only way of knowing is its rapid and accelerating rate of progress observed when one reads the scientific literature. It is salutary to recall that the current explosive growth and, one might almost say, pre-eminence of biology (e.g., Kafatos, 1998) dates from Watson and Crick's characterization of DNA, which occurred in 1953—not that much earlier than the beginnings of EBS. This makes biology a useful model for EBS which has not developed very much in comparison. In fact, I would argue that it has been on a plateau, or even declining for the past 15 or 20 years. One reason is the lack of agreement in the field even about simple things like terms, concepts, and definitions. I have previously referred to the impact on chemistry of such agreement in the 18th c. (Rapoport, 1990a). It is also the case that most sciences do this at regular intervals. Attempts to try and do this in EBS have met with no success.

Following science as an ongoing process also reveals a striking contrast between the rapid development of new fields, which are often at the interface between existing fields whereas EBS, although interdisciplinary from the beginning, has not established links with fields other than those it began with. Yet there are quite a few new fields relevant to EBS, although such relevance can only be seen at some *higher level of abstraction* which presupposes explicit agreement about terms, concept, and some conceptual framework. To mention just a few: Cognitive science, neuroscience and cognitive neuroscience, brain science, psychobiology, integrative biology, evolutionary science and paleoanthropology, genetics, sociobiology, evolutionary psychology,[7] the "new archaeology", artificial intelligence, and computational approaches generally.

Some of these fields overlap with, and have even begun in small ways to influence, social "science," for example, in terms of a revived interest in human nature or human universals (e.g., Brown, 1991; Goldsmith, 1991; Barkow et al., 1992; Horgan, 1995 among many other; see also later). In general, however, social "sciences" have rejected such findings partly on ideological grounds and partly as part of the *zeitgeist*. That is currently anti-science and leads to a rejection of science as a model for social and behavioral science, EBS, etc. Even worse is the currently fashionable assault on notions such as truth, objectivity, and rationality themselves (e.g., Gross and Levitt, 1994; Sokal & Bricmont, 1998; Nagel, 1998; Koertge, 1998). As a result, these fields and their findings have been ignored in favor of phenomenology, narrative, discourse analysis, deconstruction, hermeneutics, feminism, post-modernism, various political ideologies, "new age," mysticism, Eastern religions, social constructivism, and relativism generally (cf. Laudan, 1990; Norris, 1997). The effects on the social and behavioral "sciences," with which EBS largely still identifies, have been dire.

Computational approaches generally seem to have great potential, including computational philosophy of science (e.g., Thagard, 1988) and computational approaches to discovery (e.g., Langley et al., 1987; Langley & Jones, 1988). The latter is significant for another reason. According to the "received view" of science (e.g., Suppe, 1977), i.e. that of the logical empiricists, one could only deal with the context of justification but not the context of discovery. This was because there was a logic of the former but not of the latter. Even opponents of the logical empiricists, like Karl Popper, argued that discovery was a matter for psychologists not philosophers of science. Now discovery is a topic both for the philosophy of science and metascience (e.g., Nickles, 1980a, b; Grmek et al., 1981; Holland et al., 1986; Schaffner, 1993; Kantorovich, 1993). This was important in my discussion of explanation above, since the "received view" of expla-

---

[7] Most of these are relevant primarily to question 1 (and to a lesser extent, question 3) of EBS discussed in part 2 of this paper.

nation (Hempel's D-N and I-S models) are, in fact, *logical* arguments and, as indicated above, no longer generally accepted.

In any case, an explicit commitment to science leads to a number of metaphysical and metatheoretical assumptions which need to be made explicit. In fact, explicitness is essential throughout (e.g., Rapoport, 1990b); although agreement about assumptions, definitions, terms, and concepts is much to be preferred, at a minimum explicitness enables specifics of disagreement to be identified.

One such disagreement concerns the debate between what are called internalist and externalist accounts of the development of science. Internalists maintain that such development is due to factors within the sciences themselves. It is, they argue, caused by the growth of knowledge, clarification of problems and questions which arise from earlier work, resolution of debates through the rigorous testing of alternatives, clarification of concepts, filling in of gaps, the development of new methods and techniques and, in some cases, instruments. Very important is the nature of the world itself, or the particular part of it being studied. Externalists tend to attribute changes and development in science (if any; in general they are skeptical about development) to social, economic, ideological, political, and other such factors.

Although, in any case, the question should really be about the relative roles of these two sets of variables, my own reading of the literature leads me to accept the internalist position regarding knowledge, approaches, methods, concepts, theory building, testing, and modification and the like. At the same time, it seems clear that the choice of topics of research may sometimes be initiated by external factors like the ones listed (often through their effects on funding). How they are addressed and tackled, the answers obtained and the like are, however, due to internal factors as well as the constraints of the domain being studied, i.e. the world out there. The impact of new methods and instruments seems somewhat ambiguous with regards to these two types of factors. In any case, I will not be dealing with these.

In EBS so far it would appear that internal factors have possibly played a smaller role than is usual—and that may, in fact, be one of the major problems with the field. There has been little, if any, impact of the logic of the discipline (mainly because there is, as yet, no discipline). The field is disorganized, there are no agreed upon concepts, terms are used loosely and without shared meaning, there is no explanatory theory. However, internalist factors have played a role in the case of individuals—for example, myself. As a result of the field itself, however, rather than cumulative evolution there has been an accumulation, 30 years' worth of fragmented work (as has been argued in connection with sociology (e.g., Willer & Willer, 1973)). The fact that EBS is a highly fragmented field emphasizes the importance of unification (see Part 2).

Accepting the internalist account clearly implies a rejection of solipsism (which underlies the "social construction of reality" view) and also the acceptance of some form of realism (e.g., Norris, 1997; Newton, 1997; Miller, 1987).[8] The world is refractory and imposes itself on theories so that, as a result, the "theory-ladenness" of empirical data is not a major problem. This, however, brings us back to the metaphysical and

---

[8] Note that most scientists are "naïve realists." Among philosophers of science not only are there antirealists but also many varieties of realism. I must confess to tending toward a fairly naïve form of realism. Every time one flies at 37,000 feet and claims to be a social constructionist one is (as someone once said) a hypocrite. Another very convincing argument for realism is, for example, television, which utilizes electrons to create images. As Hacking (1983) pointed out, if they can be used they are real.

metatheoretical assumptions within which any discussion of explanatory theory must occur. These were mentioned above and now need to be made more explicit.[9] I will list a number of them but only elaborate (albeit briefly) two.

I will approach these assumptions as a sequential process, although there are no grounds for thinking that it works that way. The first step, after a commitment to science, involves clarifying what it is. The second is to consider broad, general philosophical presuppositions (which will be discussed below). The third step deals with some more specific presuppositions concerning domain definition, pattern seeking and identification, problem identification, the potential roles of deduction, induction and abduction, and the like. Only then, as step four, can one raise questions about explanatory theory: What is it? What does it do? Is it a generic concept or domain specific? How is theory constructed? How is it modified, and so on. Then follow questions about how one evaluates theory (i.e. the criteria for "good theory" (which will also be discussed below)). The fifth step concerns the nature of explanation (one answer having been sketched in this paper), its relation to prediction (symmetric or asymmetric) as well as the role of postdiction/retroduction. In step six one considers the distinctions (if any) and/or relations between theories and models, the types of models and the relations between models and mechanisms. Seventh, one considers where theory fits into larger frameworks (such as metaphysical blueprints, research programs, networks of theories, etc.) as well as the components of theory (postulates, concepts, constructs, and the like) and how they relate to theory. Step eight addresses the relation of theory to evidence, including confirmation versus falsification, the link of theory to empirical data and the "problem of the theory ladenness of evidence." The latter does not seem to be as serious as philosophers seem to think. Theoretical expectations are constantly violated in "science in use," for example, almost weekly in *Science*.[10] Step nine marks a transition to EBS as a scientific domain and raises the question of whether the study of humans can be "scientific" (i.e. whether the social and behavioral sciences are inherently different to the natural sciences).

An example of general philosophical presuppositions (based mainly on Nash, 1963) includes: a real world exists, is simple enough so that it can be understood and it imposes its structure on what is discovered (cf. Rosch & Lloyd, 1978). This principle of intelligibility (which Einstein called a miracle) is essential: It may prove wrong (and will then fail) but, if one assumes the opposite, that becomes a self-fulfilling prophecy and inevitably leads to failure. From this principle follow certain methodological consequences which I will not discuss here, but also the congruence between mind and world (and hence the possibility of a naturalized epistemology). The principle of continuity also follows, which implies the possibility (and, as I have long argued, the necessity) of separating wholes into parts and, having studied them, reconstructing the wholes. Many more assumptions of that type could be added, but in this sketch this brief list at least gives the "flavor" of these kinds of assumptions.

The nature of theory partly emerges from the list of attributes of "good theory." I began with a list of five desirable attributes proposed by Kuhn (cited in Newton-Smith, 1981, pp. 112–113). These I, Gary Moore and several generations of doctoral students developed to about fifteen. I have since added some while combining (or

---

[9] The ones discussed here I use in my contributions to our doctoral theories course. I suspect that the ones in the book on theory will not necessarily be the same.

[10] As just one example see *Science* Vol. 281, No. 5374, September 18, 1998, where 3 examples are found in just two pages (1765 and 1767).

removing) others. As in the case of my whole project I have since discovered that others in the field have also done that (McMullin, 1993).

The original five were: predictive accuracy [which is generally taken not to apply to evolutionary theory], consistency [internal and external; empirical; of logic; with other theories and metaphysical assumptions]; breadth of scope [sometimes labeled comprehensiveness or extensivity]; simplicity [there is some debate whether this is the same or different to parsimony and elegance]; fertility. Those added since include elegance and/or parsimony [see note above]; explanatory power [which implies problem solving ability, utility, etc.], public discussibility [i.e. explicitness]; testability, criticizability and/or falsifiability and confirmability (at least in principle); progress [which is a historical, *post facto* criterion]; smoothness, responsiveness or adaptability (i.e. how it responds to new findings, problems, etc.) [some argue that a theory must not be *too* smooth in this sense]; inter-theory support; communicability (ease of teaching and learning); compressibility (the ability to unify much information); empirical corroboration; beauty [there is much debate about the relation of this attribute to elegance].

Note that this is an open ended list; that many of the attributes are difficult to define and are subjective and difficult to deal with (e.g., "elegance," "beauty"); many (in addition to progress) involve time and can only be judged in retrospect after some (often significant) period of time; the relative importance of the various attributes is not established; how they are "measured" and how they may relate to one another is far from clear. In spite of all these (and other potential) problems at least some of these criteria (or something like them) are used by scientists to evaluate, judge, and accept or reject theories and models (judging by what they write). Such criteria also help to define the state of the field of inquiry by enabling at least provisional relative judgements among fields. Note also that fit to empirical data, is an important and necessary, but not sufficient criterion. After all positivism and phenomenological physics both fit empirical data but do not explain, being non-theoretical or even anti-theoretical.

Consider *compressibility*. It is clearly important since its absence (and lack of cumulativeness) paradoxically leads to the fact that there often is too much material in EBS—we are in fact "drowning" in empirical material which can become counterproductive (cf. discussion of housing later). Contrast this with the personal example in my (1990b, p. 89) and the many examples that can be found in the literature. To cite just one, the physicist Feynman reduced all of classical physics to nine equations occupying one-half page (Gribbin & Gribbin, 1997). Many other attributes are related to this, such as communicability, elegance, and *simplicity*. The above example helps clarify what is meant by that latter term—simpler means shorter, easier to remember and use. At the same time "simplicity" as a concept is surprisingly difficult to define, nor easy to use as a criterion applied to theory (e.g., Slobodkin, 1992). One can suggest that a simple theory has fewer terms than a more complex one (and this can apply both to concepts and objects (ontological parts)), i.e. there are fewer discernible parts or partitions (Slobodkin, 1992). Moreover, one recognizes simplicity when one sees it, and the search for simplicity also underlies the search for patterns as the starting point, as what needs explaining and understanding.

In concluding Part 1 of this paper, I would like to draw attention to the principle of continuity mentioned above. That leads to the possibility and importance, indeed necessity, of dismantling domains. In other words, it leads to a rejection of holism. Much clearly depends on what is meant by "holism," showing once again the need for explicit definitions. In science, for example, it is often used in the sense of discovering (or uncovering) and then taking into consideration all relevant variables, studying them and their interactions, their effects, what mechanisms are involved and then reintegrating or resynthesizing at higher (sometimes much higher) levels of abstraction. This kind of "holism" is essential. This also leads to the view that much of the discussion in EBS, being too concrete and specific, is misconceived both at the level of terms (space, architecture, the elderly, etc.) and specific fields (anthropology, psychology, etc.) rather than concepts and findings that can cut across fields.

If, however, "holism" means studying everything together, that is impossible (Phillips, 1976; Efron, 1984). Various examples will be given of how I use dismantling in my work (see also Bechtel and Richardson, 1993).

3. PART 2

3.1. Toward an Explanatory Theory of EBR

In this part I will present a sketch of a framework for EBS, based on a synthesis of some of my work, as an example that I believe fits into the framework of Part 1. Before doing so, however, I would argue that the strength of any conceptual or interpretive framework depends on its theoretical grounding, on explicit arguments about why certain concepts or data are relevant to particular questions, explicit definitions and so on. In other words, one needs a coherent theoretical perspective and explicit (even if provisional) model of how the world works before one starts presenting and analyzing data. Given the current state of EBS, explicitness and, hence, "transparency" of one's argument and its components are essential. This enables the argument to be criticized, lacunae to be discovered and modifications and improvements to be made.

As I suggested elsewhere (Rapoport, 1990b, pp. 9–11, begun at the same time as the theory project) an important first step is domain definition which includes both subject matter and important questions about it (Shapere, 1977, 1984). In my view, what I have called the three basic questions of EBR are the simplest, briefest, most fundamental and all-embracing such definition—and, hence, the most useful.

1. What biosocial, psychological, and cultural characteristics of human beings (as individuals, members of a species and of various groupings) influence (and, in design, should influence) which characteristics of the built environments.

2. What effects do which aspects of which environments have on which groups of people, under what circumstances and why.

3. Given these two-way interactions between people and environments, there must be mechanisms linking them; what are these mechanisms.

These meet Shapere's requirements for domain definition. First, they describe the body of material of concern, each subsuming a very extensive cluster of more specific material, so that they can be dismantled to any degree of specificity and precision; second, they define the important questions, again at a very high level of generality—too broad to be useful as stated, but again capable of being dismantled in line with my discussion in Part 1 and below, and Jacob's (1982) argument that, in science, small questions lead to big answers whereas large questions lead to small [or no] answers. Moreover, this formulation is open-ended allowing new developments, either in EBS or any other existing or new discipline to be incorporated and, hence, to modify (add, subtract or combine) specifics.

Since, as already suggested, "holism" (in the non-scientific sense) is impossible, dismantling is a *constant*, a standard technique or approach, applicable throughout in order to clarify and operationalize concepts. Dismantling typically produces as its first product what one might call a lexicon of potential items (Rapoport, 1990d). These, in many cases, can be further dismantled (or "unpacked"), so that there are scales and levels of dismantling. Such lexicons require empirical studies to determine which components are used in different situations, or by different groups—or are useful for specific questions or problems. One can also try to identify the relative importance (i.e. ranking) of those items. Some lexicons can be applied to different subdomains (for example, process and product characteristics to different types of environments (e.g., vernacular versus high style) (Rapoport, 1990c), to tradition (Rapoport, 1989a) to spontaneous settlements (1988b). Others, like the result of dismantling culture, environment (to be discussed later) or ambience (Rapoport, 1992a, pp. 276–280) have different uses in unification and hence different types of broad applicability. Recall that for scientific holism re-synthesis at higher levels of abstraction is necessary.

As just one example of how new information (which arrived while I was writing this paper) fits easily into such lexicons, consider question 1. At the level of species characteristics it has been known since 1975 that humans and chimpanzees share 98%–99% of their DNA, so that humans have been described as "the third chimpanzee" (Gibbons, 1998, p. 1432).[11] Currently work is beginning at trying to identify the specific reasons, at the level of genes, that make humans so different. Eventually, and it should not be too long, it should be possible to characterize what makes us human (Gibbons, 1998). Some of the other aspects of questions 1, such as culture, rule systems, and social variables, will be briefly clarified later.

An additional important aspect of research on human origins and evolution, gene-culture coevolution, etc., and the disciplines that study them is that the extreme cultural relativism in anthropology is beginning to be questioned. Anthropology has tended to emphasize cultural differences and often to treat humans as *tabula rasa*. This is slowly giving way to a more nuanced view, so that constancies, universals, and human nature are being recognized (e.g., Barkow et al., 1992; Brown, 1991; Goldsmith, 1991; Horgan, 1995, among many others). One knows from answers to question 1 that culture will play some role. In line with the previous discussion of the increasingly empirical nature of ontology, and rather than making further assumptions, and speaking of "cultural theories" (Moore, 1987, pp. 1375–1376 (or any others, for that matter)), the question becomes an empirical one: What are the relative contributions of cultural versus other variables and of both human universals and cultural differences (or specifics) in different situations in different domains (in our case the built environment) or subdomains (see Rapoport, 1975b, Fig. 1, p. 146; 1990b, Fig. 3.12, p. 111).[12] More specifically one can ask whether cultural variables are the only ones, the primary ones influencing other variables to varying degrees, playing different roles in different situations or being relatively unimportant vis-à-vis other variables (including universals). Answers cannot be given *a priori* but need to be discovered.

Regarding question 2, I have tried to show elsewhere (e.g., Rapoport, 1983a, 1985) that the choice of particular settings (and the rejection of others) is the principal influence of the environment on people. Choice also modifies (or moderates) more specific effects. These in turn, can also be identified, studied, new findings incorporated,

---

[11] There are two species of chimpanzees, the well-known *Pan Troglodytes* and the less-well known Bonobo (*Pan Paniscus*).

[12] As is the case with many of my analytical diagrams there are many, gradually evolving, versions in various publications. The latest version of some of these (including this one) will be shown below.

etc. Also, through the concepts of reduced competence and criticality, various groups and their attributes allow some predictions about the magnitude of effects of the environment on people.

Regarding question 3, the attempt to list possible mechanisms (essential for explanation) suggests that there seem to be a limited number. A first list (or lexicon) would include:

Physiology (comfort, adaptation, etc.)
Anatomy (e.g., ergonomics)
Perception (including "aesthetics"—see later)
Cognition
Affect
Evaluation (preference and choice)
Meaning
Supportiveness (physiological, anatomical, psychological, behavioral, social, cultural, etc.)
Some of the components of culture (see later)

Clearly, some of these variables interact and work together. For example: affect, meaning, and evaluation; affect and perception; ergonomics, comfort, and culture, and so on. Also clearly, the list (or lexicon) can, once again, be expanded if necessary, i.e. if new mechanisms are discovered. Moreover, the mechanisms listed can be further dismantled (for example, meaning into different types and levels, the ideal images that result, etc. (Rapoport, 1988c, 1990d, f)). Also, ever more specific and concrete mechanisms can be incorporated, often on the basis of new findings (and from newly relevant fields). One example is the relevance of frame/script theory, from computational science, for meaning (Rapoport, 1990d, pp. 236–238). Among the many new findings incorporated as I was writing this (weekly, as each new issue of *Science* arrived) consider an example related to cognition. Much early work in environmental cognition concerned mental maps. Just recently a specific *neural mechanism* of mental maps has been discovered (Kentros et al., 1998). Thus, as cognitive neuroscience and related fields gradually discover such mechanisms in different species, these findings could be linked to work on mental maps in psychology, geography, and EBS, and also species and group differences and constancies in such mechanisms, and hence the possible role of culture in the use of such maps in humans. This would provide understanding both through the identification of specific mechanisms and through unification. The latter will be helped by the ongoing and frequent discovery of neural mechanisms for way-finding, memory, learning, etc., as well as for many aspects of perception (e.g., see special issue on cognitive neuroscience, *Science*, Vol. 275, No. 5306 (14 March, 1997), pp. 1570–1610). There is even work on non-visual perception at the level of mechanisms.

Given the attributes of the three questions briefly discussed above, specific questions can be formulated, including those that, as is common, involve more than one variable within a question and those involving more than one of the basic questions. Also, other definitions or formulations of the domain can also be derived from them. As a result, this formulation meets the widely held view that the fewer assumptions the better, so that minimal ontological and epistemological assumptions are to be preferred. Thus one can derive the formulation by Moore et al. (1985) without introducing "settings," "user groups", or "socio-behavioral phenomena" as ontological assumptions. As will be seen shortly, "settings" emerge naturally from the need to operationalize "environment." In the same way, "user groups" and activities ("socio-behavioral phenomena") emerge from question 1 (and also question 3) through "culture" which partly defines humans. As a result, "culture" also emerges naturally as a relevant variable (also playing a role in question 2). It is also the case that EBS can only deal with groups or aggregates (although it also needs to be able to transcend those (Rapoport, 1997a)). This does not mean that individuals cannot be studied—in fact typically they are. It does mean that one needs to generalize, as one does in science generally.[13] The need for generalization immediately, naturally, and directly, leads to at least three very important consequences. One is the question of which attributes define relevant groups—an under-researched topic in EBS.[14] The second is that one knows immediately that "culture" will play some role, since one of its purposes is precisely to distinguish among groups within a single species (and which have been called "pseudo-species") (Rapoport, 1986a). Of course, as is always the case, "culture" needs to be dismantled in order to become useful (see below; cf. Rapoport, 1993a, Fig. 1, p. 16). The third consequence is the obvious need for the largest and most varied body of evidence. This immediately leads both to the need for unification and synthesis through comparative work of all kinds. It also suggests how the required body of evidence is to be obtained, with major conceptual and methodological implications. This involves what I have called a four-fold expansion of the body of material used—to include all environments, all cultures, the whole environment, and the full span of history (Rapoport, 1990a, b, esp. pp. 11–19, 1993a). I have since come across the same argument regarding psychiatry—the need to use the largest and most diverse body of evidence which, it is argued, is a "cornerstone of science" (Kleinman & Cohen, 1997).

Starting with the three questions also leads naturally to the need for an operational definition of environment, because they implicitly require it. Since "holism" in its non-scientific sense is not an option, this definition needs to be done through dismantling and made explicit. Thus this aspect of ontology emerges naturally through this process.[15] I have found four complementary definitions of "environment" to be most useful (although others no doubt exist). The specific dismantling or combination of dismantlings used varies with the topic or question being studied.

The environment[16] can be conceptualized (a) as the organization of space, time, meaning, and communication; (b) as the cultural landscape; (c) as a system of settings within which systems of activities (including their latent aspects) take place; (d) as composed of fixed, semi-fixed, and non-fixed elements.

The advantage of these four complementary operational definitions/dismantlings stems from the fact that they essentially specify the ontology of the environment. Also, a large number of consequences follow immediately and naturally, forming both linear and lateral linkages among themselves and with other fields and their theories. To give

---

[13] Individuals and individual variations play a role, e.g., as one of the "filters" in evaluation (Rapoport, 1992b, Fig. 1, p. 37) and in terms of individual diversity within groups (Rapoport, 1990f, Fig. 2.2, p. 12; 1990c, Fig. 4.6, p. 85). Individual variations can be accommodated through open-ended design, i.e. in research application (Rapoport, 1995c, Paper 33, pp. 533–562).

[14] This is just one small example of how conceptual frameworks, let alone theory, lead to the identification of important research areas, topics and questions (Rapoport, 1973, 1975a).

[15] This also applies to other aspects of ontology. Also, attempts to "naturalize" epistemology emerge naturally from question 1.

[16] Note that there is a whole other environmental domain with which I am not concerned here—that dealing with resources, pollution, climate change, etc. (cf. Rapoport, 1995c, Paper 30, p. 476).

just a few examples, the first not only naturally introduces time, meaning and human activities, interactions, (the organization of communication) and hence boundaries, cues, transitions, rule systems, etc., as part of, but separable from space organization. It also covers scales from regions to furnishings of interiors. It also naturally incorporates the nature of the rules, ideals, and criteria used in such organization, which vary (e.g., with culture). It also relates directly to the nature of design through the choice model of design (e.g., Rapoport, 1977, pp. 15–18, 1990c, Fig. 4.9, p. 88). This model, in turn, can be applied to *all* design (traditional, contemporary, vernacular, high-style, etc.) what varies are the specifics—time-frames involved, the nature of the criteria used, who applies these (i.e. makes decisions), the order of application (reflecting ideals, priorities, constraints, and their nature, the rules involved, etc.). Reference to rules draws attention to the fact that this concept also applies across many subdomains and thus establishes many lateral connections and major unification. Thus it helps define groups (which share rules); rules also guide behavior and play a central role in lifestyles and activity systems; they are reflected in the choice model of design; settings work through rules communicated by cues (see below), they also explain how cultural landscapes come to be.

This leads to the second conceptualization, environment as cultural landscape, which expresses the organization of space, time, meaning, and communication through the particular application of rule systems based on shared ideals, images, schemata, and the rule systems through which they are expressed. This, in turn, throws light on the nature of design, since cultural landscapes are typically not designed in the traditional sense of the term, yet the choice model of design applies, and they have identity and ambience (the latter concept being very useful once made operational through dismantling (Rapoport, 1992a, pp. 276–280). Cultural landscapes also naturally combine built and "natural" components, leading to both differences and similarities and making imperative the need to study those together. This reveals the importance of relationships among elements, as does the need to study high-style and vernacular together (Rapoport, 1992b). This in turn bears on the nature of meaning, since both types of environments (and things more generally) take on meaning through contrast. The use of cultural landscapes also makes relevant (and becomes relevant to) a large geographical literature.

The third, systems of settings, links settlements and dwellings with all kinds of outdoor and other settings. It introduces linkages, separations, and transitions among settings, temporal factors (settings vary over time as well as in space), the role of cues in settings (hence meaning and variability) as well as the control of boundaries and penetration. The control of boundaries leads to various transitions and domains, for example, from the most private to the most public (privacy also being an aspect of the organization of communication). Transitions tend to vary more (in the dwelling/settlement system) than either of the components (Rapoport, 1977, 1994a; Lawrence, 1986; Pellow, 1996). This set of concepts also makes possible the use of different bodies of literature. It can also be applied at different scales from rooms to regions (e.g., domains of sacredness) again allowing major unification. In addition, like cultural landscapes, systems of settings naturally cross scales from regions to rooms, make built and "natural" environments one, thus dealing with similarities and differences, emphasizing the need to study those (and also high-style and vernacular) together. Since systems of activities are part of the formulation, humans and their behaviors form a natural and inevitable part. The further dismantling of activities (and functions) into four categories from the most manifest and instrumental to the most latent (see below) provides a very useful way to study such activities and the role of meaning as a part of functions and activities, rather than something added on.

Finally the fact the environment consists of fixed, semi-fixed, and non-fixed elements immediately includes much of material culture generally (and thus very useful links to archaeology). It also includes, through the non-fixed elements the inhabitants and their lifestyles, dress and hairstyles, and hence links with behavior, interaction and communication, activity systems, rule systems, social organization, and also values, ideals, and the like—i.e. the other formulations. In addition it links settlements and buildings with all kinds of outdoor and indoor furnishings at many scales.

There is another way of showing the unification this approach makes possible despite (or possibly because) the different levels of abstraction and complexity of these formulations. The most fundamental and abstract formulation (the environment as the organization of space, time, meaning, and communication) is expressed physically as cultural landscapes at various scales from the region (from where it comes through geography) through townscape (urbanscape), enthnoscape, housescape, wirescape, and smaller or more specific components. Cultural landscapes are composed of systems of settings within which systems of activities take place (playing a role in producing the particular ambience). The cultural landscape, the elements comprising settings and their cues, and activity systems are made up of fixed and semi-fixed elements and both created by, and occupied by, non-fixed elements (mainly people). In this sense the four formulations are not only complementary but closely linked, partly because the different conceptualizations incorporate and use similar and related components. Other ways of explicitly combining and relating these are clearly possible without losing sight of the components and the simplicity of this ontology. This very large and rich web of relationships among these conceptualizations provides a very high level of unification (and even more potential unification, and hence explanation) and, less obviously, also possibilities of explanation through the mechanisms of question 3.

The above discussion introduced the differing levels of abstraction and complexity of these formulations. These can be further clarified. The organization of space, time meaning and communication is the most abstract, followed by systems of settings, cultural landscapes, and fixed, semi-fixed, and non-fixed elements as the most concrete. This, then, also ranks them from the most complex to the simplest. This is because the first needs much further dismantling (for example, many varieties of space can be easily listed (Rapoport, 1970, 1977, 1993a)). Moreover, all of the four variables have large and rich bodies of literature in various fields, enabling much further dismantling. Also many different rules of organization can be identified, making this first conceptualizations complex. Systems of settings are less abstract and simpler since it is a *single* "thing" with a single definition. Methods for uncovering settings exist and can be developed, are not too difficult to use and can be inferred through observation (e.g., Liu's, 1994 doctoral dissertation). Cultural landscapes are much more concrete, being directly perceptible, for example through their ambience, although that is still a complex, multisensory term and a large potential lexicon can be generated (e.g., Rapoport, 1992a, pp. 276–280). Fixed, semi-fixed, and non-fixed elements are directly perceptible visually and can also easily be captured by photographs, drawings, and inventories (from which, for example, settings can be inferred (Liu, 1994)). They are thus the most concrete as well as the simplest.

The discussion of the conceptualizations is usefully ended by pointing out that each of them can be related to a variety of disciplines, including biology, evolutionary science, territoriality, ecology, landscape ecology, geography (including the new field of "chronogeography"—the geography of time), non-verbal communication, artificial intelligence and computational approaches generally, technology (for example, both transport and telephones, computers, etc., are forms of communication), multisensory perception, archaeology, and others (cf. Rapoport, 1994a).

It may be useful briefly to elaborate on the concept of setting, partly as an example of how other concepts can be handled and partly because, unfortunately, it seems to be being replaced by "place." For example, my "system of settings" corresponds to "multiplace analysis" (Bonnes et al., 1990; Bonaiuto & Bonnes, 1996). The concept of "place" can be criticized mainly on conceptual and logical grounds (Rapoport, 1993b, 1994b).[17] It can also be criticized on empirical grounds because settings, unlike places, can be discovered, i.e. they are empirically accessible. At the same time, as will be seen later, much place-related (and other apparently incompatible work) can be used and fits into the framework being discussed (see also Rapoport, 1990a).

In relation to the discussion in Part 1 on scientific realism, I take settings to exist and to be part of the world, i.e. part of our ontology. In fact, Barker's notion of "setting" is one of the most useful and powerful concepts in EBS. It also lends itself to clarification, elaboration, and development. For example, Barker, his students, and followers largely neglected cross-cultural concerns, although they were present in their comparisons between the US and UK (cf. Liu, 1994). They also tended to neglect the role of time, i.e. how the same spaces become different settings over time (Rapoport, 1990d; Hufford, 1986; Liu, 1994). They ignored the crucially important matter of mechanisms, i.e. how settings actually work, how they are able to communicate the appropriate behaviors to people. My answer concerned the meaning of cues (fixed, semi-fixed, and non-fixed) which, if noticeable (clear) and culturally appropriate communicate the nature of boundaries, the control of entrance and penetration and, by defining the situation and the rules that operate, lead to appropriate behavior making co-action possible (Rapoport, 1990d).

Support for this hypothesized mechanism came from work in AI and cognitive science (an example of lateral links) and the development there of frame/script theory (see references to Minsky, Schank and Abelson; Abelson, Mandler, etc., in Rapoport, 1990d, pp. 235–239). This, in itself, is already an example of unification because, as Kruse (1988) points out, frame/script theory involves concepts from both psychology and linguistics. One way in which this might also work in the case of semi-fixed cues (and my argument about their importance) is partly supported by Norman's (1988) work in cognitive science.

My argument also involved a number of other lateral linkages, for example to Goffman's "role setting," Rosenthal's work on experimenter effects, to the "dramaturgical analogy" as developed by Perinbanayagam and others (all cited in Rapoport, 1990d). The idea of a *generic* use of setting (and hence the possible use of methods other than Barker's) was also reinforced and enriched by the work of Kaminsky (1987).

All of this development also had the effect of reintroducing the physical milieu into settings. Ironically, over time, Barker's initial emphasis on the milieu (which was the basis for his work in the first place) weakened, and most subsequent work emphasized social variables. In fact, in EBS generally, there tends to be (much) more on behav-

---

[17] An almost identical argument can be made about the increasing use of "home" (Rapoport, 1995d). Although this does not belong in the present discussion, the point needs to be made that, as in the case of "place" it adds an unnecessary (and logically untenable) concept. In addition, of course, generally the fewer concepts in a theory the better. "Setting" and "dwelling" are the terms of preference (being more neutral). The whole discussion, both of "place" and "home" boils down to their affective and other impacts on people.

ior than on the environment, which needs to be redressed. Of course, there is even more neglect of mechanisms and unification and, hence, explanatory theory.

The importance of rules in settings and elsewhere and consequently their relevance in many subdomains, leading to unification, has already been pointed out. One further point is that all kinds of rules can be investigated further, unifying many periods, cultures and environments. Among the rules one can study are written, unwritten, economic, legal, ideological, building, health and safety regulations, sumptuary laws, religious, etc. For example, one can include geomancy as a particular type of rule system. Such systems (e.g., Yoon, 1986, 1991, 1992a; cf. Nemeth, 1987 among many others (see also examples and references in Rapoport, 1993c, 1992b)),[18] including their absence or nonsystematic nature (Yoon, 1992b) can be compared cross-culturally (e.g., Hakim, 1986, 1994). Further unification occurs because rule systems, as already discussed, play a role in the various conceptualizations of environment. Moreover, Yoon's (1991, 1992b, 1994) notion of "geomentality" neglects the role of schemata and ideal images. Linking these provides another point of potential unification and intertheory and interfield support, as well as a suggestion about potential mechanisms involved in "geomentality." The effect of specific geomentalities on, for example, gardens can be further linked to other studies of gardens (cf. references in Rapoport, 1969, 1977, 1990d) and to cultural landscapes generally.

Geomantic and other similar systems (e.g., Rapoport, 1993c) can also be used as "model systems," corresponding to those used in biomedical research to show the role of rules generally including, for example, ideology (e.g., Nemeth, 1987; Cavalcanti, 1994). They make good model systems because, as I will discuss further later, being more "extreme" they offer clearer examples.

This elaboration and development of the concept of settings and the role of rules also has the important consequence of establishing strong links between settings and cues in them (as well as other conceptualizations of environment) and the major topic of meaning as a mechanism which, through rules systems, makes settings work. I would argue that meanings are a human universal—people expect the environment to communicate meaning and rules, indicating appropriate behavior and making co-action possible. If the environment fails to do so (either because the cues are too subtle to be noticed or, if noticed, inappropriate or incomprehensible) people will *impose* meaning. This may be a possible way of resolving the disagreement between me (Rapoport, 1990d) and Bonta's (1975) analysis of Mies's Barcelona Pavillion.[19]

However, once again, "meaning" is too global (i.e. "holistic") a concept and thus needs to be dismantled, resulting in the notion of *different levels of meaning*. This helps to explain how different meaning work, i.e. one can specify the mechanisms available (a "lexicon") and those actually used. One can also begin to make predictions, e.g., about how meanings differ in different types of environments (e.g., Rapoport, 1988c, 1990d Epilogue, 1994a). These predictions become hypotheses that can be tested over time and cross-culturally, with lateral links to other fields. Thus in archaeology/ anthropology support regarding my high-and-middle level meanings is provided by Blanton (1994, esp. pp. 8–13, cf. Binford, 1962) who does not, however, deal with low-

---

[18] When I came across Yoon's work in 1998, I discovered that he did not know Nemeth's work. I was able to relate these two, and the others being discussed—i.e. establish later links.

[19] This disagreement may also be related to the differences between critics (and designers) and users. Both these hypotheses seem worth investigating.

level meanings. There are also supporting links with political science (Goodsell, 1988) and psychology (Gibson, 1950, 1968) and others. One can, in fact, relate a number of these and continue the process of unification.

This becomes clear when one compares (Rapoport, 1994a, Table 1, p. 471) with a new diagram, (Fig. 1) showing both a further lateral link as well as how diagrams that I use can, sometimes over decades change, grow, and develop, unifying more material but without losing their fundamental point (some other examples, e.g., the dismantling of "culture," will be discussed later).

Having shown how meaning and rule systems relate to settings and enable many other linkages, thus leading to unification, I will now turn to some other modifications of the concept of settings which are of help in developing an explanatory theory of EBR.

One theme in the analysis of behavior settings has been the large numbers involved, both at the urban and architectural scales (see references to Bechtel and Le Compte and Willems in Rapoport, 1977, which has been confirmed cross-culturally by Liu (1994). This did not, however, seem to lead to the idea that settings form *systems* (Rapoport, 1977, 1986b, 1990f)). Yet setting do form systems and what happens (or does not happen) in some settings influences the use of others. Moreover, the nature of the settings in the system, the proximities and separations among them, the order in which they are used are all most important questions (Rapoport, 1986b, 1990f). An important issue is the extent of the system, which needs to be identified rather than assumed (Rapoport, 1969, references to Hartmann, Coates and Sanoff, Michelson, Van der Ryn and Silverstein and many others in Rapoport (1977; cf. Rapoport, 1986b, 1990f). Thus these and other studies are subsumed by systems of settings unifying them. This need to identify the extent of the system has been generalized, through another lateral link in a very different context, as "progressive contextualization" (Vayda, 1983). This, in turn, not only relates to people's home ranges and helps explain the functioning of regions, settlements, and neighborhoods and their differential use by various groups. It also immediately and fundamentally changes the definition of housing (e.g., Rapoport, 1980, 1990f, Fig. 2.5, p. 16, 1994a, Fig. 2, p. 464) as that subsystem of settings (within the larger system) within which a particular subsystem of activities takes place. This has major implications for the study of housing as well as leading to unification across scales through the activity system that occurs in them.

Housing also provides a particularly striking example of the need for theory. As already suggested, there is too much information, numerous disconnected pieces of empirical research which, in effect, are counterproductive. As pointed out in Part 1, a major role theory plays is to subsume vast amounts of data in easily remembered and used formats (known as compressibility). Even a conceptual framework can help, although not as much as theory, and at least it helps organize material. As has been said with regard to chemistry: One needs "unifying concepts capable of bringing order to the chaos of facts," and the only way "to thin the thicket" to obtain a viable forest is through their [the facts'] *significance*, for which one needs "consensus" [What in Part 1 I called agreement about such unifying concepts.] (Huisgen, 1994, p. 7, p. 13). Part of this is equivalent to "internal housekeeping" to: Determine what is solidly established, [cf. Boulding, 1980], what are [and need to be] current research questions [and how they fit into the framework], to decide what is speculation and what is just plain non-sense pretending to be science (Wilson, 1996, p. 704).

The redefinition of housing not only shows that it must be considered in relation to streets, open spaces, other settings, and neighborhoods, as well as rules and lifestyles

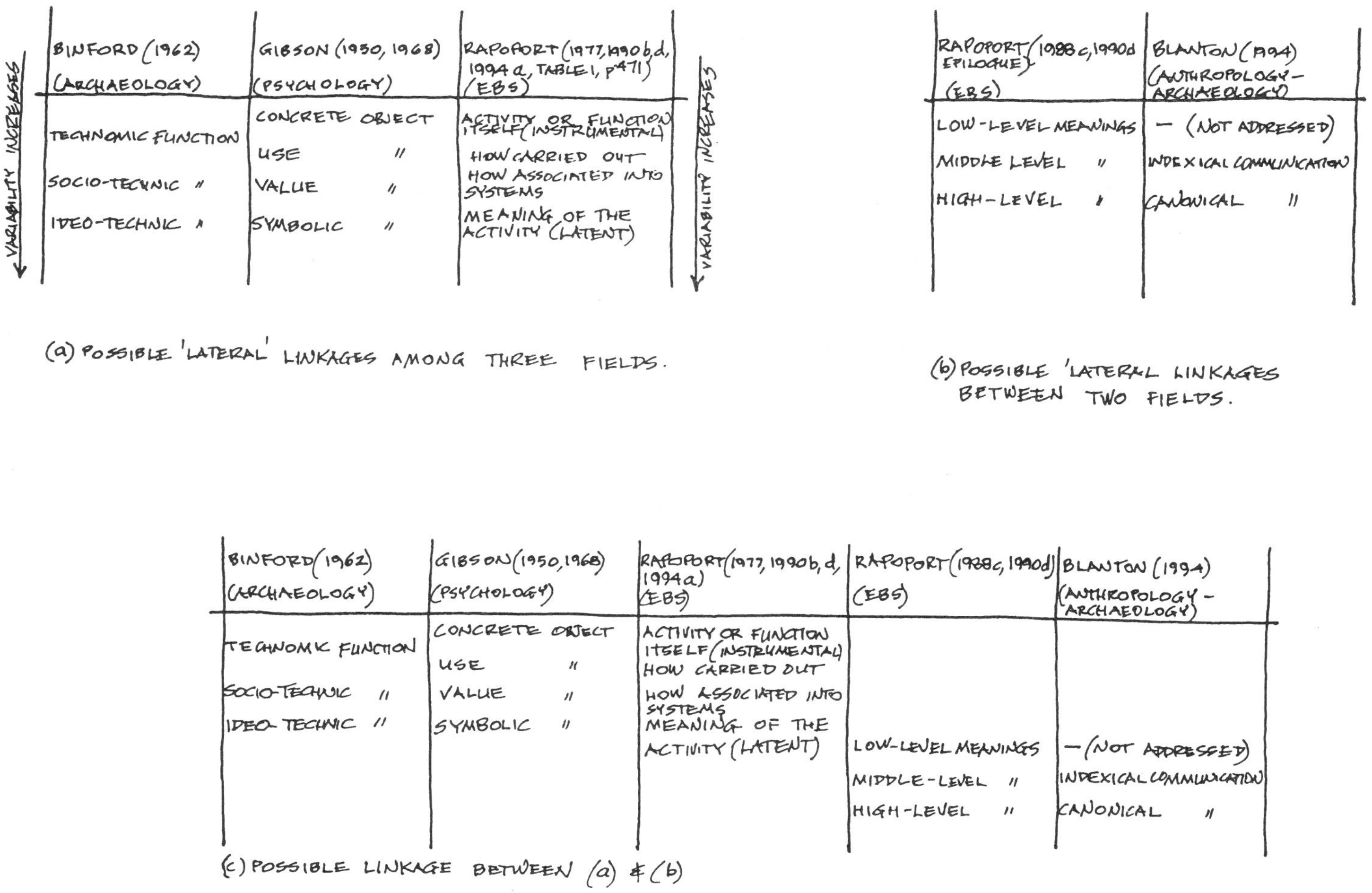

Figure 1. Example of possible 'lateral' linkages among fields leading to unification & growth.

about what behavior is appropriate in which setting (Baumgartner, 1988) thus also influencing acceptability and definitions of crowding, privacy, etc. This definition of housing also serves as an example of the operational definition of the environment as a system of settings within which systems of activities occur. This also helps in explaining the role and nature of neighborhoods (Rapoport, 1986b, 1997b), settlement patterns, transportation routes, etc. Since that is one of four operational definitions of environment this, in turn, relates housing to the other three definitions and through those to time, communication, ambience, its components and relative importance, meaning (and, hence, latent functions, status, ethnicity, and the like) and the role of fixed, semi-fixed, and non-fixed elements in communicating such meanings, and so on. Again one finds a major and "natural" unification, as well as understanding so that, for example, meanings such as identity, will vary with culture and will tend to be communicated, although through different fixed, semi-fixed, and non-fixed cues. If only the latter are used, then the communication is not through buildings or furnishings (Rapoport, 1981, 1990d).

The unification through definitions of environment and the three basic questions also "naturally" includes the concept of culture. Once again, since wholes cannot be studied, and "culture" is not a thing but a concept (or definition), one cannot deal with culture-environment relations at that level of generality and abstraction; it must be dismantled to make it operational. Over a period of 20 years I have gradually elaborated a way of doing this, starting with Rapoport (1977) through a number of iterations (some unpublished) (including Rapoport, 1993a, Fig. 1, p. 16; 1994a, Fig. 3, p. 476; 1995e, Fig. 2, p. 16) through a presentation in Seoul, Korea in August 1998 a further development of which will be illustrated here. This particular way (with which some anthropologists disagree (e.g., Cooper & Rodman, 1995)) I find extremely useful. Undoubtedly, like other models, it will be further modified and developed.

This particular dismantling (see Fig. 2) leads to much more manageable, smaller, and specific questions about how certain features of the environment (as discussed above) can be related to more concrete social variable (roles, family, kinship, social networks, institutions, status, etc.) and also to more specific expressions of culture, such as world views, values, ideals, images and schemata, norms and rules, lifestyles and activity systems. As I have discussed in detail elsewhere, many of these can be further dismantled—for example lifestyle and activities (e.g., Rapoport, 1977, 1981, 1985, 1990f). It should be noted that social variables are treated separately (as opposed to speaking of socio-cultural variables). Also, culture is treated as an ideational concept from which both more concrete and more specific variables can be derived.

This not only enables one relatively easily to relate aspects of culture to components of environment, but also helps one to define relevant group, lifestyle, as I have argued elsewhere, often being the most useful. I have also pointed out the availability of over 50 lifestyle profiles in the US marketing literature (Rapoport, 1985, 1995c Paper 30, pp. 471–488. Two recent newspaper stories provide an interesting example, where lifestyle and especially family structure relate the housing needs (at the level of the dwelling) of two apparently very different groups (Polygamous Mormons and

Figure 2. Dismantling of "culture" and relating its expressions to the built environment (the width of arrows corresponds approximately to the hypothetical feasibility & ease of relating the various elements (based on Rapoport 1977, Fig. 1.9, p. 20, 1993a, Fig. 1., p. 16, 1994a, Fig. 3, p. 476, 1995e, Fig. 2, p. 5 and unpublished & new components).

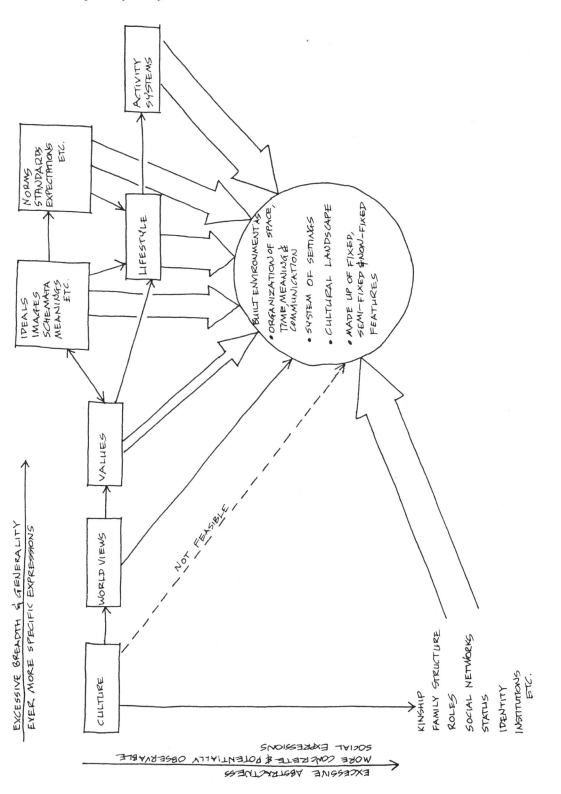

Hasidim) (Berger, 1997; Sontag, 1998; Williams, 1997). One can, in fact, plot these on Fig. 2. Moreover, differences can be identified at the level of the neighborhood where the needs differ, and the differences can be explained. This then also relates to the redefinition of housing—the specific subsystem needs to be discovered. Once again, this shows unification and the ability of the framework easily to accommodate new data (see also the Conclusion).

The expression of culture in terms of world views, values, ideals and images, and norms and rules leads to a single evaluation model (again elaborated and developed for some time) which can be applied in many situations, including the numerous housing and settlement types one finds when the body of evidence is increased and made more diverse. It also accommodates the role of latent functions, such as meaning and therefore *wants* (as opposed to "needs"). See Fig. 3.

Note that this model is also related to choice, that is a result of evaluation and preference. We have already seen that this also plays a role in question 2, since the most important effect of environment on people is through choice, i.e. habitat selection (Rapoport, 1983a). This importance of choice pervades EBS; it applies to understanding "aesthetics," environmental quality, housing and other systems of settings and also design conceived as a choice process (e.g., Rapoport, 1977, pp. 15–20). The latter, as already seen, itself can be used to understand all forms of design and also cultural landscapes. Thus choice is another very important concept which emerges from the analysis, can be applied very widely in EBS and can thus play and important role in unification.

In all these cases, while ideals play an important role, they may vary in their specifics. People have (an often implicit) image of ideal people leading ideal lives in ideal environments, which can be discovered. This allows the ideals of different groups to be identified while using the same model. Further unification becomes possible when it is realized that there are at least two major extreme possibilities (with a range in between)—people may prefer the familiar (traditional) or the unfamiliar (novelty) which provide potential links to the literature on sensation-seeking people.

This type of work not only clarifies cultural specifics and differences but, because ideals are always involved, also universals and constancies (discussed earlier). These become evident at least partly through extensive comparative work. The evaluation model also helps explain changes over time, both as a result of changing ideals and the reduction or loosening of constraints. The latter allows universals, if any, to emerge. One such example is the emergence and growth in popularity of "suburban" housing in Japan, France, Italy, Russia, Mexico, Indonesia, Thailand, Africa, and elsewhere as soon as some constraints (in this case economic) were loosened. The preference for narrow lot houses in Milwaukee, and the Worcester, MA, triple-decker (as opposed to row houses or apartments) can also be understood in terms of the ideal image of the detached house, even if the spacing was extremely narrow (Rapoport, 1990d, p. 33). These examples, and others (e.g., Rapoport, 1990d, pp. 160–167) even allow predictions to be made. For example, I was surprised during a trip to Korea in 1992 to discover that apartments were preferred to detached, single-family dwellings. I "predicted" that this was likely to change. Six years later, in July 1998 I discovered, both from a poster at the IAPS 15 conference in Eindhoven, and during a return visit to Korea in August 1998, that this was changing and that preference was now for the detached house. One might suggest, as a hypothesis, that this preference, for a majority (although not all) is as widespread as that suggested for Savannah landscapes. Both of these hypotheses, of course, need empirical testing.

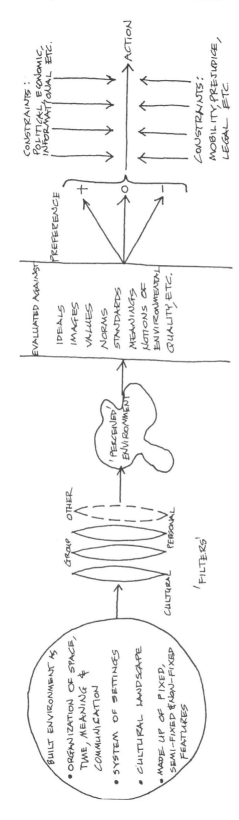

Figure 3. Model of evaluative process (which, with minor modifications, can also be applied to other subdomains).

The evaluation model can also be used and applied to larger systems, such as neighborhoods, cultural landscapes, cities, regions, countries, and so on, and to all settings within those larger systems—educational, work, streets, parks, movement, etc., in all cases based on lexicons of attributes of these entities.[20] One could, of course continue and the network, synthesis, and unification would continue to "grow naturally" (by analogy to Rapoport, 1990b, Fig. 3.10, p. 96; cf. Fig. 3.1, p. 67, Fig. 3.2, p. 67). This growth would be both internal and through intertheory and interfield support. Moreover, by identifying mechanisms (question 3) one could begin to make the arrows in the various models much more concrete and meaningful, thus helping to explain EBR in both ways discussed in Part 1. Also this process also simplifies the field—there is a beginning of compressibility.

One example is the topic of environmental aesthetics on which there has been some interesting work done during the past few years (by Nasar, Stamps, and others). However, as I argued at EDRA 26 it is not really needed as a separate subdomain. It really concerns the perceptual aspects of environmental quality which, in its profile form (Rapoport, 1995c, Paper 30), can include it among all other aspects. By looking at environmental aesthetics in this way no special treatment is necessary. Moreover, in terms of question 3 one realizes that mechanisms other than perception are involved, the most important being meaning. Together these also relate to affect and evaluation (and, hence, preference) (e.g., Nasar, 1998); the same evaluation model discussed above can be used.

Another result of unification and the "natural" emergence of concepts from the three basic questions is that there is no need to introduce the notion of "special user groups." It is sufficient to refer to groups (which form part of the human characteristics in question 1) all of which have sets of attributes.[21] The typical categories used, such as children, youth, elderly, women, disabled, homeless, cognitively impaired, urban poor of the developing countries and the like are not needed, except possibly in *applied* research and *research applications*. They may also be useful as "model systems," corresponding to those used in biomedical research, in which EBR are seen more clearly, being more extreme and at higher criticality (as already discussed earlier). In any case, all these aggregates and/or groups are then defined not by age, or economic status or other imposed etic categories, but through lists of attributes that follow from the 3 basic questions and the dismantling of culture (ideals, social variables, lifestyle, and all the others already discussed).

The discussion in Part 2 makes clear that it is possible to use the same concepts, models, mechanisms, and so on in many situations, cases, subdomains, etc., and a few examples of this approach and way of thinking have been given. This also applies to the diagrams which I regard as very important and which, as already pointed out, show development over long periods of time. They can be used in all the classes that I teach, in different books, papers, and invited topics with which I am initially unfamiliar (cf. Rapoport, 1990a). This is deliberate and part of unification and theory development. The fewer elements one needs, and the more widely they can be applied, the better. As already suggested, it should be possible to make the lines and arrows much more causal,

---

[20] Note that we are not dealing with "grand" theory. As has often been pointed out in the metascience literature, most explanatory theories are inherently of the "middle range."

[21] Note that these are often actually aggregates, i.e. defined through imposed etics rather than through emics (as groups are), or at least through derived etics.

by relating them to mechanisms. It should also be possible, in such diagrams, to begin to distinguish between mediators and moderators (Evans & Lepore, 1997) and to make other useful and relevant distinctions and refinements. These diagrams, which stand for networks of concepts and emerging theory, can grow, develop, and change—and finally be discarded for ones that better reflect knowledge and more developed explanatory theory.

4. CONCLUSION

In the brief, summary outline sketches in Parts 1 and 2 of this paper much has been left out, and much left undeveloped. There has been no attempt to establish a one to one correspondence or congruence between Parts 1 and 2, nor have I shown how Part 2 follows directly from Part 1. In this paper the correspondence is global. However, a careful reading and thinking about this correspondence will, I believe, show that it exists. At the same time, while suggesting that there are outlines of an emergent theory, I did not develop or construct an actual theory, nor discuss how one might evaluate it or test it.[22] At the same time, I think that enough has been said in Part 2 to show, even in this sketchy form, how much can be derived just from the domain definition. There is a high level of unification and the identification of mechanisms also follows, promising the possibility of an explanatory theory of EBR. Much more unification already exists but has not been discussed, much more fleshing out of mechanisms and many more linkages to the existing literature in many fields is possible, promising further unification. In addition to leaving out many linkages already made, I did not attempt to make new ones. To give just one more example, with recent developments in design methods research, unification between that field and EBS becomes feasible (e.g., Heath, 1984; Oxman et al., 1995; Purcell et al., 1993; the work of Ömer Akin etc.; cf. Rapoport, 1990a on the divergence of these two fields.)

Note that the EBS framework discussed is such that as new work becomes available, whether in journals, conference presentations, poster sessions, proceedings, books, preprints, and readings in other fields (including "science in use") it always seems to fit easily, without "distorting" the framework. Thus, for example, there are references in this paper from sources that arrived as it was being written and many more could have been incorporated. Within the context of theory building, the first question in all these cases needs to be: does it fit and if so where, how and why and if not—why? I have already pointed out that this applies to different classes I teach and to new topics I am asked to address (cf. Rapoport, 1990a). This is often the case even if the approach taken is one I have criticized and rejected (e.g., material using semiotics, phenomenology (as hypotheses), "home" or "place" (e.g., work using "multiplace analysis" rather than systems of settings) and others. I gave a number of examples in the epilogue to Rapoport (1990d).

I also argued there that, in many cases, some of these terms are just used because they are fashionable (as is the obligatory attack on "positivism") and nothing is lost (and much often gained) by eliminating them. The question is really how the work was

---

[22] I expect that at least some of that, and possibly much more will be done in the proposed book, although even then I suspect that it will remain at a metatheoretical level (although more specific); theory development is a collective endeavor.

done, what the findings are, how they can be interpreted in terms of concepts such as those in Part 2 and whether they will survive scientific testing, fit into the unifying framework and whether mechanisms can be identified. In the long run, of course, it is likely that some material will not fit in. Then both it and the framework will need to be evaluated and one or the other modified or even discarded.

Because the framework (and, hence, the emergent theory) deals with EBR rather than environmental-design, architecture, buildings, and the like, it is relevant to, and can be used by, other fields (or subfields) in which EBR plays any role. Those can, and do, in turn help EBS—interfield and intertheory support become a two-way street. Among examples are archaeology, geography, human ecology, parts of cognitive science, human evolution, and others.

Understanding (through explanatory theory) is the only way validly to attempt to change the world, i.e. applied research. For example, the student "experiments" described in Rapoport (1989b, 1990a, fn. 11, p. 99) show that one can achieve explicit objectives even with minimal means (and even make predictions, which such objectives are) if one understands the relevant EBR. Without such understanding it often becomes almost (or completely) impossible even with maximum means. As a result, this paper has emphasized pure, basic research at the meta-level. I have argued, among other things, that science works by identifying problems and asking (and answering) "why" and "how" questions (which explanation is partly about). I have, however, mentioned both applied research and research application, without discussing whether they (and particularly the latter) are subfields, or merely the possible use of bodies of knowledge, or the identification of problems and questions at a *more specific* level and trying to answer them (as well as asking "what" questions: What should be done).

I have argued elsewhere (Rapoport, 1990a) that one cannot apply something that does not exist. Nor can one apply something inconsistent—and, at the moment, without explanatory theory, without even agreement on domain definition, on concepts and terms, on philosophical presuppositions, EBS is highly inconsistent. It will, therefore, continue to be impossible (or, at least, extremely difficult) to apply EBS until it develops all these attributes. For those pure, basic scientific research is essential and, hence, is the only basis for application.

This implies that there also is the need for philosophical analysis not only of science and theory development, but of application.[23] Consider the topic of "appropriate technology" in design and development. What does it mean? What attributes define it? Appropriate for whom? How does one know it is appropriate? And so on. Similar analyses are needed for "sustainable urban development" (cf. Rapoport, 1994c) and "good" or "better" environments (cf. Rapoport, 1983b).

In any case, all three (basic research, applied research, and research application) are necessary and useful. Applications (in any field—technology, medicine, or design) are ways of solving a different class of problems—more specific and concrete societal problems. My examples have been from science not only because I view EBS as the science of EBR, but because design, seen properly, is research application of EBR theory (Rapoport, 1989b, 1995a, b). Thus design needs to be seen differently. The term "design," at least in English, is used extremely broadly, and I have suggested elsewhere that analogies with material science, biotechnology, bioengineering, protein engineer-

---

[23] There has, in fact, recently developed the field of philosophy of technology, distinct from but related to, philosophy of science.

ing, and the like would be much more appropriate and relevant than analogies with art (Rapoport, 1995a, b). In many social "sciences," and often in EBS, research application (and sometimes applied research) are the closest thing there is to experiments.

Design needs to be seen differently, as a problem identifying and solving field, setting clear, explicit and justified goals and objectives as hypotheses based on the best available knowledge. The next step involves identifying mechanisms and, hence, deriving lexicons of means to achieve these objectives, i.e. predictable results, implementing these followed by evaluation (not "criticism"). If done in this way, i.e. properly, design or the application of any body of knowledge and then evaluating the outcomes is equivalent to experimental (or, at least, quasi-experimental) science, or at least not that different from it. It is the only way of finding out what works and what does not—and why. It is also the only way to get rid of the many meaningless statements so popular in the social "sciences," philosophy and even in EBS (cf. Gross and Levitt, 1994; Sokal & Bricmont, 1998; Nagel, 1998; Koertge, 1998).[24]

Unless one knows what a thing is supposed to do, and on the basis of what evidence, one cannot judge whether it does it well or not, whether it is successful or unsuccessful, and why (which requires the process of design described above). Under these circumstances one cannot either make a better environment or learn either from successes or failures. It also bears on the question of scientific realism—if the use of concepts and mechanisms, i.e. explanatory theory achieves expected results, then they are real (by analogy to Hacking, 1983). We have come full circle. As someone once said: "There is nothing as practical as a good theory."

REFERENCES

Agnew, B. (1998, June 5). Biology by design: From software to skin. *Science, 280*(5369), 1517.

Bailar III, J. C. (1997, July 25). "Assessing assessments," Review of M. Hunt *How science takes stock (a history of metaanalysis)*. New York: Russell Sage Foundation, *Science, 277*(5325), 528–529.

Barkow, J. H., Cosmides, L., & Tooby, J. (Eds.). (1992). *The adapted mind (evolutionary psychology and the generation of culture)*, New York, Oxford University Press.

Barnes, E. (1992). "Explanatory unification and scientific understanding," in D. Hull et al. (Eds.) *PSA 1992*, East Lansing, MI, Philosophy of Science Association, vol. 1, pp. 3–12.

Baumgartner, M. (1988). *The moral order of the suburb.* New York, Oxford University Press.

Bechtel, W. (Ed.). (1986). *Integrating scientific disciplines.* Dordrecht, Martinus Nijhoff.

Bechtel, W., & R. C. Richardson (1993). *Discovering complexity (decomposition and localization as strategies in scientific research).* Princeton, NJ: Princeton University Press.

Berger, J. (1997, Jan. 13). Growing pains for a rural Hasidic enclave. *New York Times.*

Binford, L. R. (1962). Archaeology as anthropology. *American Antiquity, 28*(2), 217–226.

Blanton, R. E. (1994). *Houses and households: A comparative study.* New York: Plenum.

Bonaiuto, M., & Bonnes, M. (1996). Multiplace analysis of the urban environment: A comparison between a large and a small Italian city. *Environment and Behavior, 28*(6), 699–747.

Bonnes, M., Mannetti, I., Secchiaroli, G., & Tanucci, G. (1990). The city as a multi-place system: An analysis of people-urban environment transactions. *Journal of Environmental Psychology, 10,* 37–65.

Bonta, J. P. (1975). *An anatomy of architectural interpretation: A semiotic review of the criticism of Mies Van Der Rohe's Barcelona Pavillion,* Barcelona, Gustavo Gili.

Boulding, K. E. (1980, Feb. 22). Science: Our common heritage. *Science, 207*(4433), 831–836.

Brown, D. E. (1991). *Human universals.* Philadelphia, Temple University Press.

---

[24] Many of the points Nagel (1998) makes, including those about Kuhn and Feyerabend, and some additional ones about Kuhn, I make every year in our doctoral theories class.

Cavalcanti, M. (1994). *Urban reconstruction and autocratic regimes: A case study of Bucharest*, Ph. D. Dissertation, Oxford Brookes University (unpublished).

Churchland P. M. (1990). On the nature of theories: A neurocomputational perspective. In C. W. Savage (Ed.), *Scientific theories* (pp. 59–101). Minneapolis: University of Minnesota Press.

Churchland, P. M. (1989). *A neurocomputational perspective (the nature of mind and the structure of science)*. Cambridge, MA: MIT Press (Bradford Books).

Churchland, P. S. (1988). *Neurophilosophy (Toward a unified science of the mind-brain)*, Cambridge, MA: MIT Press (Bradford Books).

Cooper, M., & Rodman, M. (1995). Culture and spatial boundaries: Cooperative and non-profit housing in Canada. *Architecture and Behavior, 11*(2), 123–138.

Darden, L., & Maull, N. (1977). Interfield theories. *Philosophy of Science, 44,* 43–64.

Efron, E. (1984) *The apocalyptics (Cancer and the big lie)*. New York: Simon and Schuster.

Evans, G. W., & Lepore, S. J. (1997). Moderating and mediating processes in environment-behavior research. In G. T. Moore, & R. W. Marans (Eds.), *Advances in environment, behavior, and design, vol. 4* (pp. 256–285). New York: Plenum.

Frankel, H. (1980). Hess's development of his seafloor spreading hypothesis. In T. Nickles (Ed.), *Scientific discovery: Case studies* (pp. 345–366). Dordrecht: Reidel.

Gibbons, A. (1998, Sept. 4). Which of our genes make us human? *Science, 281*(5383), 1432–1434.

Gibson, J. J. (1950). *The perception of the visual world*. Boston: Houghton-Mifflin.

Gibson, J. J. (1968). *The senses considered as perceptual systems.* London: Allen and Unwin.

Gibson, J. J. (1979). *The ecological approach to visual perception.* Boston: Houghton-Mifflin.

Giere, R. N. (Ed.). (1992). *Cognitive Models of Science, vol. 15.* Minnesota Studies in the Philosophy of Science, Minneapolis, University of Minnesota Press.

Goldsmith, T. H. (1991). *The biological roots of human nature (forging links between evolution and behavior)*. New York: Oxford University Press.

Goodsell, C. T. (1988). *The social meaning of civic space: Studying political authority through architecture.* Lawrence, KS: University Press of Kansas.

Gribbin, J., & Gribbin, M. (1997). *Richard Feynman (A life in science).* New York: Dutton.

Grmek, M. D., Cohen, R. S., & Cimino, G. (Eds.). (1981). *On scientific discovery.* Dordrecht, Reidel.

Gross, P. R., & Levitt, N. (1994). *Higher Superstition: The Academic Left and Its Quarrels with Science.* Baltimore: Johns Hopkins University Press.

Hacking, I. (1983). *Representing and intervening.* Cambridge: Cambridge University Press.

Hahlweg, K., & Hooker, C. A. (Eds.). (1989). *Issues in evolutionary epistemology*. Albany: SUNY Press.

Hakim, B. S. (1986). *Arabic-Islamic cities (Building and planning principles).* London: KPI.

Hakim, B. S. (1994, Summer). The 'urf' and its role in diversifying the architecture of traditional Islamic cities, *Journal of Architectural and Planning Research, 11*(2), 108–127.

He, T-C., Sparks, A. B., Rago, C., Hermeking, H., Zawel, L., da Costa, L. T., Morin, P. J., Vogelstein, B., & Kinzler, K. W. (1998, Sept. 4). Identification of c-MYC as a target of the APC pathway, *Science, 281*(5382), 1509–1512.

Heath, T. (1984). *Method in architecture.* Chichester, UK: John Wiley.

Hempel, C. G. (1965). *Aspects of scientific explanation and other essays in the philosophy of science.* New York: Free Press.

Himsworth, H. (1986). *Scientific knowledge and philosophic thought.* Baltimore: Johns Hopkins University Press.

Holland, J. H., Holyoak, K. J., Nisbet, R. E., & Thagard, P. (1986). *Induction (Processes of inference, learning, and discovery).* Cambridge, MA: MIT Press.

Horgan, J. (1995, Oct). The new social Darwinists. *Scientific American, 273*(4), 174–181.

Hufford, M. (1986). *One space, many places (Folklife and land use in New Jersey's pineland national reserve).* Washington, DC: American Folklife Center (Library of Congress).

Huisgen, R. (1994). *The adventure playground of mechanisms and novel reactions* (Volume in Profiles, Pathways, and Dreams; J. I. Seeman Series Editor). Washington, DC: American Chemical Society.

Humphreys, P. (1989). *The chances of explanation (Ccausal explanation in the social, medical, and physical sciences).* Princeton, NJ: Princeton University Press.

Jacob, F. (1982). *The possible and the actual.* New York: Pantheon Books.

Kafatos, F. C. (1998, May 29). Challenges for European biology. *Science, 280*(5368), 1327 (Editorial).

Kaiser, J. (1997, March 14). New institute seen as brains behind big boost in spending. *Science, 275*(5306), 1562–1563. (This is a special issue on cognitive neuroscience).

Kalgutkar, A. S., Crews, B. C., Rowlinson, S. W., Garner, C., Seibert, K., & Marnett, L. J. (1998, May 22). Aspirin-like molecules that covalently inactivate Cyclooxygenase-2. *Science, 280*(5367), 1268–1270.

Kaminsky, G. (1987). Cognitive bases of situation processing and behavior setting participation. In G. R. Semeni, & B. Krahé (Eds.), *Issues in contemporary German social psychology* (pp. 218–240). Beverly Hills: CA, Sage.

Kantorovich, A. (1993). *Scientific discovery (Logic and tinkering)*. Albany: SUNY Press.

Kaplan, A. (1964). *The conduct of inquiry*. New York: Harper and Row.

Kentros, C., Hargreaves, F., Hawkins, R. D., Kandel, E. R., Shapiro, M., & Muller, R. V. (1998, June 26). Abolition of long-term stability of new hippocampal place cell maps by NMDA receptor blockade. *Science, 280*(5372), 2121–2126.

Kitcher, P., & Salmon, W. C. (Eds.). (1989). *Scientific explanation, vol. 13.* Minnesota Studies in the Philosophy of Science, Minneapolis, University of Minnesota Press.

Kleinman, A., & Cohen, M. (1997, March). Psychiatry's global challenge. *Scientific American, 276*(3), 86–89.

Koertge, N. (Ed.). (1998). *A house built on sand: Exposing postmodernist myths about science.* New York: Oxford University Press.

Kornblith, H. (Ed.). (1985). *Naturalizing epistemology.* Cambridge, MA: MIT Press.

Kruse, L. (1988). Behavior settings, cognitive scripts, linguistic frames. In N. Prak et al. (Eds.) *IAPS 10* (pp. 106–109). Delft: The Netherlands.

Kuhn, T. (1965). *The structure of scientific revolutions* (1st ed.). Chicago: University of Chicago Press.

Lang, J. (1987). *Creating architectural theory.* New York: Van-Nostrand Reinhold.

Langley, P., Simon, H. A., Bradshaw, G. L., & Zytkow, J. M. (1987). *Scientific discovery (Computational explorations of the scientific process).* Cambridge, MA: MIT Press.

Langley, P., & Jones, R. (1988). A computational model of scientific insight. In R. J. Sternberg (Ed.), *The meaning of creativity* (pp. 177–201). Cambridge: Cambridge University Press.

Laudan, L. (1990). *Science and relativism (Some key controversies in the philosophy of science)* Chicago: University of Chicago Press.

Lawrence, R. J. (1986). *Le seuil franchi . . .* Geneva: Georg.

Liu, C-W. (1994). *From old town to new city: A study of behavior settings and meanings of streets in Taiwan.* Ph. D. Dissertation, Department of Architecture, University of Wisconsin-Milwaukee (2 vols.).

Lynch, K. (1981). *A theory of good city form.* Cambridge, MA: MIT Press.

McMullin, E. (1993) Rationality and paradigm change in science. In P. Horwich (Ed.) *World changes (Thomas Kuhn and the nature of science)* (pp. 55–78). Cambridge, MA: MIT Press (Bradford Books).

Menard, H. W. (1986). *The ocean of truth (A personal history of global tectonics).* Princeton, NJ: Princeton University Press.

Miller, R. W. (1987). *Fact and method (Explanation, confirmation, and reality in the natural and the social sciences).* Princeton: Princeton University Press.

Moore, G. T. (1987). Environment and behavior research in North America: History, developments, and unresolved issues. In D. Stokols & I. Altman (Eds.), *Handbook of environmental psychology, Vol. 2* (pp. 1359–1410). New York: Wiley.

Moore, G. T., Tuttle, D. P., & Howell, S. C. (1985). *Environmental design research directions.* New York: Praeger.

Morris. R. (1983). *Dismantling the universe (The nature of scientific discovery).* New York: Simon and Schuster.

Nagel, T. (1998, Oct. 12). The sleep of reason, review of A. Sokal and J. Bricmont. *Fashionable nonsense,* New York, Picador, *The New Republic, 4369,* 32–38.

Nasar, J. L. (1998). *The evaluative image of the city.* Thousand Oaks, CA: Sage.

Nash, L. K. (1963). *The nature of the natural sciences.* Boston: Little, Brown, and Co.

Nemeth, D. J. (1987). *The architecture of ideology (Neo-confucian imprinting on Cheju Island, Korea), Vol. 26.* Berkeley, CA: University of California Press, Geography.

Newton, R. G. (1997). *The truth of science (Physical theories and reality).* Cambridge, MA: Harvard University Press.

Newton-Smith, W. H. (1981). *The rationality of science.* Boston: Routledge and Kegan Paul.

Nickles, T. (Ed.). (1980a). *Scientific discovery, logic, and rationality.* Dordrecht: Reidel.

Nickles, T. (Ed.). (1980b). *Scientific discovery: Case studies.* Dordrecht: Reidel.

Norman, D. A. (1988). *The psychology of everyday things.* New York: Basic Books.

Norris, C. (1997). *Against relativism (Philosophy of science, deconstruction, and critical theory).* Oxford: Blackwell.

Nye, M. J. (1993). From chemical philosophy to theoretical chemistry (Dynamics of matter and dynamics of disciplines, 1800–1950). Berkeley: University of California Press.

Oxman, R. M. et al. (Eds.). (1995). *Design research in The Netherlands.* Faculty of Architecture, Planning and Building Science: Eindhoven University of Technology.

Papineau, D. (1993). *Philosophical naturalism*. Oxford: Blackwell.

Pellow, D. (Ed.) (1996) *Setting boundaries (The anthropology of spatial and social organization)*. Westport, CT: Bergin and Garvey.

Pennisi, E. (1998a, May 22). Building a better aspirin, *Science, 280*(5367), 1191–1192.

Pennisi, E. (1998b, Sept. 4). How a growth control path takes a wrong turn to cancer, *Science, 281*(5382), 1438–1441.

Phillips, D. C. (1976). *Holistic thought in social science*. Stanford, CA: Stanford University Press.

Prieditis, A. (Ed.). (1988). *Analogica: The first workshop on analogical reasoning*. Los Altos, CA: Morgan-Kaufmann.

Purcell, A. T., Williams, P., Gero, J. S., & Colbron, B. (1993). Fixation effects: Do they exist in design problem solving?, *Environment and Planning B: Planning and Design, 20*, 333–345.

Rapoport, A. (1969). *House form and culture*. Englewood Cliffs, NJ: Prentice-Hall.

Rapoport, A. (1970). The study of spatial quality, *Journal of Aesthetic Education, 4*(4), 81–95.

Rapoport, A. (1973) An approach to the construction of man-environment theory. In W. F. E. Preiser (Ed.), *Environment Design and Research (EDRA4), Vol. 2* (pp. 124–135). Stroudsburg, PA: Dowden, Hutchinson, and Ross.

Rapoport, A. (1975a) Aging-environment theory: A summary. In P. Windley et al. (Eds.), *Theory development and aging* (pp. 263–281). Washington, DC: Gerontolgical Society.

Rapoport, A. (1975b). An 'anthropological' approach to environmental design research. In B. Honikman (Ed.), *Responding to social change (EDRA 6)* (pp. 145–151). Stroudsburg, PA: Darden, Hutchinson, and Ross.

Rapoport, A. (1977). *Human aspects of urban form*. Oxford: Pergamon Press.

Rapoport, A. (1980). Towards a cross-culturally valid definition of housing. In R. R. Stough, & A. Wandersman (Eds.), *Optimizing environments (Research, practice, and theory)* (pp. 310–316). (EDRA 11). Washington, DC: EDRA.

Rapoport, A. (1981). Identity and environment: A cross-cultural perspective. In J. S. Duncan (Ed.) *Housing and identity: Cross-cultural perspectives* (pp. 6–35). London: Croom-Helm.

Rapoport, A. (1983a). The effect on environment on behavior. In J. B. Calhoun (Ed.), *Environment and population (Problems of adaptation)* (pp. 200–201). New York: Praeger.

Rapoport, A. (1983b). Debating architectural alternatives, *RIBA Transactions 3, 2*(1) (20th c. Series), 105–109.

Rapoport, A. (1985) Thinking about home environments: A conceptual framework. In I. Altman, and C. M. Werner (Eds*.) Home environments (Vol. 8 of Human behavior and environment)* (pp. 255–286). New York: Plenum

Rapoport, A. (1986a). Culture and built form–a reconsideration. In D. G. Saile (Ed.), *Architecture in cultural change (Essays in built form and culture research)* (pp. 157–175). Lawrence: University of Kansas.

Rapoport, A. (1986b) The use and design of open spaces in urban neighborhoods. In D. Frick (Ed.), *The quality of urban life: Social, psychological, and physical conditions* (pp. 159–175). Berlin: de Gruyter.

Rapoport, A. (1988a). Review of J. Lang Creating architectural theory, New York, Van-Nostrand Reinhold, 1987. *Design Book Review 15* (Fall), 69–70.

Rapoport, A. (1988b). Spontaneous settlements as vernacular design. In C. V. Patton (Ed.), *Spontaneous shelter* (pp. 51–77). Philadelphia: Temple University Press.

Rapoport, A. (1988c). Levels of meaning in the built environment. In F. Poyatos (Ed.), *Cross-cultural perspectives in non-verbal communication* (pp. 317–326). Toronto: C. J. Hogrefe.

Rapoport, A. (1989a) On the attributes of tradition. In J. P. Bourdier, & N. Alsayyad (Eds.) *Dwellings, settlements, and tradition* (pp. 77–105). Lanham, MD: University Press of America.

Rapoport, A. (1989b). A different view of design. *The University of Tennessee Journal of Architecture, 11*, 28–32.

Rapoport, A. (1990a). Science and the failure of architecture. In I. Altman, & K. Christensen (Eds.), *Environment and behavior studies: emergence of intellectual traditions, Vol. 11 of Human Behavior and Environment*, 79–109. New York: Plenum.

Rapoport, A. (1990b). *History and precedent in environmental design*. New York: Plenum.

Rapoport, A. (1990c). Defining vernacular design. In M. Turan (Ed.), *Vernacular architecture (Paradigms of environmental response)* (pp. 67–101). Aldershot, UK: Avebury.

Rapoport, A. (1990d). *The meaning of the built environment*. Tucson: University of Arizona Press (Revised edition).

Rapoport, A. (1990e). Levels of meaning and types of environments, Y. Yoshitake et al. (Eds.) *Current issues in environment-behavior research, Proceedings of 2nd Japan—US Seminar, Kyoto 1990* (pp. 135–147). Tokyo: University of Tokyo.

Rapoport, A. (1990f). Systems of activities and systems of settings. In S. Kent (Ed.) *Domestic architecture and the use of space* (pp. 9–20). Cambridge: Cambridge University Press.

Rapoport, A. (1992a). On regions and regionalism. In N. C. Markovich et al. (Eds.), *Pueblo style and regional architecture* (pp. 272–294). New York, Van Nostrand-Reinhold (Paperback Edition).

Rapoport, A. (1992b, Spring). On cultural landscapes. *Traditional Dwellings and Settlements Review, 3*(2), 33–47.

Rapoport, A. (1993a). *Cross-cultural studies and urban form*. College Park, MD: Urban Studies and Planning Program, University of Maryland.

Rapoport, A. (1993b) A critical look at the concept 'place'. In R. P. B. Singh (Ed.), *The spirit and power of place: Human environment and sacrality (Essays dedicated to Yi-Fu Tuan)* (pp. 31–45). Varanasi, Banaras Hindu University (National Geographical Society of India).

Rapoport, A. (1993c). On the nature of capitals and their physical expression. In J. Taylor et al. (Eds.), *Capital cities (International perspectives)* (pp. 31–67). Ottawa: Carleton University Press.

Rapoport, A. (1994a). Spatial organization and the built environment. In T. Ingold (Ed.), *Companion encyclopedia of anthropology: Humanity, culture, and social life* (pp. 460–502). London: Routledge.

Rapoport, A. (1994b). A critical look at the concept 'place'. *National Geographic* Journal *of India, 40*, 31–45.

Rapoport, A. (1994c). *Sustainability, meaning, and traditional environments*. Berkeley, CA, University of California, Center for Environmental Design Research, Traditional Dwellings and Settlements Working Paper Series, vol. 75/IASTE (pp. 75–94).

Rapoport, A. (1995a). Rethinking design. In E. Schaur (Ed.), *Building with intelligence* (pp. 236–242). ILF, University of Stuttgart, IL 41.

Rapoport, A. (1995b). On the nature of design, *Practices, 3/4*(Spring), 32–43.

Rapoport, A. (1995c). *Thirty-three papers in environmental behavior research*. Newcastle, UK: Urban International Press.

Rapoport, A. (1995d) A critical look at the concept 'home'. In D. Benjamin (Ed.), *The Home: Words. Interpretations, Meanings, and Environments* (pp. 25–52). Aldershot, UK: Avebury.

Rapoport, A. (1995e). Response to the theme of EDP 95. In M. Deobhakta (Ed.), *Education of a design professional (Synthesis of tradition and modernity for a sustainable society)* (CAA/WSE Conference Document) (pp. 11–18). Mumbai. India: Rudra Sansthapan Publications.

Rapoport, A. (1997a). Theory in environment-behavior studies: Transcending times, settings, and groups. In S. Wapner et al. (Eds.), *Handbook of Japan-United States environment-behavior research (Toward a transactional approach) (pp. 399–421)*. New York: Plenum.

Rapoport, A. (1997b). The nature and role of neighborhoods. In M. Shokoohy (Ed.), *Urban design studies, 3*, 93–118. London: School of Architecture and Landscape, University of Greenwich.

Rosch, E., & Lloyd, B. B. (Eds.). (1978). *Cognition and categorization*. Hillsdale, NJ: Erlbaum.

Rubinstein, R. A., Laughlin, C. D. Jr., & McManus, J. (1984). *Science as cognitive process (Toward an empirical philosophy of science)*. Philadelphia: University of Pennsylvania Press.

Salmon, W. C. (1990). *Four decades of scientific explanation*. Minneapolis: University of Minnesota Press.

Salmon, W. C. (1998). *Causality and explanation*. New York: Oxford University Press.

Schaffner, K. F. (1993). *Discovery and explanation in biology and medicine*. Chicago: University of Chicago Press.

Sellers, P. J., Dickinson, R. E., Randall, D. A., Betts, A. K., Hall, F. G., Berry, J. A., Collatzi, G. Denning, A. S., Mooney, H. A., Nobre, C. A., Sato, N., Field, C. B., & Henderson, A. (1997, Jan. 17). Modeling the exchanges of energy, water, and carbon between continents and the atmosphere. *Science, 275*(5398), 502–509.

Shapere, D. (1977). Scientific theories and their domains. In F. Suppe (Ed.), *The structure of scientific theories (2nd ed.)* (pp. 518–573). Urbana: University of Illinois Press.

Shapere, D. (1984). *Reason and the search for knowledge*. Dordrecht: Reidel.

Shrager, J., & Langley, P. (Eds.). (1990). *Computational models of scientific discovery and theory formation*. San Mateo, CA: Morgan Kaufman.

Slobodkin, L. R. (1992). *Simplicity and complexity in games of the intellect*. Cambridge, MA: Harvard University Press.

Sokal, A., & Bricmont, J. (1998). *Fashionable nonsense: Postmodern philosophers' abuse of science*. New York: Picador.

Sontag, D. (1998). Devout community burgeons in Brooklyn, *New York Times* (Jan 7).

Suppe, F. (Ed.). (1977). *The structure of scientific theories (2nd ed.)*. Urbana: University of Illinois Press.

Suppe, F. (1989). *The semantic conception of theories and scientific realism*. Urbana: University of Illinois Press.

Taubes, G. (1997, Jan. 17) Center seeks synthesis to make ecology more useful. *Science, 275*(5298), 310–311.

Thagard, P. (1988). *Computational philosophy of science.* Cambridge, MA: MIT Press (Bradford Books).

Thagard, P. (1989, Sept). Explaining coherence. *Behavioral and brain sciences, 12*(3), 435–502.

Vayda, A. P. (1983, Sept). Progressive contextualization, *Human Ecology, 11*(3), 265–282.

Weber, R. L. (Ed.). (1982). *More random walks in science.* London: British Institute of Physics.

Whewell, W. (1840). *The philosophy of inductive science.* London: Parker.

Willer, D., & Willer, J. (1973). *Systematic empiricism: Critique of a pseudoscience.* Englewood Cliffs, NJ: Prentice-Hall.

Williams, F. (1997, Dec. 11). A house, 10 wives: Polygamy in suburbia. *New York Times.*

Wilson, E. O. (1998). *Consilience: The unity of knowledge.* New York: Knopf.

Wilson, T. P. (1996, Nov. 1). Letter to the editor. *Science, 274*(5288), 703–704.

Yoon, H-K. (1986). The nature and origin of Chinese geomancy. *Eratosthène, 1,* 88–102.

Yoon, H-K. (1991). On geomentality. *Geo Journal, 25*(4), 387–392.

Yoon, H-K. (1992a). The expression of landforms in Chinese geomantic maps, *The Cartographic Journal, 24,* 12–15.

Yoon, H-K. (1992b). Maori identity and Maori geomentality. In D. Hooson (Ed.), *Geography and National Identity* (pp. 293–310). Oxford: Blackwell.

Yoon, H-K. (1994). Two different geomentalities, two different gardens: The French and Japanese cases. *Geo Journal, 33*(4), 471–477.

11

LINKING BUILT ENVIRONMENTS TO EVERYDAY LIFE

Assumptions, Logic, and Specifications

William Michelson

Department of Sociology
University of Toronto
Toronto, Ontario, Canada M5T 1P9
and Faculty of Arts and Sciences
University of Toronto
Toronto, Ontario, Canada M5S 3G3

1. INTRODUCTION

Imagine two circles. One is small. The other is big. Within each circle is a body of knowledge. Outside each circle is what's not known. The larger the amount of what's known, the greater the perimeter of the circle. Therefore, the more that is known, the greater exposure there is to what's not known. In other words, the more you learn, the more you realize how much more there is yet to learn. This simple perspective[1] illustrates the theme of this paper on the study of built environments. The pursuit of empirical research in this field brings out a greater awareness of conceptual and methodological challenges yet to be overcome.

This paper is consciously personal, to give "full and open disclosure" (Altman, 1997, p. 433) of the assumptions embedded in my work—and the challenges they have presented. My abiding interest within environment-behavior research is the extent and ways in which built environments impact upon people's everyday lives. I shall attempt to describe at first the outlook with which I started with this interest, the conceptual state of the field at that time, and then the increasing challenges with which I had to deal as I proceeded. Only after about thirty-five years in this pursuit do I start to get a feeling of gaining an initial grasp on the reality sought.

---

[1] It must be noted clearly that this outlook is not original, but goes back to ancient Greece.

Theoretical Perspectives in Environment-Behavior Research, edited by Wapner *et al.*
Kluwer Academic / Plenum Publishers, New York, 2000.

2. LINKING PEOPLE AND ENVIRONMENTS

2.1. Initial Outlook

In 1959–1960, as an undergraduate student of Sociology, I had the opportunity to study for a year in Denmark.[2] The housing, urban development, and ways of life I encountered in Scandinavia that year left me with curiosity about the potential for built environments at various levels to enhance the quality of everyday life. Taking my first course in Urban Sociology the next year, I found the traditional substance of the field extremely interesting; but it did not address this curiosity. No linkage was made between built environments and everyday life. I attempted to delve further as a graduate student, through a fortunate combination of course selection, research assistance, and independent writing. This interest was considered so unusual that my faculty advisor, despite being a specialist in interdisciplinary perspectives, asked me several times if I was still in his program. Nonetheless, I persevered. I carried out dissertation research examining the nature and extent of linkages between people's value orientations and their preferences for aspects of urban form (Michelson, 1965).[3] This was the beginning of my attempts to link built environments to people's lives. The choice was made to study people's subcultures with regard to built environment because it seemed obvious to me that people of different subcultures put different demands on their urban space. It was also the beginning of further needs to specify this kind of relationship more accurately and completely, insofar as environmental choice was found related to consistent rationales in this sector of life but not to general value orientations.

The experience of this dissertation left me with the reaction that what the field needed was not simply more and better research of the same kind but rather a more thoroughgoing conceptualization of how built environments (referred to then as urban environment) could be linked to people and their everyday lives. At that time, there was not an absence of theories dealing with environment and behavior. Lewin (1936) led the way with his field theory; but his theory was one dealing with what appears to be the path of least resistance, not clearly articulated variables for either environment or behavior. Barker's behavior setting theory (1968) linked a number of considerations; but they were considered component aspects of a single phenomenon, not independent of each other, and behavior settings, in any case, are just a limited set of everyday environments and are limited to relatively microscopic scales. The Ekistics grid makes cross-classifications of knowledge at different scales of environment and human group, but contains no dynamic elements of the linkages on its own.[4] The proxemics perspective put forward by Hall (1966) does contain a dynamic linkage between human spacing for specific types of behavior and largely micro environmental settings; but its scope is relatively limited to cultural spacing norms and relevant aspects of built environment such as furniture arrangements, however important these might be at times. The mainstream theoretical approach to urban environment and behavior was still that of the Chicago school of Human Ecology, which did not treat space as a tangible, dynamic

---

[2] I am grateful to Edward A. Tiryakian and Princeton University for their support for this opportunity.
[3] Encouragement for this academic leap from traditional substance came from Charles Tilly (instrumentally, as supervisor), Martin Meyerson, Florence Kluckhohn, and Talcott Parsons.
[4] See the full run of Ekistics, the journal.

variable, only an epiphenomenon of economic land value combining and separating groups of people and land-uses (c.f. Michelson, 1970). Indeed, a major sociological analyst of the time counselled that, "We must root out of our thinking . . . the assumption that the physical form of our communities has social consequences" (Glazer, 1965, pp. 57–59).

A major field of environment-behavior research by sociologists at that time, on the impact(s) of high density on people, was almost totally uninformed by any theoretical linkage. A number of studies exhibited correlations of varying strengths between such measures of density and of supposed outcomes as could be found through available statistics (usually not at the individual level but at the aggregate level, a problem in itself); but there was seldom any theoretical justification as to why the particular pair of variables were linked. The exceptions were animal studies in which glandular malfunctioning appeared to accompany the onset of pressures of higher densities with fixed resources; but there were limitations as to how well animal studies could apply to human life and its expanded forms of communication (c.f. Calhoun, 1963).

I thus had a firm assumption, counter to Glazer's, but found it necessary to seek linkages between people and their urban environments with more explicit dynamics.

2.2. Marginal Progress

I attempted to establish a linkage between people and built environments in a book (Michelson, 1970)[5] which centered around the concept of congruence. It adopted the systems approach of the sociological theorist, Talcott Parsons, expanding his three systems (social, personal, and cultural) to include a fourth, the environmental system. The original systems were linked to each other as representing mutually limiting factors. The cultural system of a society, for example, would bring with it limitations on the nature and functioning of phenomena in the social system—and vice versa. I tried to suggest in this book that built environments play the same type of role with respect to the other systems—to limit which expressions of social, personal, and cultural systems might easily take place within the spatial parameters of built environments. This kind of relationship is not deterministic. It points rather to the need to understand which system characteristics co-exist more easily and hence might be considered more compatibly for purposes of analysis and practice. Some person-environment relationships are congruent, and others are incongruent. Do high residential densities, for example, foster the interaction patterns of some cultures but not others? Such a conceptual view might more accurately be called possibilistic, neither deterministic or even probabilistic. The book started with an attempt to characterize the unique parameters of built environments and then turned to evidence from then-current literature to link particular aspects of built environment to congruent aspects of the three main systems covered by the social sciences.

But while this conceptualization made it more possible to link factors having to do with built environments with characterizations of human groups—independent phenomena with interacting relationships—what I realized then is that it did not

---

[5] A translation to Japanese was subsequently made by Shoji Ekuan.

contribute more than marginal progress because it did not provide a dynamic explanation of how these factors impacted on one another. The factors may have been linked, but not in a way which gave understanding of any kind of causal sequence or necessary association. Further reflection led to the realization that a third type of phenomenon provides more meaningful linkage between the social science systems and built environments, behavior (c.f. Michelson, 1987). It is behavior which actually occurs within built environments. While precipitated by factors in the social, personal, and cultural systems, behavior is given opportunity to occur or is constrained by the existence and design of particular built environments. Behavior is a mediating factor in the linkage between people and their built environments. Life style was seen as a contemporary, though somewhat imprecise, concept with which to view the appropriateness and impacts of built environments among diverse cultures and groups (Michelson & Reed, 1974).

3. APPROPRIATE MEASURES

Therefore, when formulating subsequent research plans to test what differences housing types make in people's everyday lives, it became essential to gather explicit data on people's behavior. Not just a single kind of behavior, but the wide range of what people do during their everyday lives. Not just in a single place, but in the package of contexts which constitute people's habitats. However, now that I was sufficiently aware of both the kinds of linkages which joined people and environments, as well as the critical place of behavior in this regard, another unexpected problem came into focus. For all the study of particular behaviors in the social sciences (e.g., children's play, crime, courtship, learning), little emphasis had been put on an integrative measure of everyday behavior. For all the study of behavior settings (in built environments), measures of behavior occurring both within and between the wide range of public and private settings were notably absent. Once this became evident, it was crucial to devise such measures, in which people's characteristics, behavior, and physical setting were measured but not synonymous.

One of the complexities of doing research on built environment is that both the substance of theory and the data suggested as appropriate by theoretical substance are highly variable. They vary by the type of human group, by the purpose of the unit(s) of built environment involved, and by the scale of one or both of them. In other words, there are different dynamics and behaviors involved when the focus is on a nuclear family than on a crowd at a sporting event. Behaviors to be observed in schools are not the same as those in hospitals, even though both deal with many people and large buildings. And phenomena appropriate to residential kitchens differ markedly from those concerning hotel restaurant kitchens. It is no wonder that no single theory provides the necessary person-environment linkages to all questions. And then, because theoretical conceptualizations are so variable, so are the phenomena which must be measured in a given study. While behavior is surely a vital focus for measurement, it is not self-evident that appropriate measures for the types of behavior relevant for a given study exist. Certainly, no single measure fits all purposes. In my studies of housing type (Michelson, 1977, 1993a,b, 1994), I have found it helpful to use a variety of methods to study behavior within and between settings. Again, this was not fully anticipated when the research thrust was first contemplated.

The research technique which most fully addressed everyday behavior(s)—and which had to be adapted to built environment research—is called the time-budget. Collected from people either as diaries or through interview techniques, time-budgets are an account of what the person did during a fixed time period. This is usually a single day, though sometimes it covers several days or even a week. The time-budgets are a chronological account of what a respondent did from start to finish during the period covered. They take the form of a matrix of information, in which each successive episode is a row, while different aspects of the episode of behavior form the columns: the overt nature of the activity (e.g., sleeping, sewing, working, playing), when it started, when it finished, what secondary activity may have been happening simultaneously, who else was with the respondent, where it occurred, and, often, subjective aspects or outcomes of the activity. With this technique, the wide-range of everyday behavior (save for secret or embarrassing activities) are covered, as are the various contexts used by respondents. Furthermore, each episode of behavior is grounded in its social and physical contexts, as well as in the total pattern of behavior constituting the day.

Time-budget research has been practiced throughout most of the century just ending. However, most usages have been for macro planning purposes such as industrial location and transportation planning, or just for descriptive comparisons of aggregate behavior in different nations. Nonetheless, the link at the episode level between behavior and place could be refined to refer to very specific places pertinent to housing and design considerations. For example, the comparison of families living in high rise apartments to those living in single-family homes showed a number of differences.[6] The apartment dwellers spent much more time outside their apartments, with less active pursuits once home. The house dwellers were not only engaged more actively but had this take the form of entertaining more often; they also spent much more time in home maintenance, as might be expected. A later study, on the impacts of experimental design characteristics in several Swedish housing projects (compared to conventional control areas) indicated that such social objectives as enhanced social contact and reduced difficulty in child care were more likely to be fulfilled in residential areas which incorporated innovations in both physical design and social organization.[7] Reference in the time-budget protocols to location and more specifically to particular kinds of rooms and spaces enabled the testing of design hypotheses having to do with intended behaviors. It also enabled the gathering of information on the use of spaces not readily open to observation and on the simultaneous activities of the various family members and co-residents of residential areas.

This technique is helpful also for the study of macroscopic environments such as cities and metropolitan areas. For example, in a third study, on the daily lives of employed mothers and their families,[8] the interface of domestic responsibilities with the purposes, means, and extent of daily travel were shown to be extremely important in the understanding of stress among employed women with children (Michelson, 1985, 1986). These measures underscored the importance of domestic division of labor, urban land-use and transportation systems, and flexibility at the workplace.

Nonetheless, just as the time-budget measures provide data which are unparal-

---

[6] This study was supported by grants from Canada Mortgage and Housing Corporation and from Canada Council.

[7] Conducted in association with Professor Birgit Krantz and the Department of Building Functions Analysis, University of Lund, Sweden, this study was funded by the Swedish Building Research Council.

[8] Support for this study came from a donation by the Federal Ministry of Health and Welfare, Government of Canada, and Grant CA-11-0024 from the U.S. Department of Transportation, Urban Mass Transportation Administration.

leled in their scope and integration, observational techniques enable data whose qualitative nature is not evident from the time-budgets. For example, in the Swedish experimental housing study, we were able to study the nature of activity and participants in social contacts found in the common spaces of eight housing projects; the contributions of student research personnel enabled simultaneous systematic observations at fixed days and times. This kind of observation gave more insight into who got together and what they did in spaces with particular design objectives. For example, the time-budgets told us that the indoor common spaces in one housing area were largely used (during the winter) by children. The observations indicated that a combination of loud play by the children in relatively unstructured spaces and low temperatures found there created a vacuum that adults did not rush to fill.

This study also demonstrated the value of conventional survey methods, to fill gaps that the former two research approaches left. The survey was able to elicit information on people's residential histories, motivations, and subjective evaluations which do now flow from the more objectively-narrow time-budgets or the behavioral and qualitative directions of systematic observation.

Thus, appropriate measures are at once a challenge and a basis for rewarding innovation in the diverse person-environment research context, in which the use of customary, proven methods cannot be assumed to apply to new research projects.

4. NON-CONTEXTUAL FACTORS

For better or for worse, though, even crystal clear results from carefully-chosen measures do not always provide sufficient information for understanding the phenomenon studied. Factors outside the person-environment context may take one or more roles in explanation. For example, in the high-rise apartment study, even the most comprehensive data on people's behavior in different housing types did not deal with causality. A typical theoretical dilemma revolves around whether the specific residential site provides such opportunities that common behavior is fostered among its residents or whether, to the contrary, only people already practicing a particular set of behaviors were induced to move to the site. The former might be appropriately called situated behavior (Michelson, 1977), while the latter has been called self-selection (Bell, 1958). It took a longitudinal design in our study, not just appropriate cross-sectional measures, to indicate that both processes accounted, differentially, for the behavior differences separating high-rise apartments from single-family homes.

Moreover, even then, the relevance of the findings required reference to additional non-contextual factors. We had assumed that people not interested in or doing the modal behaviors in specific residential settings would be less happy there than those happily engaged in congruent activities. Residential satisfaction was found related (through additional questions fortuitously placed in the conventional survey part of data gathering) not primarily to environment-behavior congruence but rather to the extent that families felt that they could eventually achieve their residential aspirations. People living in apartments, for example, typically hoped to move in the future to very different environments; their satisfaction with the apartments depended on whether they felt they had a realistic chance to move out of them. Very similar people varied in their housing satisfaction, dependent on not feeling trapped forever in their current housing. This finding brought up the importance of considerations completely outside our data set—and unanticipated at the start. The nature and cost of different housing types in the housing market was essential in the understanding the person-environment dynamic. It led to the recommendation that public policy focus on the provision of new single-family houses with less space, fewer amenities, and, of paramount importance, lower costs than had become the norm at the time.

5. ABSOLUTE VALUES

Again, learning some things about the person-environment relationship leads to the realization that there are other aspects still not clear. In the case of my studies, it was the realization that even with attempts to conceptualize and document the extent and reasons for relationships between built environments, behavior, and different types of people, the absolute value of this information is not clear. Long ago, Abraham Maslow (1954) made clear that what people value most varies with their socio-economic status (c.f. Rainwater, 1966). People with nothing appreciate the most elemental aspects of housing. People with everything still find reasons for dissatisfaction and the need to change. The question of how much the environment-behavior linkage actually matters for people is typically external to empirical research. Most researchers are motivated by the assumption that the satisfaction of user needs is an end in itself. This, too, needs examination.

The Swedish experimental housing study finally provided some evidence on this matter, helped by the cultural situation that multiple-family housing is not overshadowed by aspirations primarily for single-family housing there. In one of our case studies, we found that persons pursuing the modal behaviors found in their housing area were much more likely to expect to remain there in the future. Stayers were typically active socializers and pursued a range of active leisure activities. Potential future movers were, to the contrary, almost entirely dedicated to spending time in passive, private leisure activities. This provides some suggestion, despite non-contextual factors which may impinge at any time or place, that environment-behavior fit may indeed matter.

6. CONCLUDING REMARKS

Hopefully, our circles continue to enlarge. This should be a source of encouragement. But it must not be a source of complacency. It brings with it the need to attend more and more to more and more which was previously unseen. It is a potential source of satisfying and engrossing research.

REFERENCES

Altman, A. (1997). Environment and behavior studies: A discipline? Not a discipline? Becoming a discipline? In S. Wapner, J. Demick, T. Yamamoto, & T. Takahashi (Eds.), *Handbook of Japan-United States environment-behavior research* (pp. 423–434). NY: Plenum.
Barker, R. (1968). *Ecological psychology*, Stanford, CA: Stanford University Press.
Bell, W. (1958). Social choice, life styles, and suburban residence. In W. M. Dobriner (Ed.), *The suburban community* (pp. 225–247). NY: Putnam.
Calhoun, J. B. (1963). Population density and social pathology. In L. J. Duhl (Ed.), *The urban condition* (pp. 33–43). NY: Basic Books.
Glazer, N. (1965). Slum dwellings do not make a slum. *New York Times Magazine* (pp. 55ff.), Nov. 21.

Hall, E. T. (1966). *The hidden dimension*. Garden City, NY: Doubleday.

Lewin, K. (1936). *Principles of topological psychology*. NY: McGraw-Hill.

Maslow, A. (1954). *Motivation and personality*. NY: Harper and Row.

Michelson, W. (1965). *Value orientations and urban form*. Cambridge, MA: Harvard University, Ph.D. thesis.

Michelson, W. (1970). *Man and his urban environment: A sociological approach*. Reading, MA: Addison-Wesley. (translated to Japanese by Shoji Ekuan, Hozansha, 1975).

Michelson, W. (1977). *Environmental choice, human behavior, and residential satisfaction*. NY: Oxford University Press.

Michelson, W. (1985). *From sun to sun: Daily obligations and community structure in the lives of employed women and their families*. Totawa, NJ: Rowman & Allanheld.

Michelson, W. (1986). Divergent convergence: the daily routines of employed spouses as a public affairs agenda. *Public Affairs Report, 26*(4): entire issue for August, 1995.

Michelson, W. (1987). Congruence: the evolution of a contextual concept. In W. van Vliet, H. Choldin, W. Michelson, & D. Popenoe (Eds.), *Housing and neighborhoods* (pp. 19–28). NY: Greenwood Press.

Michelson, W. (1993a). The behavioral dynamics of social engineering: lessons for family housing. In E. Arias (Ed.), *The meaning and use of housing* (pp. 303–325). Aldershot: Avebury.

Michelson, W. (1993b). Grounding time-use in microspace: empirical results. *Social Indicators Research, 30*, 121–137.

Michelson, W. (1994). Measuring new objectives in suburban housing. In M. Baldassare (Ed.), *Research in Community Sociology, 4*, 253–269. Greenwich, CT: JAI Press.

Michelson, W., & P. Reed (1974). Life style in environmental research. In C. Beattie, & S. Crysdale (Eds.), *Sociology Canada: Readings* (pp. 406–419). Toronto: Butterworths.

Rainwater, L. (1966). Fear and the house-as-haven in the lower class. *Journal of the American Institute of Planners, 32*, 23–31.

12

A HYPOTHETICAL MODEL OF ENVIRONMENTAL PERCEPTION

Ambient Vision and Layout of Surfaces in the Environment

Ryuzo Ohno

Department of Built Environment
Interdisciplinary Graduate School of Science and Engineering
Tokyo Institute of Technology
Yokohama, 226-8502, Japan

1. INTRODUCTION

This paper reexamines a hypothetical model of environmental perception under-lying my research development since 1978 when I started to study visual perception of texture in an environmental context. While studying the perception of texture in the environment, it became clear that perception of discrete elemental features (objects) and continuous environmental features (textures) are quite different: texture, which is perceived without focal attention, creates a context in which attended objects stand out, and contributes to enhance the subtle ambience or feelings of the environment (Ohno, 1980; Ohno & Komuro, 1984).

Beyond the study on texture, my research interest has extended to our uncon-scious process of perception for continuous environmental features in general (Ohno, 1985). Unlike most psychological research, the research emphasis lies not in the psy-chological process itself but in the source of information in the physical environment, more particularly in the method to describe relevant physical features which are poten-tially sensible.

Although discussions in this paper deal with visual perception, the dichotomy here is a part of the comprehensive model of multi-modal perception in the environ-ment, in which our senses are distinguished between two basic modes: subject centered (autocentric) and object centered (allocentric). The former concerns people's feeling and pleasure whereas the latter is concerned with objectification and understanding, and involves attention and directionality (Rapoport, 1977). This differentiation among

Theoretical Perspectives in Environment-Behavior Research, edited by Wapner *et al.*
Kluwer Academic / Plenum Publishers, New York, 2000.

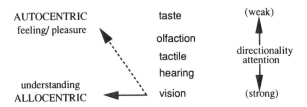

Figure 1. Differentiation among senses.

senses is shown in Fig. 1 where vision is most allocentric, but there is also autocentricity in vision which is emphasized in this paper.

2. A HYPOTHETICAL MODEL OF ENVIRONMENTAL PERCEPTION

The hypothetical model of environmental perception which has served as a foundation for my studies was constructed by combining two theories: one concerned perceptual process, namely the parallel processing theory, and the other concerned description of the environmental features based on Gibson's theory (1966, 1979).

2.1. Parallel Processing Theory: Focal and Ambient Vision

My attention to the parallel processing theory was initiated by Ittelson's (1973) discussion on the nature of environmental perception as contrasted with object perception. Among other informational content of the environment he pointed out that environments include peripheral as well as central information and provide more information than can possibly be processed. In fact, environments include enormous amounts of information from which we gain necessary information very fast and without appreciable effort. How do we deal with this mass of incoming information from the environment?

Computer scientists think that the quick and effortless performance of our vision is likely to have a computational basis using a parallel algorithm, and parallel processing may be essential for competent vision systems because the architecture of serial processing is too inefficient to deliver the massive amount of computation required (Ballard, Hilton, & Sejnowski, 1983).

The parallel processing in the visual system has been also suggested by physiological and psychological studies. It was suggested that parallel visual pathways in animals, which concurrently analyze different properties of the visual scene, subserved separate visual functions (e.g., Trevarthen, 1968). Two roles of vision were distinguished: one, ambient vision, involved in orienting the animal in space and guiding its larger movements, the other, focal vision, used for the detailed examination and identification of objects. Evidence of similar parallel pathways has also been demonstrated in humans (Bassi, 1989).

Although ambient and focal vision are different in their functions, they are considered to interact with each other to obtain perceptual synthesis. On the interplay between the two channels of visual analysis, Julesz and Bergen (1983) regarded ambient vision as a preattentive visual system. Based on findings of experiments on texture discrimination, they suggested the existence of a separate preattentive visual system that cannot process complex forms, yet can, almost instantaneously and without conscious effort or scrutiny, detect differences in a few local conspicuous features regardless of where they occur. Preattentive vision, therefore, was believed to serve as an early warning system by pointing out those loci that should be attended to. With this interplay of two visual systems, we can pick up wanted information from a wide area of the environment with limited attentional effort.

When we treat perception as an information processing system, the two visions can be characterized by the difference of strategies for controlling the flow of information. Focal vision eliminates unwanted information by selective attention while enhancing the elements attended to. Ambient vision deals with broader areas, with scattered or unconscious attention, and provides quick global impressions.

Although the above discussions were extracted from experiments under special laboratory situations, a hypothetical model of environmental perception can be formulated based upon the dual mode of vision. The dichotomy of the two visions is schematically represented in Fig. 2, in which focal vision is characterized by an active visual line fixated on an object, while ambient vision is characterized by numerous radiant visual lines converging on the station point, which I call "visual radiation."

2.2. A Source of Ambient Information: Continuous Environmental Surfaces

In order to make the preceding discussion feasible for environmental design, we have to know the relevance of the physical features in the environment to focal and ambient information. We cannot simply determine that a certain element always provides focal information, since that depends on human action and mental state, namely arousal, motivation and/or expectation. We can however estimate the relation by considering common behavior in a certain places. Considering usual situations, we can tell what kind of physical features tend to provide focal or ambient information.

We can assume that the continuous surfaces which surround us are usually processed by ambient vision, while discrete objects scattered in the visual field are processed by focal vision.

The author's attention to surfaces in the environment has stemmed from Gibson's discussions (Gibson, 1966) on the environment as a source of stimulation in which he showed a new approach as contrasted with traditional object-oriented one. The essential difference between the two approaches seem to be derived from different conceptions of the basic units which convey stimulus information in the visual field, in other words, the differences in ways to abstract the visual field. Landwehr (1984) clearly

Figure 2. A schematic representation of the two visions.

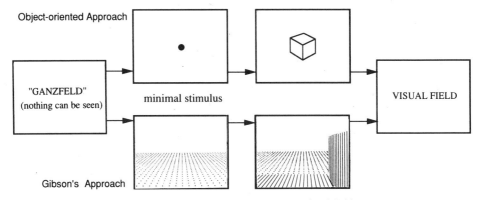

Figure 3. Two approaches for filling up visual field.

demonstrated this difference as shown in Fig. 3, in which the homogeneous "Ganzfeld" where nothing is seen is supposed to be filled up in two different ways. Traditional theory begins with a minimal point and then extends to lines, figures, and abstract solids. Gibson's theory on the other hand begins with a textured horizontal plane as a minimal stimulus information, which may be associated with the plane of the earth's surface, and extends various layouts of textured surfaces, or "ambient optic arrays" (Gibson, 1979).

Both of these different approaches can be accepted if we consider that each of the two approaches puts emphasis on one side or the other of the two aspects of the visual system. The object-oriented approach may explain more about perception of individual elements by focal vision, whereas Gibson's theory may explain more about perception of larger fields by ambient vision. As a hypothetical model, the differences of the two visions discussed in the previous section and the relevant source of information are summarized in Table 1.

Table 1. A Hypothetical Model of Environmental Perception: A Dichotomy of the Two Visions

	Focal vision	Ambient vision
Perceiver's attitude	Focal attention/conscious/active	Scattered attention/unconscious/passive
Visual pathway	Visual cortex	Superior colliculus*
Nature of information processing	Time consuming process	Instant process
	Detailed information per area of visual field	Limited information per area of visual field
	Perceptual selection	Perceptual integration
Outcome/function	Recognition of objects	Body posture/locomotion
		Attention evocation/orientation
	Understanding	Global impression/feeling
Source of information	Discrete elemental features (objects)	Continuous environmental features (surfaces)

*The superior colliculus has recently been found as a second projection area in the brain.

3. DEVELOPMENT OF A METHOD FOR MEASURING AMBIENT VISUAL INFORMATION

Ambient visual information has rarely been taken into account as a environmental variable by researchers and designers since it is not easy to manipulate and describe the continuous surfaces surrounding people by conventional methods. Thus it is worth having an alternative tool to describe ambient visual information of the environment. A computer program was, therefore, developed to provide a visual representation and statistical analysis of the ambient visual information. In order to develop a computer program, the preceding discussion about the nature of ambient vision has to be extended to postulate hypotheses which serve as design criteria of the computer program.

The basic units which convey ambient visual information were postulated to be areas of visible surfaces divided up by differences in meaning for basic human behavior, or differences in their "affordance". They are, for instance, surfaces of pavement, water, grass, trees, building, and sky. Pavement affords walking but water surface cannot, and trees afford going through or under but building wall cannot. The way of dividing environmental surfaces is similar to Thiel's (1997) concept of "basic pattern areas" in his "notation" and is virtually made following his ideas, but it does not distinguish surfaces of the same affordance regardless of their texture or color.

Ambient vision deals with broad areas, with scattered attention and provides a quick global impression. This "instant" process which integrates information from broad areas implies that ambient vision performs a simple statistical analysis. The program was, therefore, designed to assess visual information from all directions around the perceiver and conduct some statistical calculation for integration although accuracy of details was not required.

The program, in practice, assesses surrounding scenes by numerous scanning lines radiated from a station point in all directions with equal density, and records the array of visible surfaces of various components and the distance between the surfaces and the station point. The concept of spatial volume, a mean length of visual lines, is similar to Benedict's (1979) two-dimensional "Isovists", although in this present case three-dimensional volume are measured. Having this data, it then calculates various measures which are expected to describe various psychological impacts of "visual radiation" from the surfaces surrounding the perceiver.

4. SOME EMPIRICAL STUDIES

The followings are brief reviews of some empirical studies which examine quantitative relations between measured ambient visual information obtained by the program and a variety of human responses.

4.1. Sensory Information and Behavior in the Japanese Gardens

This study (Ohno, Hata, & Kondo, 1997) examined the hypotheses that people's behavior commonly changes at certain places in the Japanese garden, and can be explained by the sensory information in the environment. It has been often mentioned that Japanese gardens have been designed so as to control visitors experience, particularly the vistas, as they move along the garden paths. If we can learn from these sophisticated skills of landscape design, we could naturally direct people's attention to something we want to be viewed.

In a typical circuit-style garden in Kobe City, each of the twenty-one participants was asked to stroll at will along the main circuit path. The participant's behavior was recorded on videotape by a TV camera from a position about 5 meters behind them. From the videotape each participant's motion (viewing directions) and walking pace were observed and recorded at every 0.5 meter consecutive points along the path. In order to analyze participants' behavior with objective data, the changes of sensory input latent in the environment as people moved along the path were measured by a set of personal computer programs (Ohno & Kondo, 1994).

By comparing the results of the experiment and the measurement, it was found that participants commonly change their viewing direction where ambient vision detects a sudden change in surrounding scenes. It occurred when a participant was moving from enclosed space to open space, and passing over water on a bridge. An asymmetrical distribution of participant's viewing direction suggested that people tend to extend their attention to open areas where vistas are larger. Although influences by other senses (hearing, kinesthetic) are acknowledged, the measurement of ambient visual information in the environment was found to predict these behaviors.

4.2. Evaluation of Landscape of Housing Neighborhood and Ambient Visual Information

An impression or feeling of landscapes was attempted to explain by ambient visual information measured by the program (Ohno, 1991). Nine simulated housing neighborhoods which were systematically different in density of vegetation were created and scale models (1/250) of the sites were made. A sequence of scenes along a typical path in each of the housing sites was presented to forty-three participants by a series of slides taken by an endoscope at ten consecutive points each ten meters apart. The participants were asked to rate the impressions of each simulated landscape using a rating sheet which contained ten bipolar adjective pairs. The program was applied to eight consecutive points along the path, and profiles of the numerical measures, namely visible area of components and the spatial volume, were obtained. From these data, averages of each measure and the coefficient of sequential variation were calculated.

The relation between the subjective judgment of "natural—artificial" and the visible area of greenery, suggests that greenery softens the negative impression of "artificial," but the positive impression of "natural" would not be enhanced by an increase of greenery beyond 15 percent. As for the sequential variation, changes in visible spatial volume was found to relate with the impressions of "unique—featureless" and "pleasant—boring." This study, in which subjects rated sequential landscapes of different housing neighborhoods, revealed that the global impressions or feelings of a place can be well explained by some of the measures obtained by the program.

4.3. Site Planning of Multifamily Housing Considering Residents' Mutual Visual Interactions

In order to develop a tool for the site planning of multifamily housing as related to the residents' psychological responses, the program was modified to measure the amount of visual interaction, or the probability of being seen, at a given point in a proposed environment (Ohno & Takeyama, 1994). While negative aspect in the visual interactions causes a problem of visual invasion of privacy, there is also the positive aspect of "natural surveillance", related to feelings of safety (Newman, 1972). The basic hypothesis for this study is that the residents' perceptions and attitudes concerning privacy and security from crime are a function of the amount of "visual radiation" from the surrounding buildings. The amount of "visual radiation" at a given station point is postulated to depend on the visible area of surrounding surfaces which potentially have residents' eyes, namely the surrounding buildings' facade with windows.

An empirical study was conducted using a questionnaire which asked the respondents (1) to rate how much they cared about neighbors' visual invasion of privacy on a 6 point scale, and (2) to mark the area on the housing site map where they do not let children play and they avoid themselves for safety reasons. It was revealed that the residents' sense of privacy was related to the measures of "visual radiation" to each individual apartment from surrounding buildings and paths, and the outdoor spaces in the housing site where residents feel unsafe could be predicted by lack of residents' eyes measured by "visual radiation" from surrounding buildings.

5. CONCLUSION

The three empirical studies generally support the validity of the measurement of ambient visual information latent in the disposition of environmental surfaces in the visual field, and they clarified some numerical relations between the measures obtained by the program and human responses at various levels. In this paper I put more emphasis on direct, ongoing perception rather than on cognitive aspects of perception. Since the latter aspects, more related with focal information, are receiving considerable attention by conventional environment-behavior studies, I stress the former aspects which tend to be ignored. The study of ambient perception may, I believe, make up for the deficiency of traditional object-oriented approach and contribute to a more complete theory of environmental perception.

Finally, I would like to draw attention to ambient information unconsciously received by nonvisual senses. Considering the changes in our built environments which tend to cause increasing dependence on artificial elemental features, we should stress the importance of the ambient features of natural and traditional environments which enrich our sensory experience.

REFERENCES

Ballard, D. H., Hilton, G. E., & Sejnowski, T. J. (1983). Parallel visual computation, *Nature*, *306*, 21–26.
Bassi, C. J. (1989). Parallel processing in the human visual system. In M. Wall, & A. A. Sadun (Eds.), *New methods of sensory visual testing* (pp. 1–13). New York: Springer-Verlag.
Benedict, M. L. (1979). To take hold of space: Isovists and isovists fields. *Environment and Planning B*, *6*, 47–65.
Gibson, J. J. (1966). *The senses Considered as perceptual systems*. Boston: Houghton-Mifflin.
Gibson, J. J. (1979). *The ecological approach to visual perception*. Boston: Houghton-Mifflin.
Ittelson, W. H. (1973). Environmental perception and contemporary perceptual theory. In W. H. Ittelson (Ed.), *Environment and Cognition* (pp. 1–19). New York: Seminar Press.
Julesz, B., & Bergen, J. R. (1983). Textons, the fundamental elements in preattentive vision and perception of textures, *The Bell System Technical Journal*, *60*(6), 1619–1645.

Landwehr, K. (1984). *On the minimal stimulus information for something to be seen.* Paper presented at the XXIII International Congress of Psychology.

Newman, O. (1972). *Defensible space, crime prevention through urban design.* New York: Macmillan Pub.

Ohno, R. (1980). Visual perception of texture: development of a scale of the perceived surface roughness of building materials, *Proceeding of EDRA, 11*, 193–200.

Ohno, R. (1985). Notion of duality in the visual system and its implication for environmental design. *Cross cultural research in environment and behavior: Proceedings of the Second US-Japan Seminar*, 209–217.

Ohno, R. (1991). Ambient vision of the environmental perception: describing ambient visual information, *Proceeding of EDRA, 22*, 237–252.

Ohno, R., Hata, T., & Kondo, M. (1997). Experiencing Japanese gardens: sensory information and behavior. In S. Wapner, J. Demick, T. Yamamoto, & T. Takahashi, (Eds.) (1997). et al. (Eds.), *Handbook of Japan-United States environment-behavior research* (pp.163–182). New York: Plenum.

Ohno, R., & Kondo, M. (1994). Measurement of the multi-sensory information for describing sequential experience in the environment: An application to the Japanese circuit-style garden. In S. J. Neary et al. (Eds.), *The urban experience: A people-environment perspective* (pp.425–437). London: E & FN Spon.

Ohno, R., & Komuro, K. (1984). Texture perception versus object perception in the environment, *Proceedings of 8th IAPS*, 17–21.

Ohno, R., & Takeyama, K. (1994). Measurement and graphic representation of the residents' mutual visual interactions for the site planning of multi-family housing, *Proceedings of the 6th International Conference on Engineering Computer Graphics and Descriptive Geometry*, 568–571

Rapoport, A. (1977). *Human aspects of urban form* (pp. 195–201). New York: Pergamon Press.

Thiel, P. (1997). *People, paths, and purpose: notations for a participatory envirotecture.* Seattle: U.W. Press.

Trevarthen, C. B. (1968). Two mechanisms of vision in Primates, *Psychologische Forschung, 31*, 299–337.

13

A WAY OF SEEING PEOPLE AND PLACE

Phenomenology in Environment-Behavior Research

David Seamon

Department of Architecture
Kansas State University
Manhattan, KS 66506-2901

1. INTRODUCTION

In simplest terms, phenomenology is the interpretive study of human experience. The aim is to examine and clarify human situations, events, meanings, and experiences "as they spontaneously occur in the course of daily life" (von Eckartsberg, 1998a, p. 3). The goal is "a rigorous description of human life as it is lived and reflected upon in all of its first-person concreteness, urgency, and ambiguity" (Pollio, Henley, & Thompson, 1996, p. 5).

This preliminary definition, however, is oversimplified and does not capture the full manner or range of phenomenological inquiry. Herbert Spiegelberg, the eminent phenomenological philosopher and historian of the phenomenological movement, declared that there are as many styles of phenomenology as there are phenomenologists (Spiegelberg 1982, p. 2)—a situation that makes it difficult to articulate a thorough and accurate picture of the tradition.

In this article, I can only claim to present my understanding of phenomenology and its significance for environment-behavior research. As a phenomenological geographer in a department of architecture, my main teaching and research emphases relate to the nature of environmental behavior and experience, especially in terms of the built environment. I am particularly interested in why places are important for people and how architecture and environmental design can be a vehicle for place making. Empirical phenomenological studies with which I have been involved range from the use of my own personal experience to understand the nature of a particular place (Seamon, 1992) to the interpretation of photography and imaginative literature as a way to understand essential experiential qualities of the person-environment relationship (Seamon,

Theoretical Perspectives in Environment-Behavior Research, edited by Wapner *et al.*
Kluwer Academic / Plenum Publishers, New York, 2000.

157

1990a, 1993). I have also written on the ways that the phenomenological approach might be used to interpret architecture and to contribute to better environmental design (Coates & Seamon, 1993; Lin & Seamon, 1994; Seamon, 1990b, 1991, 1993b, 1994; Seamon & Mugerauer, 1985).

In demonstrating in this chapter the value of phenomenology to environment-behavior research, I occasionally draw on my own studies but give most attention to phenomenological work done by other researchers, since the breadth of phenomenological possibilities is considerable, and my work indicates only a small portion of the potential whole. Throughout the chapter, most of the studies to which I refer are explicitly phenomenological, though occasionally I incorporate work that is implicitly phenomenological in that either the authors choose not to involve the tradition directly (e.g., Brill, 1993; Pocius, 1991; Tuan, 1993) or are unaware that their approach, methods, and results parallel a phenomenological perspective (e.g., Krapfel, 1989; Lipton, 1990; Walkey, 1993). I justify the inclusion of these studies because they present aspects of human life and experience in new ways by identifying generalizable qualities and patterns that arise from everyday human life and experience—for example, qualities of the built environment that contribute to a sense of place, order, and beauty (e.g., Alexander, 1987, 1993; Alexander, Ishikawa, & Silverstein, 1977; Silverstein, 1993b; Rattner, 1993).[1]

Specifically, I discuss the following themes in this chapter:

- the nature of phenomenology;
- key assumptions of a phenomenological approach;
- the methodology of empirical phenomenological research;
- trustworthiness and phenomenological research;
- phenomenology and environmental design.

2. THE NATURE OF PHENOMENOLOGY

Phenomenology is a critical, descriptive science that is related, in method and philosophical outlook, to other interpretive traditions that include existentialism and hermeneutics (Stewart & Mukunis, 1990). Phenomenology includes different conceptual approaches that range from the transcendental or "pure" phenomenology of philosopher Edmund Husserl to the hermeneutic phenomenology of philosopher Paul Ricoeur to the existential phenomenology of philosophers Martin Heidegger and Maurice Merleau-Ponty (Spiegelberg, 1982). In using the term here, I refer to a way of knowing that seeks to describe the underlying, essential qualities of human experience and the world in which that experience happens (Burch, 1989; Pollio, 1997; Valle, 1998; van Manen, 1990).

I therefore define phenomenology as the exploration and description of phenomena, where *phenomena* refers to things or experiences as human beings experience them. Any object, event, situation or experience that a person can see, hear, touch,

---

[1] In this article, I largely highlight research of the last ten years. For discussions of earlier phenomenological work relating to environment-behavior research, see Seamon, 1982; Seamon, 1987; Seamon, 1989.

smell, taste, feel, intuit, know, understand, or live through is a legitimate topic for phenomenological investigation. There can be a phenomenology of light, of color, of architecture, of landscape, of place, of home, of travel, of seeing, of learning, of blindness, of jealousy, of change, of relationship, of friendship, of power, of economy, of sociability, and so forth. All of these things are phenomena because human beings can experience, encounter, or live through them in some way.

The ultimate aim of phenomenological research, however, is not idiosyncratic descriptions of the phenomenon, though such descriptions are often an important starting point for existential phenomenology. Rather, the aim is to use these descriptions as a groundstone from which to discover underlying commonalities that mark the essential core of the phenomenon. In other words, the phenomenologist pays attention to specific instances of the phenomenon with the hope that these instances, in time, will point toward more general qualities and characteristics that accurately describe the essential nature of the phenomenon as it has presence and meaning in the concrete lives and experiences of human beings.

3. SOME CORE ASSUMPTIONS OF A PHENOMENOLOGICAL APPROACH

In the last several years, there has appeared a growing number of works that discuss the relation of phenomenology to the scholarly and professional worlds in general terms (Burch, 1989, 1990, 1991; Embree, 1997; Stewart & Mukunis, 1990) and to specific disciplines—e.g, anthropology (Jackson, 1996; Weiner, 1991); art (Berleant, 1991; Davis, 1989; Jones, 1989); education (van Manen, 1990); environmental design (Berleant, 1992; Condon, 1991; Corner, 1990; Dovey, 1993; Mugerauer, 1994; Howett, 1993; Vesely, 1988); geography (Cloke, Philo, & Sadler, 1991, chap. 3; Relph, 1989a, b, 1990; Seamon, 1997); psychology (Moustakis, 1994; Pollio Henley, & Thompson, 1996; Valle, 1998); philosophy (Casey, 1993, 1997); and natural science (Bortoft, 1996; Heelan, 1983; Jones, 1989; Riegner, 1993; Seamon & Zajonc, 1998).

In much of this work, commentators have placed phenomenology within the wider conceptual and methodological rubric of *qualitative inquiry* (Cloke et al., 1991; Lincoln & Guba, 1985; Low, 1987). For example, Patton (1990, pp. 66–91) associates phenomenology with such other qualitatively-oriented theories and orientations as ethnography, heuristic inquiry, ethnomethodology, symbolic interactionism, and ecological psychology. Patton argues that, in broadest terms, all these perspectives present variations on "grounded theory" (e.g., Glaser and Strauss, 1967)—in other words, perspectives assuming "methods that take the researcher into and close to the real world so that the results and findings are 'grounded' in the empirical world" (Patton, 1990, p. 67). This perspective approaches theory inductively, in contrast to "theory generated by logical deduction from a priori assumptions" (ibid., p. 66).

Patton's identification of phenomenology with qualitative orientations is certainly acceptable, though it is also important to realize that these various qualitative perspectives involve as many differences as similarities, thus, for example, ethnographic inquiry typically studies a *particular* person or group in a *particular* place in time; in contrast, a phenomenological study might begin with a similar real-world situation but would then use that specific instance as a foundation for identifying deeper, more generalizable patterns, structures, and meanings. Similarly, both symbolic interactionism and phenomenology examine the kinds of symbols and understandings that give meaning to a particular group or society's way of living and experiencing. The perspective of the symbolic interactionalist, however, most typically emphasizes the more explicit, cognitively-derived layers of meaning whereas a phenomenological perspective defines meaning in a broader way that includes bodily, visceral, intuitive, emotional, and transpersonal dimensions.

Phenomenology, therefore, can be identified as one style of qualitative inquiry but involving a particular conceptual and methodological foundation. Here, I highlight two broad assumptions that, at least for me, mark the essential core of a phenomenological approach:

- Person and world as intimately part and parcel;
- Phenomenology as a radical empiricism.

I emphasize these two broad assumptions because the first relates to the particular subject matter of phenomenology, while the second relates to the means by which that subject matter is to be understood. I hope discussion of these two assumptions gives the reader a better sense of what makes phenomenology distinctive and how this distinctiveness can offer a valuable tool for environment-behavior research.

3.1. Person and World Intimately Part and Parcel

A central focus of phenomenology is the way people exist in relation to their world. In *Being and Time*, Heidegger (1962) argued that, in conventional philosophy and psychology, the relationship between person and world has been reduced to either an idealist or realist perspective. In an idealist view, the world is a function of a person who acts on the world through consciousness and, therefore, actively knows and shapes his or her world. In contrast, a realist view sees the person as a function of the world in that the world acts on the person and he or she reacts. Heidegger claimed that both perspectives are out of touch with the nature of human life because they assume a separation and directional relationship between person and world that does not exist in the world of actual lived experience.

Instead, Heidegger argued that people do not exist apart from the world but, rather, are intimately caught up in and immersed. There is, in other words, an "undissolvable unity" between people and world (Stewart & Mickunas, 1990, p. 9). This situation—always given, never escapable—is what Heidegger called *Dasein*, or *being-in-the-world*. It is impossible to ask whether person makes world or world makes person because both exist always together and can only be correctly interpreted in terms of the holistic relationship, being-in-world (Pocock, 1989; Relph, 1989a; Seamon, 1990a).

In this sense, phenomenology supplants the idealist and realist divisions between person and world with a conception in which the two are *indivisible*—a person-world whole that is one rather than two. A major phenomenological challenge is to describe this person-world intimacy in a way that legitimately escapes any subject-object dichotomy.

One broad theme that phenomenologists have developed to overcome this dichotomy is *intentionality*—the argument that human experience and consciousness necessarily involve some aspect of the world as their object, which, reciprocally, provides the context for the meaning of experience and consciousness (Stewart & Mickunas, 1990, p. 90; Pollio, 1997, p. 7). In examining peoples' intentional relationships with their worlds, environment-behavior researchers using phenomenology have typically drawn on two central notions that I review here—*lifeworld*, and *place*. These notions are significant for a phenomenological approach to environment-behavior research because each refers to a phenomenon that, in its very constitution, holds people and world always together and also says much about the physical, spatial, and environmental aspects of human life and events.

3.1.1. Lifeworld. The lifeworld refers to the tacit context, tenor, and pace of daily life to which normally people give no reflective attention. The lifeworld includes both the routine and the unusual, the mundane and the surprising. Whether an experience is ordinary or extraordinary, however, the lifeworld in which the experience happens is normally out of sight. Typically, human beings do not make their experiences in the lifeworld an object of conscious awareness. Rather, these experiences *just happen*, and people do not consider how they happen, whether they could happen differently, or of what larger experiential structures they might be a part. One of my earliest phenomenological efforts was a book-length study that sought to identify the underlying geographical aspects of the lifeworld, which I explored in terms of three existential themes—*movement*, *rest*, and *encounter* (Seamon, 1979).

One research focus relating to the lifeworld in recent phenomenological research is its perceptual taken-for-grantedness (Abrams, 1996; Pocock, 1993). For example, partly influenced by the seminal works on the acoustic dimensions of the lifeworld by Schafer (1977) and Berendt (1985), there have been phenomenological studies of the multimodal ways in which the senses contribute to human awareness and understanding (Jarvilouma, 1994; Pocock, 1993; Porteous, 1990; Schonhammer, 1989; Tuan, 1993). Other phenomenological researchers have considered how particular circumstances relating to the environment or to the person lead to particular lifeworld experiences, thus Behnke (1990) and Rehorick (1986) examined the experience of earthquakes phenomenologically, while Hill (1985) explored the lifeworld of the blind person and Toombs (1992a, 1995a, 1995b) drew upon her own experience of chronic progressive multiple sclerosis to provide a phenomenological explication of the human experience of disability.

One insightful study relating to material aspects of the lifeworld is Palaasma's architectural examination of how the design aesthetic of Modernist-style buildings largely emphasized intellect and vision and how a more comprehensive architecture would accommodate an environmental experience of all the senses as well as the feelings (Pallasmaa, 1996). Another study linking lifeworld with the physical environment is Nogué i Font's efforts at a phenomenology of landscape (Nogué i Font, 1993). He attempted to describe the essential landscape character of *Garroxta*, a Catalonian region in the Pyrenees foothills north of Barcelona. In developing a phenomenology of this region, Nogué i Font conducted in-depth interviews with five groups of people familiar with Garroxta in various ways—farmers, landscape painters, tourists, hikers, and recently-arrived residents who were formerly urbanites.

In this study, Nogué i Font addressed a central phenomenological question: Can there be a phenomenology of landscape in its own right, or does there exist only a phenomenology of that landscape as particular individuals and groups experience and know it? He concluded that both phenomenologies exist, and one does not exclude the other. In describing the meanings of Garroxta for the farmers and painters, for example, Nogué i Font found that, in some ways, the landscape has significantly contrasting meanings for the two groups. In spite of these differences, however, both farmers and painters spoke of certain physical elements and experienced qualities that mark the uniqueness of Garroxta as a "thing in itself." For example, both groups saw the region as a wild, tangled landscape of gorges, precipices, and forests that invoke a sense of respect and endurance.

3.1.2. Place. One significant dimension of the lifeworld is the human experience of place, which, in spite of criticism from non-phenomenologists (e.g., Rapoport, 1993), continues to be a major focus of phenomenological work in environment-behavior research (Barnes, 1992; Boschetti, 1993; Chaffin, 1989; Hester, 1993; Hufford, 1986; Oldenburg, 1989; Pocius, 1991; Porteous, 1989; Relph, 1993; Seamon, 1992, 1993a; Sherry, 1998; Smith, 1989; Weiner, 1991).

In philosophy, Casey (1993, 1997) has written two book-length accounts that argue for place as a central ontological structure founding human experience: "place, by virtue of its unencompassability by anything other than itself, is at once the limit and the condition of all that exists . . . [P]lace serves as the *condition* of all existing things . . . To be is to be in place" (1994, pp. 15–16). Drawing on Merleau-Ponty (1962), Casey emphasized that place is a central ontological structure of being-in-the world partly because of our existence as *embodied* beings. We are "bound by body to be in place" (1993, p. 104), thus, for example, the very physical form of the human body immediately regularizes our world in terms of here-there, near-far, up-down, above-below, and right-left. Similarly, the pre-cognitive intelligence of the body expressed through action—what Merleau-Ponty (1962) called "body subject"—embodies the person in a prereflective stratum of taken-for-granted bodily gestures, movements, and routines (Ediger, 1994; Hill, 1985; Seamon, 1979; Toombs, 1992a, 1995a, 1995b).

The broad philosophical discussions of Relph (1976, 1990, 1993, 1996) continue to be a significant conceptual guide for empirical phenomenologies of place (Boschetti, 1990, 1993, 1996; Chaffin, 1989; Masucci, 1992; Million, 1992; Seamon, 1993). Perhaps the most comprehensive example is provided by Million (1992), who examined phenomenologically the experience of five rural Canadian families forced to leave their ranches because of the construction of a reservoir dam in southern Alberta. Drawing on Relph's notions of insideness and outsideness (Relph, 1976), Million sought to identify the central lived-qualities of what she called *involuntary displacement*—the families' experience of forced relocation and resettlement. Using in-depth interviews with the families as her descriptive base, she demonstrated how place is prior to involuntary displacement with the result that this experience can be understood metaphorically as a forced journey marked by eight stages.

Becoming uneasy, struggling to stay, and *having to accept* emerge in Million's study (1992) as the first three stages of involuntary displacement whereby the families realize that they must leave their home place. The process then moves into *securing a settlement* and *searching for the new*—two stages that mark a "living in between"—i.e., a middle phase of a forced journey and a time when the families feel farthest away from place. Finally, with *starting over, unsettling reminders,* and *wanting to settle,* the families move into three stages of a rebuilding phase. Million's study is significant because it examined the foundations of place experience for one group of people and delineates the lived stages in the process of losing place and attempting to resettle.[2]

3.2. A Radical Empiricism

If one key phenomenological assumption is the intimate connectedness between person and world, a second assumption relates to what I call "radical empiricism"— the particular manner in which this person-world connectedness is to be studied. In using this descriptive phrase, I attempt to encapsulate the heart of phenomenological method by indicating a way of study whereby the researcher seeks to be open to the phenomenon and to allow it to show itself in its fullness and complexity *through her own direct involvement and understanding*. In that this style of study arises through firsthand, grounded contact with the phenomenon as it is experienced by the researcher, the approach can be called *empirical*, though the term is used much differently than by positivist scientists who refer to data that are materially identifiable and mathematically recordable.

If, in other words, phenomenological method can be called empirical, it must be identified as *radically* so, since understanding arises directly from the researcher's personal sensibility and awareness rather than from the usual secondhand constructions of positivist science—e.g., *a priori* theory and concepts, hypotheses, predetermined methodological procedures, statistical measures of correlation, and the like. More precisely, one can make the following claims about phenomenological method as a radical empiricism:

1. The study must involve the researcher's direct contact with the phenomenon. If the phenomenologist studies a person or group's experience, then she must encounter that experience as directly as possible. Methodological possibilities include the researcher's participating in the experience, her conducting in-depth interviews with the person or group having the experience, or her carefully watching and describing the situation supporting or related to the experience.

If the phenomenon being studied is some artifactual text—for example, photographs, a novel, music, or a landscape—the researcher must find ways to immerse herself in the text so that she becomes as familiar as possible with it. Thus, she might carefully study the text and thoroughly record her experience and understanding. She might ask other parties to respond to the text and provide their insights and awareness. Or she might study other commentator's understandings of the text—for example, reading reviews of the novel or studying all critical commentaries on the author or artist in question.

---

[2] Closely related to the theme of place are three other significant phenomenological topics: first, *sacred space* (Barnes, 1992; Brenneman & Brenneman, 1995; Chidester & Linenthal, 1995; Cooper Marcus, 1993; Eliade, 1961; Lane, 1988; Lin, 1991; Lin & Seamon, 1994; Merant, 1989; Muguerauer, 1994, chap. 4; Whone, 1990; Wu, 1993); second, *home and at-homeness* (Bachelard, 1964; Barbey, 1989; Bollnow, 1961; Boschetti, 1990, 1993, 1995; Casey, 1993; Chawla, 1994, 1995; Cooper Marcus, 1995; Day, 1996; Dovey, 1985; Graumann, 1989; Heidegger, 1971; Koop, 1993; LeStrange, 1998; Mugerauer, 1994; Norris, 1990; Pallasmaa, 1995; Rouner, 1996; Seamon, 1993a; Shaw, 1990; Silverstein, 1994; Sinclaire, 1994; Stefanovic, 1992; Winning, 1990, 1991; Wu, 1991); and, third, work dealing with *a phenomenology of environmental ethics* (Abrams, 1996; Cheney, 1989; Foltz, 1995; Margadant-van Archen, 1990; Stefanovic, 1991; Weston, 1994).

In short, the researcher must facilitate for herself an intimacy with the phenomenon through prolonged, firsthand involvement.

2. The phenomenologist must assume that she does not know the phenomenon but wishes to. Ideally, the phenomenologist approaches the phenomenon as a beginner—in fact, phenomenology is often defined as a "science of beginnings" (Stewart & Mukunas, 1990, p. 5). Whereas, in positivist research, the student typically begins her inquiry *knowing* what she does not know, the phenomenologist, does *not* know what she doesn't know. The phenomenon is an uncharted territory that the student attempts to explore.

The phenomenologist must therefore always adapt her methods to the nature and circumstances of the phenomenon. A set of procedures that work for one phenomenological problem may be unsuitable elsewhere. In this sense, the central instrument of deciphering the phenomenon is the *phenomenological researcher herself*. She must be directed yet flexible in the face of the phenomenon.

In short, the phenomenologist has no clear sense of what she will find or how discoveries will proceed. The skill, perceptiveness, and dedication of the researcher are the engine for phenomenological research and presuppose any specific methodological procedures.

3. Since the researcher as human instrument is the heart of phenomenological method, the specific research methods she uses should readily portray human experience in experiential terms. The best phenomenological methods, therefore, are those that allow human experience to arise in a rich, unstructured, multidimensional way. If the interview format seems the best way to gather an account of the phenomenon, then the researcher must be open to respondents and adapt her questions, tone, and interest to both respondents' commentaries and to her own shifting understanding as she learns more about the phenomenon. If the researcher uses a novel, photograph or some other artifactual text to examine the phenomenon, then she must be willing to return to its parts again and again, especially if an exploration of one new part offers insights on other parts already considered.

In short, phenomenological method incorporates a certain uncertainty and spontaneity that must be accepted and transformed into possibility and pattern. The phenomenological approach to a particular phenomenon must be developed creatively and allow for a fluidity of methods and research process.

4. SPECIFIC PHENOMENOLOGICAL METHODS

Having considered, broadly, some central components of phenomenological method, I next wish to review attempts to identify specific methodological types of phenomenological research. For the most part, it has been *psychologists*—especially psychologists associated with what has come to be called the "Duquesne School of Phenomenological Psychology"—who have sought to establish reliable procedural methods for conducting empirical phenomenological research (Giorgi, Barton, & Maes, 1983; Valle, 1998; also see Moustakas, 1994).

Drawing on the designations of Duquesne phenomenological psychologist von Eckartsberg (1998a, b), I discuss two methodological approaches—what von Eckartsberg calls the *existential* and the *hermeneutic*. I also add a third approach that I call *first-person*. I describe this approach first, since it draws on the realm of experience closest to the researcher—her own lived situation.

4.1. First-Person Phenomenological Research

In first-person phenomenological inquiry, the researcher uses her own firsthand experience of the phenomenon as a basis for examining its specific characteristics and qualities (Chaffin, 1989; Lane, 1988; Shaw, 1990; Wu, 1991). For example, this approach was one of several I used in trying to understand the unique character of *Olana*—nineteenth-century American landscape painter Frederic Church's home looking out over the Hudson River (Seamon, 1992). Through being on the site and walking, looking, writing, sketching, and so forth, I attempted to empathize with and identify the architectural, environmental, and human qualities that make Olana a special place, at least for me as a representative twentieth-century visitor. Another example is the work of Violich (1985, 1998), who examined the contrasting qualities of place for several Dalmatian towns with varying spatial layouts. Using such techniques as sketching, mapping, and journal entries, he immersed himself in each place for several days and sought to " 'read' each as a whole" (Violich, 1985, p. 113).

One of the most sensitive and exhaustive uses of first-person phenomenological research is the work of Toombs (1992a, 1992b, 1995a, 1995b), who lives with multiple sclerosis, an incurable illness that affects her ability to see, to hear, to sit, and to stand. In her work, which most broadly can be described as a phenomenology of illness, she demonstrated how phenomenological notions like the lived body provide "important insights into the profound disruptions of space and time that are an integral element of changed physical capacities such as loss of mobility" (Toombs, 1995b, p. 9).

Toombs' method involved a continual dialectic between phenomenological notions as conceptually understood versus their concreteness as known directly in her own lived experience. For instance, to provide an understanding of how the disabled person's loss of mobility leads to a changed interaction with the surrounding world, Toombs recounted in detail a typical experience—her journey by airplane to a professional conference. At one point in her narrative she described airport check-in:

Once in the terminus I go to the airline check-in counter. In my battery-operated scooter I am approximately three and a half feet tall and the counter is on a level with my head. All my transactions with the person behind the counter take place at the level of my ear. The person behind the counter must stretch over it to take my tickets, and I must crane my neck and shout to be heard (ibid., p. 14).

From such lived examples, Toombs drew phenomenological generalizations—for example, she described how her loss of upright posture relates to Merleau-Ponty's broader notions of bodily intentionality and the transformation of corporeal style (Merleau-Ponty, 1962, p. 76). Thus the loss of uprightness is not confined to problems of locomotion but also involves deeper experienced dimensions like the diminishment of one's own autonomy and the tendency of able persons to treat the disabled as dependent or even subnormal.

Another way in which the first-person approach can be used in phenomenology is as a starting place from which the phenomenologist can bring to awareness "her preconceived notions and biases regarding the experience being investigated so that the researcher is less likely to impose these biases when interpreting [the phenomenon]" (Shertock, 1998, p. 162).

Provided the phenomenologist has access in her own experience to the phenomenon she plans to study, first-person research can offer clarity and insight grounded in one's own lifeworld. This understanding is derived from a world of one, however, and the researcher must find ways to involve the worlds of others. This need leads to the method of existential-phenomenological research.

4.2. Existential-Phenomenological Research

The basis for generalization in existential-phenomenological research is the specific experiences of specific individuals and groups involved in actual situations and places (von Eckartsberg, 1998a, p. 4). In the discussion of lifeworld and place research above, Million's phenomenology of involuntary displacement (Million, 1992) and Nogué i Font's phenomenology of landscape (Nogué i Font, 1993) are good examples in that the basis for generalization is the real-world experiences of the ranchers forced to relocate or the farmers and landscape painters of Garroxta. Similarly, in my work on a "geography of the lifeworld" (Seamon, 1979), I asked volunteers to participate in discussion groups and share accounts of their own environmental and place experiences (Seamon, 1979).

Phenomenological psychologists, particularly those associated with the Duquesne School, have devoted considerable effort to establishing a clear set of procedures and techniques for this style of phenomenology (see Valle, 1998). For example, von Eckartsberg (1998b) speaks of four steps in the process: (1) identifying the phenomenon in which the phenomenologist is interested; (2) gathering descriptive accounts from respondents regarding their experience of the phenomenon; (3) carefully studying the respondents' accounts with the aim of identifying any underlying commonalities and patterns; and (4) presenting results, both to the study respondents (in the form of a "debriefing" about the study in ordinary language) and to fellow researchers (in the form of scholarly presentation).

Other phenomenologists have discussed the steps in existential-phenomenological work in ways that more or less echo von Eckartsberg's four stages (e.g., Giorgi, 1985; Churchill, Lowery, McNally, & Rao, 1998; Wertz, 1984). Whatever the particular phrasing, the common assumption is that the individual descriptive accounts, when carefully studied and considered collectively, "reveal their own thematic meaning-organization if we, as researchers, remain open to their guidance and speaking, their disclosure, when we attend to them" (von Eckartsberg, 1998b, p. 29).

In claiming to generate accurate generalization, the existential-phenomenological approach makes one important assumption: that there is a certain equivalence of meaning for the respondents whose experience the researcher probes. In other words, the claim is that "people in a shared cultural and linguistic community name and identify their experience in a consistence and shared manner" (von Eckartsberg, 1998a, p. 15). Procedurally, this claim means that respondents (1) must have had the experience under investigation and (2) be able to express themselves clearly and coherently in spoken, written, or graphic fashion, depending on the particular tools used for eliciting experiential accounts. Ideally, the respondents will also feel a spontaneous interest in the research topic, since personal concern can motivate the respondent to provide the most thorough and accurate lived descriptions (Shertock, 1998, p. 162).

These requirements mean that inquiry is not carried out, as in positivist science, on a random sample of interchangeable subjects representative of the population to which findings will be generalizable. Rather, some respondents will be more appropriate than others because of their particular situation in relation to the phenomenon studied or because they seem more perceptive, thus better able to articulate their experience. Usually, in phenomenological research, "subjects" are instead called "respondents" or "co-researchers," since any generalizable understanding is a function of the sensibilities of both respondent and researcher.

In practice, there is no exact step-by-step procedure for conducting existential-phenomenological research beyond the general stages identified above. As explained earlier, the individual style of the researcher and the specific nature of the phenomenon are much more important for establishing the specific research procedure and tools of description. In her study of involuntary displacement, for example, Million (1992) spent much time locating participants who wished to share their experience and who appeared to be able to offer that sharing in a thoughtful, articulate way. She involved these participants in several in-depth interviews, the formats of which shaped and reshaped themselves as she learned more about each family's experience and the broader events of the dam construction. In addition, she lived with some of the ranch families and asked them to accompany her on "field trips" to the flooded areas that used to be their ranches. In short, Million's specific methods and procedures were auxiliary to the nature and needs of her own individual research style, her research participants, and her phenomenon of involuntary displacement.

4.3. Hermeneutic-Phenomenological Research

Most broadly, hermeneutics is the theory and practice of interpretation (Mugerauer, 1994, p. 4), particularly the interpretation of *texts*, which may be any material object or tangible expression imbued in some way with human meaning—for example, a public document, a personal journal, a poem, a song, a painting, a dance, a sculpture, a garden, and so forth. The key point hermeneutically is that the creator of the text is not typically available to comment on its making or significance, thus the hermeneutic researcher must find ways to discover meanings through the text itself. As von Eckartsberg (1998b, p. 50) describes the hermeneutical process:

One embeds oneself in the process of getting involved in the text, one begins to discern configurations of meaning, of parts and wholes and their interrelationships, one receives certain messages and glimpses of an unfolding development that beckons to be articulated and related to the total fabric of meaning. The hermeneutic approach seems to palpate its object and to make room for that object to reveal itself to our gaze and ears, to speak its own story into our understanding.

In my own phenomenological research, hermeneutic study has been important—for example, I used the New York photographs of pioneer photographer André Kertész as a way to examine the person-world relationship and urban lifeworld (Seamon, 1990a). Similarly, I drew on the novels of writer Doris Lessing to develop a phenomenology of human and place relationship (Seamon, 1993a). Overall, much of the phenomenological work relating to environment-behavior research has been hermeneutic because the aim is often an understanding people in relation to *material* environments, whether furnishings, buildings, cultural landscapes, settlement patterns, and the like (Alexander, 1987, 1993; Alexander et al., 1977; Anella, 1990; Brenneman and Brenneman, 1995; Chaffin, 1989; Chawla, 1994; Chidester & Linenthal, 1995; Condon, 1991; Harries, 1988, 1993, 1997; Lin, 1991; Lipton, 1990; Mugerauer, 1993, 1994, 1995; Norberg-Schulz, 1980, 1988, 1996; Paterson, 1991, 1993a, 1993b; Relph, 1976, 1990, 1993; Riegner, 1993; Seamon, 1991, 1993a, 1994; Silverstein, 1993b; Stefanovic, 1994; Sturm, 1990; Swentzell, 1990; Thiis-Evensen, 1987; Walkey, 1993; Wu, 1993).

One useful example of the value of a hermeneutic-phenomenological approach in environment-behavior research is the work of Norwegian architect Thiis-Evensen (1987), who proposes a universal language of architecture by focusing on the experienced qualities of *floor*, *wall*, and *roof*, which he says are "the most basic elements in architecture" (ibid., p. 8). Through a hermeneutic reading of many different buildings in different cultures and historical periods, Thiis-Evensen suggests that these three architectural elements are not arbitrary but, rather, common to all architectural styles and traditions. The essential existential ground of floor, wall, and roof, he argues, is the relationship *between inside and outside*: Just by being what they are, the floor, wall, and roof automatically create an inside in the midst of an outside, though in different ways: the floor, through *above* and *beneath*; the wall, through *within* and *around*; and the roof, through *under* and *over*.

Thiis-Evensen demonstrates that a building's relative degree of insideness or outsideness in regard to floor, wall, and roof can be clarified through *motion, weight*, and *substance*—what he calls the three "existential expressions of architecture" (ibid., p. 21). Motion relates to the sense of dynamism or inertia evoked by the architectural element—i.e., whether it seems to expand, contract, or rest in balance. Weight involves the sense of heaviness or lightness of the element and its relation to gravity. Substance refers to the material sense of the element—whether it is soft or hard, coarse or fine, warm or cold, and so forth. The result, claims Thiis-Evensen, is an intricate set of tensions between architectural elements and experience.

In his work, Thiis-Evensen assumes that architectural form and space both presuppose and contribute to various shared existential qualities—insideness-outsideness, gravity-levity, coldness-warmth, and so forth—that mark the foundation of architecture as human beings experience it (Seamon, 1991). For example, if one studies the lived qualities of stairs, one realizes that narrow stairs typically relate to privacy and make the user move up them more quickly than up wide stairs, which better express publicness and ceremonial significance. Similarly, steep stairs express struggle and strength, isolation and survival—experienced qualities that sometimes lead to the use of steep stairs as a sacred symbol, as in Mayan temples or Rome's Scala Santa. On the other hand, shallow stairs encourage a calm, comfortable pace and typically involve secular use, as, for example, Michelangelo's steps leading up to the Campidoglio of Rome's Capitoline Hill (Thiis-Evensen, 1987, pp. 89–103).

I discuss Thiis-Evensen's work at length here because it is an exceptional example of one researcher's effort to look at a text—buildings in many different times and places—and to identify a series of experiential themes that do justice to "the integrity, complexity, and essential being of the phenomenon" (von Eckartsberg, 1998b, p. 50). One test of the value of Thiis-Evensen's experiential theory is that other researchers have found his interpretation to be a useful language for examining in detail the work of specific architects and specific architectural styles (e.g., Kushwah, 1993; Lin, 1991; Lin & Seamon, 1994; Ramaswami, 1992). At the same time, it is important to emphasize that Thiis-Evensen does not claim that his way of architectural interpretation is the only way, and clearly there could be other hermeneutics of architecture that provide other ways of presenting and understanding architectural meaning (e.g., Harries, 1988, 1993, 1997; Mugerauer, 1993, 1994; Alexander, 1987, 1993). This is a key aspect of all hermeneutical work: there are many ways to interpret the text, thus interpretation is never complete but always underway.

4.4. Commingling Methods

Very often the phenomenological researcher uses the first-person, existential, and hermeneutic approaches in combination, thus, for example, Nogué i Font (1993), in his phenomenology of the Garroxta landscape, made use of interviews but also did hermeneutic readings of nineteenth-century Garroxtan photographs and the pictures of artists associated with the nineteenth-century Garroxta school of landscape painting.

One of the most sensitive examples of a phenomenological study drawing on multiple methods is Chaffin's study of one Louisiana river landscape as it evokes a sense of place and community (Chaffin, 1989). Chaffin's focus is Isle Brevelle, a 200-year-old river community on the Cane River of Louisiana's Natchitoches Parish. His conceptual vehicle to explore this place is simple but effective: to move from outside to inside, first, by presenting the region's history and geography, then by interviewing residents, and, finally, by canoeing the Cane River, which he comes to realize is the "focus of the community-at-home-and-at-large" (ibid., p. 41). As he glided by the river banks, he became aware of a rhythm of water, topography, vegetation, and human settlement:

Once on the water, the earlier feelings of alienation and intrusion were gone. I came directly in contact with a spatial rhythm. As the valley's horizon is formed by the surrounding sand hills, so the river's horizon is formed by the batture [the land that slopes up from a waterway to the top of a natural or artificial levee], silhouetted against the sky when viewed from a canoe. I had the paradoxical sensation of being both high and low at the same time; held down between the banks, yet as high as the surrounding fields.

The meanders of the once-wild current organized this experience. As I paddled around the bends, the rhythm unfolded. On the outside of the curve, I was contained by a steep bank, emphasized by red cedar sentinels. Only rooftops and cars passing along the river road hinted at a world beyond. On the inside, I was released into a riverside world of inlets, peninsulas, and undulating banks softened by black willows, some even growing directly from the water on submerged bars. . . .

As the curves changed direction, the containment and release offered by the two sides of the river altered in turn and, in "my own little world of the river," everything seemed to fit (ibid., p. 102).

In his study, Chaffin began with a hermeneutic study of the natural and cultural landscape through scientific and historical documents. He also observed the community of Isle Brevelle firsthand and recognized a strong sense of place, which he understood more fully through an existential stage of study involving interviews. Finally, through the first-person experience of canoeing on the river, he saw clearly that the river is not an edge that separates the two banks but, rather, a seam that gathers the two sides together as one place.

5. RELIABILITY AND PHENOMENOLOGICAL RESEARCH

Though phenomenological research in the human sciences has been criticized on a number of grounds,[3] perhaps the most significant concern among conventionally-trained, positivist social scientists is the issue of *trustworthiness*—in other words, what criteria can be used to establish the reliability of phenomenological descriptions and interpretations?

From a phenomenological perspective, the issue of reliability first of all involves *interpretive appropriateness*: In other words, how can there be an accurate fit between

---

[3] I have discussed a number of these criticisms elsewhere (Seamon, 1987, pp. 15–19).

experience and language, between what we know as individuals in our own lives versus how that knowledge can be accurately placed theoretically? As von Eckartsberg (1998a, p. 15) explains,

> How is it that we can say what we experience and yet always live more than we can say, so that we could always say more than we in fact do? How can we evaluate the adequacy or inadequacy of our expression in terms of its doing justice to the full lived quality of the experience described?

Beyond the issue of phenomenological interpretation's rendering experience faithfully is the potential dilemma that several phenomenologists, dealing with the same descriptive evidence, may present their interpretations differently and arrive at entirely different meanings. In an article comparing three phenomenologically-based interpretations drawing on the same descriptive evidence, Churchill and colleagues (Churchill, Lowery, McNally, & Rao, 1998) attempted to deal with this issue of interpretive relativity. They pointed out that, in conventional positivist research, reliability refers to the fact that one can establish an *equivalence* of measurement, where measurement refers to quantification according to a predetermined scale or standard (ibid., p. 64). If, however, "measurement" must be applied to the qualitative descriptions of phenomenological research, the required equivalence is much more difficult to establish: "[N]ot only is the criterion for agreement between two verbal descriptions not clearly defined, but also an agreement among judges regarding the equivalence of descriptions becomes equally difficult to establish" (ibid., p. 64).

As a way to consider the issue of reliability phenomenologically, Churchill and colleagues organized the following phenomenological experiment: They presented the same set of narrative descriptions to three researchers all trained in phenomenological method.[4] Each researcher was free to bring his or her set of concerns and questions to the descriptions. After studying the three resulting interpretations, Churchill and colleagues concluded that, though there were some differences in emphases, there was also a common thematic core.[5] This result indicates that phenomenological interpretation offers *some* degree of equivalence, since a "somewhat coherent set of themes can be gleaned from three different interpretive research results" (ibid., p. 81). On the other hand, there were also differences among the three interpretations, but these differences do not so much indicate the failure of phenomenology as a method but, rather, demonstrate the existential fact that human interpretation is always only *partial*.[6]

In this sense, reliability from a phenomenological perspective cannot be defined as some equivalence of measurement based on some predefined scale of calculation

---

[4] The description related to the current sexual practices of a young woman who had previously been the victim of a date rape.

[5] This thematic core involved a common focus on "a vacillation within the [respondent's] experience from active to passive agency, with passivity emerging precisely at those moments when a decision is called for on the [respondent's] part. Likewise, all three researchers see her as 'disowning' her body—disconnecting her 'self' from her actions when her integrity is at stake. Finally, all three see that her integrity within the situation is a function of her . . . desire for a sexual experience that is 'shared and reciprocal'" (ibid., p. 81).

[6] From a phenomenological perspective, Churchill's experiment is artificial in the sense that the researchers interpreting the lived description did not actually gather it from the respondent, thus they had no sense of the lived context out of which the description arose. In addition, these researchers were recruited after the description was already solicited, thus they had no personal interest or stake in the phenomenon being studied. It is significant that, in spite of these weaknesses, the three researchers were able to identify similar core themes.

separate from the experience and understanding of the researcher. Rather, reliability can only be had through what can be called *intersubjective corroboration*—in other words, can other interested parties find in their own life and experience, either directly or vicariously, what the phenomenologist has found in her own work? One can conclude that the conclusions of any phenomenological study are no more and no less than interpretive *possibilities* open to the public scrutiny of other interested parties. As Giorgi (cited in Churchill et al., 1998, p. 81) explains:

> Thus the chief point to be remembered with this kind of research is not so much whether another position with respect to the [original descriptions] could be adopted (this point is granted beforehand) but whether a reader, adopting the same viewpoints as articulated by the researcher, can also see what the researcher saw, whether or not he agrees with it. That is the key criterion for qualitative research.

In spite of the relativity of phenomenological trustworthiness, there have been efforts by phenomenologists to establish qualitative criteria that can help to judge the validity of phenomenological interpretation—at least in broad terms (e.g., van Manen, 1990; Polkinghorne, 1983). Polkinghorne (1983, p. 46), for example, presented four qualities to help readers judge the trustworthiness of phenomenological interpretation: *vividness*, *accuracy*, *richness*, and *elegance*. First, vividness is a quality that draws readers in, generating a sense of reality and honesty. Second, accuracy refers to believability in that readers are able to recognize the phenomenon in their own lifeworlds or they can imagine the situation vicariously. Third, richness relates to the aesthetic depth and quality of the description, so that the reader can enter the interpretation emotionally as well as intellectually. Finally, elegance points to descriptive economy and a disclosure of the phenomenon in a graceful, even poignant, way.

Using these four criteria, one can evaluate the effectiveness of specific phenomenological work—for example, the above-mentioned first-person studies of Toombs and Violich. Note that, from a conventional positivist perspective, the reliability of this work would immediately be called into question because of the issue of subjectivity and first-person interpretation: How can the reader be sure that the two researchers' understandings of their own experiences speak in any accurate way to the realm of human experience in general?

But also note that, in terms of Polkinghorne's four criteria, the issue is no longer subjectivity but, rather, the *power to convince*: Are Toombs' and Violich's first-person interpretations strong enough to engage the reader and get her to accept the researchers' conclusions?

In this regard, Toombs' first-person phenomenology of illness (Toombs, 1992a, b) succeeds in terms of all Polkinghorne's criteria: Her writing is vivid, accurate, and rich in the sense that the reader is drawn into the reality of her descriptions and can believe they relate to concrete experiences that she, the reader, can readily enter secondhand. In addition, Toombs' work is elegant because there is a clear interrelationship between real-world experiences and conceptual interpretation. In sum, the reader can imaginatively participate in Toombs' situations and conclusions. What she says "seems right" as her connections between phenomenological theory and lived experience allow the reader to "see" her situation in a thorough, heartfelt way.

On the other hand, Violich's portrait of Dalmatian towns can be judged as less trustworthy in terms of Polkinghorne's four criteria because Violich's interpretations seem too much the image of an outsider experiencing place for only a short time. He describes these towns largely in terms of physical features and human activities as they can be read publicly in outdoor social spaces. There is no sense of what these places mean for the people who live and work there. The resulting interpretation seems incomplete and lacking in the potential fullness of the places as they are everyday lifeworlds.[7]

Ultimately, the most significant test of trustworthiness for any phenomenological study is its relative power to draw the reader into the researcher's discoveries, allowing the reader to see his or her own world or the worlds of others in a new, deeper way. The best phenomenological work breaks people free from their usual recognitions and moves them along new paths of understanding.

6. MAKING BETTER WORLDS

In the end, the phenomenological enterprise is a highly personal, interpretive venture. In trying to see the phenomenon, it is very easy to see too much or too little. Looking and trying to see are very much an intuitive, spontaneous affair that involves feeling as much as thinking. In this sense, phenomenology might be described as a method to cultivate a mode of seeing that cultivates both intellectual *and* emotional sensibilities, with the result that understanding may be more whole and comprehensive.

As Thiis-Evensen's work indicates, many of the more recent phenomenological works relevant to environment-behavior research use phenomenological insights to examine design issues.[8] Because architecture and design also regularly involve a process of intuitive awareness and discovery, a phenomenological approach may be one way to rekindle designers' interest in environment-behavior research—an interest that seriously waned as architects and other designers became uncomfortable with the strong positivist stance of environment-behavior studies in the 1970s and 1980s. According to Franck (1987, p. 65), a key reason for this discomfort was the unwillingness of social scientists to "understand or accept the [more intuitive] strategies and priorities of the design professions" (ibid). Franck emphasized that one of the greatest values of phenomenology is its potential for providing a place for dialogue between designers and social scientists because it gives attention "to the essence of human experience rather than to any abstraction of that experience and because of its ability to reconcile, or perhaps to bypass completely, the positivist split between 'objective' and 'subjective'" (ibid., pp. 65–66).

In placing phenomenological work in today's broader intellectual landscape, Mugerauer (1993, pp. 94–95) points to critics on both the "right" and "left." On the "right," are the positivists, who see phenomenology as "subjective," "soft," and "anec-

---

[7] On the other hand, Violich's work is still important because it serves as one model for first-person phenomenologies of place. More such studies are needed, coupled with other ways to read place as in Million's and Chaffin's work (Million, 1992; Chaffin, 1989). Other useful models include Hufford's interpretation of the New Jersey Pinelands (Hufford, 1986), Lane's work on American sacred spaces and places (Lane, 1988), Mugerauer's hermeneutic readings of the contemporary North American landscape (Mugerauer, 1994, 1995), Pocius' in-depth study of a Newfoundland harbor village (Pocius, 1991, and Walkey's presentation of the multi-story, guild-build houses of mountainous northern Greece, western Turkey, and the adjoining Balkan states (Walkey, 1993).

[8] Examples include: Alexander, 1987, 1993, et al., 1977; Barbey, 1989; Boschetti, 1990; Brill, 1993; Coates, 1998; Coates & Seamon, 1993; Cooper Marcus, 1993; Dorward, 1990; Dovey, 1993; Hester, 1993; Howett, 1993; Mugerauer, 1993, 1994, 1995; Munro, 1991; Paterson, 1993a, b; Porteous, 1989; Seamon, 1990b; Silverstein, 1993a, b; Thiis-Evensen, 1987; Violich, 1998; Walkey, 1993.

todal." On the "left," are the post-structuralists and deconstructivists, who question phenomenology's belief in commonality, continuity, pattern, and order.[9] In phenomenology and hermeneutics, Mugerauer sees a *middle way* between the absolutism of positivism, on one hand, and the relativism of post-structuralism, on the other. This is so, says Mugerauer, because in its efforts to see and understand human experience and meaning in a kindly, open way, phenomenology strives for a balance between person and world, researcher and phenomenon, feeling and thinking, and experience and theory. This effort of balance, he believes is crucial "if we are to adequately understand, plan, and build a socially pluralistic and ecologically appropriate environment" (ibid., p. 94).

The long-term impact of phenomenology on environment-behavior research remains to be seen. The advances in the last ten years are encouraging, though the approach is still obscure among many mainstream researchers. I hope to have demonstrated in this chapter that phenomenology offers an innovative way for looking at the person-environment relationship and for identifying and understanding its complex, multidimensioned structure. I also hope to have suggested that phenomenology provides a useful conceptual language for bridging the environmental designer's more intuitive approach to understanding with the academic researcher's more intellectual approach. In this sense, phenomenology may be one useful way for the environment-behavior researcher to reconcile the difficult tensions between feeling and thinking, between understanding and designing, and between firsthand lived experience and its secondhand conceptual accounts.

REFERENCES

Abrams, D. (1996). *The spell of the sensuous.* New York: Pantheon.

Alexander, C. (1987). *A new theory of urban design.* New York: Oxford University Press.

Alexander, C. (1993). *A foreshadowing of 21st century art: The color and geometry of very early Turkish carpets.* New York: Oxford University Press.

Alexander, C., Ishikawa, S., & Silverstein, M. (1977). *A pattern language.* New York: Oxford University Press.

Anella, T. (1990). Learning from the Pueblos. In N. Markovich, W. Preiser, & F. Sturm (Eds.), *Pueblo style and regional architecture* (pp. 31–45). NY: Van Nostrand Reinhold.

Bachelard, G. (1964). *The poetics of space.* Boston: Beacon.

Barbey, G. (1989). Introduction: Towards a phenomenology of home. *Architecture and Behavior, 5,* 1–10.

Barnes, A. (1992). *Mount Wellington and the sense of place.* Master's thesis, Department of Environmental Studies, University of Tasmania, Hobart, Tasmania, Australia.

Behnke, E. (1990). Field notes: Lived place and the 1989 earthquake in northern California. *Environmental and architectural phenomenology newsletter, 1* (2), 10–14.

Berendt, J. (1985). *The third ear: On listening to the world.* New York: Henry Holt.

Berleant, A. (1991). *Art and engagement.* Philadelphia: Temple University Press.

Berleant, A. (1992). *The aesthetics of environment.* Philadelphia, PA: Temple University Press.

Bollnow, O. F. (1961). Lived-space, *Philosophy Today, 5,* 31–39.

---

[9] Post-structuralism and deconstruction have become be a significant conceptual force in social science and, especially, in architecture (Mugerauer, 1994, chap. 3). For deconstructivists, meaning, pattern, and quality are plural, diverse, and continuously shifting. The aim is relativist interpretation and "deconstruction"—the undermining and dismantling of all assumed and taken-for-granted givens, be they existential, cultural, historical, political, or aesthetic. The aim is the freedom to change and to reconstitute oneself continually. To have this shifting freedom, one must vigilantly remember that all life is a sham and so confront the unintelligible, relative nature of the world and human being (Mugerauer, 1988, p. 67). An excellent discussion of the poststructural-deconstructivist criticisms of phenomenology is Mugerauer, 1994, especially chap. 6.

Bortoft, H. (1996). *The wholeness of nature: Goethe's way toward a science of conscious participation in nature.* Hudson, New York: Lindesfarne Press.

Boschetti, M. (1990). Reflections on home: Implications for housing design for elderly persons. *Housing and society, 17* (3), 57–65.

Boschetti, M. (1993). Staying in place: Farm homes and family heritage. *Housing and society, 10,* 1–16.

Boschetti, M. (1995). Attachment to personal possessions: An interpretive study of the older person's experience, *Journal of interior design, 21,* 1–12.

Brenneman, W. L., Jr., & Brenneman, M. G. (1995). Crossing the Circle at the Holy Wells of Ireland. Charlottesville: University Press of Virginia.

Brill, M. (1993). An architecture of peril: Design for a waste isolation pilot plant, Carlsbad, New Mexico. *Environmental and Architectural Phenomenology Newsletter, 4* (3), 8–10.

Burch, R. (1989). On phenomenology and its practices [part I], *Phenomenology + Pedagogy, 7,* 187–217.

Burch, R. (1990). On phenomenology and its practices [part II], *Phenomenology + Pedagogy, 8,* 130–160.

Burch, R. (1991). On phenomenology and its practices [part III], *Phenomenology + Pedagogy, 9,* 167–193.

Casey, E. S. (1993). *Getting back into place.* Bloomington: Indiana University Press.

Casey, E. S. (1997). *The fate of place: A philosophical history.* Berkeley: University of California Press.

Chaffin, V. F. (1989). Dwelling and rhythm: The Isle Brevelle as a landscape of home. *Landscape journal, 7,* 96–106.

Chawla, L. (1994). *In the first country of places: Nature, poetry, and childhood memory.* Albany, New York: SUNY Press.

Chawla, L. (1995). Reaching home: Reflections on environmental autobiography. *Environmental and Architectural Phenomenology Newsletter, 6, 2,* 12–15.

Cheney, J. (1989). Postmodern environmental ethics: Ethics as bioregional narrative. *Environmental ethics, 11,* 117–134.

Chidester, D., & Linenthal, E. T. (Eds.). (1995). *American sacred space.* Bloomington: Indiana University Press.

Churchill, S. D., Lowery, J. E., McNally, O., & Rao, A. (1998). The question of reliability in interpretive psychological research. In R. Valle (Ed.), *Phenomenological inquiry in psychology* (pp. 63–85). New York: Plenum.

Cloke, P., Philo, C., & Sadler, D. (1991). *Approaching human geography: An introduction to contemporary theoretical debates.* New York: Guildford Press.

Coates, G. J. (1997). *Erik Asmussen, Architect.* Stockholm: Byggförget.

Coates, G. J., & Seamon, D. (1993). Promoting a foundational ecology practically through Christopher Alexander's pattern language: The example of Meadowcreek. In D. Seamon (Ed.), *Dwelling, seeing, and designing: Toward a phenomenological ecology* (pp. 331–354). Albany, New York: SUNY Press.

Condon, P. M. (1991). Radical romanticism. *Landscape journal, 10,* 3–8.

Cooper Marcus, C. (1993). Designing for a commitment to place: Lessons from the alternative community Findhorn. In D. Seamon (Ed.), *Dwelling, seeing, and designing: Toward a phenomenological ecology* (pp. 299–330). Albany, New York: SUNY Press.

Cooper Marcus, C. (1995). House as a mirror of self. Berkeley, California: Conari.

Corner, J. (1990). A discourse on theory I: "Sounding the depths"—origins, theory, and representation. *Landscape Journal, 9,* 61–78.

Davis, T. (1989). Photography and landscape studies, *Landscape journal, 8,* 1–12.

Day, M. D. (1996). *Home in the postmodern world: An existential phenomenological study.* Paper presented at the International Human Science Research Conference, Halifax, Nova Scotia.

Dorward, S. (1990). *Design for mountain communities: A landscape and architecture guide.* New York: Van Nostrand Reinhold.

Dovey, K. (1985). Home and homelessness. In I. Altman, & C. M. Werner (Eds.), *Home environments* (pp. 33–64). New York: Plenum.

Dovey, K. (1993). Putting geometry in its place: Toward a phenomenology of the design process. In D. Seamon (Ed.), *Dwelling, seeing, and designing: Toward a phenomenological ecology* (pp. 247–269). Albany, New York: SUNY Press.

Ediger, J. (1993). *A phenomenology of the listening body.* Doctoral dissertation, Institute of Communication Research, University of Illinois, Champaign-Urbana.

Eliade, M. (1961). *The sacred and the profane.* New York: Harcourt.

Embree, L. (1997). *The encyclopedia of phenomenology.* Dordrecht, the Netherlands: Kluwer.

Foltz, B. (1995). *Inhabiting the earth: Heidegger, environmental ethics, and the metaphysics of nature.* New York: Humanities Press.

Franck, K. (1987). Phenomenology, positivism, and empiricism as research strategies in environment-behavior research and design. In G. T. Moore, & E. Zube (Eds.), *Advances in environment, behavior, and design, Vol. 1* (pp. 59–67). New York: Plenum.

Giorgi, A., (Ed.). (1985). *Phenomenology and psychological research*. Pittsburgh, PA: Duquesne University Press.

Giorgi, A., Barton, A., & Maes, C., (Eds.) (1983). *Duquesne studies in phenomenological psychology, Vol. 4.* Pittsburgh: Duquesne University Press.

Glaser, B. G., & Straus, A. (1967). *The discovery of grounded theory: Strategies for qualitative research.* Chicago: Aldine.

Graumann, C. F. (1989). Towards a phenomenology of being at home. *Architecture and Behavior, 5,* 117–126.

Harries, K. (1988). The Voices of space, *Center, 4,* 34–49.

Harries, K. (1993). Thoughts on a non-arbitrary architecture. In D. Seamon (Ed.), *Dwelling, seeing, and designing: Toward a phenomenological ecology* (pp. 41–59). Albany, New York: SUNY Press.

Harries, K. (1997). *The ethical function of architecture.* Cambridge, Massachusetts: MIT Press.

Heelan, P. A. (1983). *Space-perception and the philosophy of science.* Berkeley: University of California Press.

Heidegger, M. (1962). *Being and time.* New York: Harper & Row.

Heidegger, M. (1971). *Poetry, language, and thought.* New York: Harper & Row.

Hester, R., Jr. (1993). Sacred structures and everyday life: A return to Manteo, North Carolina. In D. Seamon (Ed.), *Dwelling, seeing, and designing: Toward a phenomenological ecology* (pp. 271–297). Albany, New York: SUNY Press.

Hill, M. H. (1985). Bound to the environment: Towards a phenomenology of sightlessness. In D. Seamon, & R. Mugerauer (Eds.), *Dwelling, place, and environment* (pp. 99–111). New York: Columbia University Press.

Howett, C. (1993). "If the doors of perception were cleansed": Toward an experiential aesthetics for the designed landscape. In D. Seamon (Ed.), *Dwelling, seeing, and designing: Toward a phenomenological ecology* (pp. 61–73). Albany, New York: SUNY Press.

Hufford, M. (1986). *One space, many places: Folklife and land use in New Jersey's Pinelands National Reserve.* Washington, D.C.: American Folklife Center.

Jackson, M. (1996). *Things as they are: New directions in phenomenological anthropology.* Bloomington: University of Indiana Press.

Jarviluoma, H., (Ed.) (1994). *Soundscapes: Essays on Vroom and Moo.* Tampere, Finland: Tampere University.

Jones, E. (1989). *Reading the book of nature: A phenomenological study of creative expression in science and painting.* Athens, OH: Ohio University Press.

Koop, T. T. (1993). *The idea of home: A cross-cultural interpretation.* Doctoral dissertation, Department of Geography, University of Minnesota.

Krapfel, P. (1989). *Shifting.* Cottonwood, CA. [privately printed].

Kushwah, R. (1993). *Louis I. Kahn and the Phenomenology of Architecture: An Interpretation of the Kimbell Art Museum Using Thomas Thiis-Evensen's Theory of Architectural Archetypes.* Master's thesis, Department of Architecture, Kansas State University, Manhattan, Kansas.

Lane, B. (1988). *Places of the sacred: Geography and narrative in American spirituality.* New York: Paulist Press.

LeStrange, R. (1998). *Psyche speaking through our place attachments: Home and journey as a process of psychological development.* Doctoral dissertation, Department of Clinical Psychology, Pacifica Graduate Institute, Carpinteria, California.

Lin, Y. (1991). *LeCorbusier's Chapel at Ronchamp, Frank Lloyd Wright's Unitarian Church, and Mies van der Rohe's Chapel at IIT: A phenomenological interpretation of modern sacred architecture based on Thiis-Evensen's Archetypes in architecture.* Master's thesis, Department of Architecture, Kansas State University, Manhattan, Kansas.

Lin, Y., & Seamon, D. (1994). A Thiis-Evensen Interpretation of Two Churches by Le Corbusier and Frank Lloyd Wright. In R. M. Feldman, G. Hardie, & D. G. Saile (Eds.), *Power by design: EDRA Proceedings 24,* 130–142. Oklahoma City, Oklahoma: Environmental Design Research Association.

Lincoln, Y. S., & Guba, E. G. (1985). *Naturalistic inquiry.* Newbury Park, California: Sage.

Lipton, T. (1990). Tewa visions of space. In N. Markovich, W. Preiser, & F. Sturm (Eds.), *Pueblo style and regional architecture* (pp. 133–139). NY: Van Nostrand Reinhold.

Low, S. M. (1987). Developments in research design, data collection, and analysis: qualitative methods. In G. T. Moore, & E. Zube (Eds.), *Advances in environment, behavior, and design, Vol. 1* (pp 279–303). New York: Plenum.

Margadant-van Archen, M. (1990). Nature experience of 8-to-12-year-old children, *Phenomenology + Pedagogy, 8*, 86–94.

Masucci, M. (1992). The Chesapeake Bay Bridge: Development Symbol for Maryland's Eastern Shore. In D. G. Janelle (Ed.) *Geographical snapshots of North America* (pp. 74–77). New York: Guildford.

Merleau-Ponty, M. (1962). *The phenomenology of perception.* New York: Humanities Press.

Meurant, R. (1989). *The aesthetics of the sacred.* Whangamat, New Zealand: The Institute of Traditional Studies.

Million, M. L. (1992). *"It was home": A phenomenology of place and involuntary displacement as illustrated by the forced dislocation of five southern Alberta families in the Oldman River Dam Flood Area.* Doctoral dissertation, Saybrook Institute Graduate School and Research Center, San Francisco, California.

Moustakas, C. (1994). *Phenomenological research methods.* Newbury Park, CA: Sage.

Mugerauer, R. (1988). Derrida and beyond. *Center, 4*, 66–75.

Mugerauer, R. (1993). Toward an architectural vocabulary: The porch as a between. In D. Seamon (Ed.), *Dwelling, seeing, and designing: Toward a phenomenological ecology* (pp. 103–128). Albany, New York: SUNY Press.

Mugerauer, R. (1994). *Interpretations on behalf of place: Environmental displacements and laternative responses.* Albany, New York: SUNY Press.

Mugerauer, R. (1995). *Interpreting environments: Tradition, deconstruction, hermeneutics.* Austin: University of Texas Press.

Munro, K. A. (1991). *Planning for place: Phenomenological insights in urban design.* Master's thesis, School of Community and Regional Planning, University of British Columbia, Vancouver, British Columbia.

Nogué i Font, J. (1993). Toward a phenomenology of landscape and landscape experience: An example from Catalonia. In D. Seamon (Ed.), *Dwelling, seeing, and designing: Toward a phenomenological ecology* (pp. 159–180). Albany, New York: SUNY Press.

Norberg-Schulz, C. (1980). *Genius loci: Toward a phenomenology of architecture.* New York: Rizzoli.

Norberg-Schulz, C. (1988). *Architecture: Meaning and place.* New York: Rizzoli.

Norberg-Schulz, C. (1996). *Nightlands: Nordic Building.* Cambridge: MIT Press.

Norris, C. (1990). Stories of paradise: What is home when we have left it? *Phenomenology + Pedagogy, 8*, 237–244.

Oldenburg, R. (1989). *The great good place.* New York: Paragon House.

Pallasmaa, J. (1995). Identity, intimacy, and domicile: A Phenomenology of home. In D. N. Benjamin (Ed.), *The home: Words, interpretations, meanings, and environments* (pp. 33–40). London: Avery.

Pallasmaa, J. (1996). *The eyes of the skin: Architecture and the senses.* London: Academy Editions.

Paterson, D. D. (1991). Fostering the avant-garde within. *Landscape Journal, 10*, 27–36.

Paterson, D. D. (1993a). Dualities and dialectics in the experience of landscape. *Design + Values, CELA Conference Proceedings, Vol. 4* (pp. 147–166). Washington, D. C.: Council of Educators in Landscape Architecture.

Paterson, D. D. (1993b). Design, language, and the preposition: On the importance of knowing one's position in place. *Trames, Vol. 8* (pp. 74–86). Quebec: Faculté de l'aménagement, Université de Montréal.

Patton, M. Q. (1990). *Qualitative evaluation and research methods* (2nd ed.). Newbury Park, CA: Sage.

Pocius, G. L. (1991). *A Place to belong: Community Order and everyday space in Calvert, Newfoundland.* Athens: University of Georgia Press.

Pocock, D. C. D. (1989). Humankind-environment: Musings on the role of the hyphen. In F. W. Boal, & D. N. Livingston (Eds.), *The behavioural environment* (pp. 82–90). London: Routledge.

Pocock, D. C. D. (1993). The senses in focus. *Area, 25*, 11–16.

Polkinghorne, D. (1983). *Methodology for the human sciences.* Albany: SUNY Press.

Pollio, H. R., Henley, T. B., & Thompson, C. J. (1996). *The phenomenology of everyday life.* New York: Cambridge University Press.

Porteous, J. D. (1989). *Planned to death: The annihilation of a place called Howdendyke.* Toronto: University of Toronto Press.

Porteous, J. D. (1990). *Landscapes of the mind: Worlds of sense and metaphor.* Toronto: University of Toronto Press).

Ramaswami, M. (1992). *Toward a phenomenology of wood: Interpreting the Yoshimura House, a Japanese vernacular dwelling, through Thiis-Evensen's architectural archetypes.* Master's thesis, Department of Architecture, Kansas State University, Manhattan, Kansas.

Rapoport, A. (1993). A critical look at the concept "place." *National Geographic Journal of India, 40*, 31–45.

Rattner, D. M. (1993). Moldings: The atomic units of classical architecture. *Traditional Building, 6* (4) 4, 72–73.

Rehorick, D. (1986). Shaking the foundation of lifeworld: A phenomenological account of an earthquake expeience. *Human Studies, 9*, 379–391.

Relph, E. (1976). *Place and placelessness.* London: Pion.

Relph, E. (1989a). A curiously unbalanced condition of the powers of the mind: Realism and the ecology of environmental experience. In F. W. Boal, & D. N. Livingston (Eds.), *The Behavioral environment: Essays in reflection, application, and re-evaluation.* London: Routledge.

Relph, E. (1989b). Responsive methods, geographical imagination, and the study of landscapes. In A. Kobayashi, & S. MacKenzie (Eds.), *Remaking human geography* (pp. 149–163). Boston: Unwin Hyman.

Relph, E. (1990). Geographical imagination. *National geographical journal of India, 36*, 1–9.

Relph, E. (1993). Modernity and the reclamation of place. In D. Seamon (Ed.), *Dwelling, seeing, and designing: Toward an phenomenological ecology* (pp. 25–40). Albany, NY: SUNY Press.

Relph, E. (1996). Reflections on Place and placelessness. *Environmental and Architectural Phenomenology Newsletter, 7* (3), 14–16.

Riegner, M. (1993). Toward a holistic understanding of place: Reading a landscape through its flora and fauna. In D. Seamon (Ed.), *Dwelling, seeing, and designing: Toward a phenomenological ecology* (pp. 181–215). Albany, New York: SUNY Press.

Rouner, L. S., (Ed.) (1996). *Longing for home.* Notre Dame, Indiana: Notre Dame University Press.

Schafer, M. (1977). *The tuning of the world.* New York: Knopf.

Schönhammer, R. (1989). The walkman and the primary world of the senses. *Phenomenology + Pedagogy, 7*, 127–144.

Seamon, D. (1979). *A geography of the lifeworld.* New York: St. Martin's.

Seamon, D. (1982). The phenomenological contribution to environmental psychology. *Journal of Environmental Psychology, 2*, 119–140.

Seamon, D. (1987). Phenomenology and environment-behavior research. In G. T. Moore, & E. Zube (Eds.), *Advances in environment, behavior, and design, Vol. 1* (pp 3–27). New York: Plenum.

Seamon, D. (1989). Humanistic and phenomenological advances in environmental design. *Humanistic Psychologist, 17*, 280–293.

Seamon, D. (1990a). Awareness and reunion: A phenomenology of the person-environment relationship as portrayed in the New York photographs of André Kertész. In L. Zonn (Ed.) *Place images in the mdia* (pp. 87–107). Totowa, New Jersey: Roman and Littlefield.

Seamon, D. (1990b). Using pattern language to identify sense of place: American landscape painter Frederic Church's Olana as a test case. In, R. Selby (Ed.), *Coming of age: Proceedings, EDRA*, 171–179. Oklahoma City, Oklahoma.

Seamon, D. (1991). Toward a phenomenology of the architectural lifeworld. In J. Hancock, & W. Miller (Eds.), Architecture: back . . . to . . . life. *Proceedings of the 79th Annual Meeting of the Association of Collegiate Schools of Architecture* (pp. 3–7). Washington, D. C.: ACSA Press.

Seamon, D. (1992). A Diary interpretation of place: Artist Frederic Church's Olana. In D. G. Jannelle (Ed.), *Geographical snapshots of North America* (pp. 78–82). New York: Guilford Press.

Seamon, D. (1993a). Different worlds coming together: A phenomenology of relationship as portrayed in Doris Lessing's diaries of Jane Somers. In D. Seamon (Ed.), *Dwelling, seeing, and designing: Toward a phenomenological ecology* (pp. 219–246). Albany, NY: SUNY Press.

Seamon, D., (Ed.) (1993b). *Dwelling, seeing, and building: Toward a phenomenological ecology.* Albany, NY: SUNY Press.

Seamon, D. (1994). The life of the place: A phenomenological reading of Bill Hiller's space syntax. *Nordisk arkitekturforskning [Nordic Journal of Architectural Research], 7*, 35–48.

Seamon, D. (1997). [Phenomenology and] behavioral geography. In L. Embree (Ed.), *Encyclopedia of phenomenology* (pp. 53–56). Dordrecht, the Netherlands: Kluwer.

Seamon, D., & Mugerauer, R., (Eds.) (1985). *Dwelling, place, and environment: Towards a phenomenology of person and world.* New York: Columbia University Press.

Seamon, D., & Zajonc, A. (1998). *Goethe's way of science: Toward a phenomenology of nature.* Albany, NY: SUNY Press.

Shaw, S. (1990). Returning home. *Phenomenology + Pedagogy, 8*, 224–236.

Sherry, J. F., Jr., (Ed.). (1998). *Servicescapes: The concept of place in contemporary markets.* Chicago: NTC/Contemporary Publishing Co.

Shertock, T. (1998). Latin American women's experience of feeling able to move toward and accomplish a meaniful and challenging goal. In R. Valle (Ed.), *Phenomenological inquiry in psychology* (pp. 157–174). New York: Plenum.

Silverstein, M. (1993a). Mind and the world: The interplay of theory and practice. *Architecture California, 15*, 2, 20–28.

Silverstein, M. (1993b). The first roof: Interpreting a spatial pattern. In D. Seamon (Ed.), *Dwelling, seeing, and designing: Toward a phenomenological ecology* (pp. 77–101). Albany, New York: SUNY Press.

Silverstein, M. (1994). Is place a journey? *Environmental and Architectural Phenomenology Newsletter. 5* (1), 12–15.

Sinclaire, C. (1994). *Looking for home: A phenomenological study of home in the classroom.* Albany, NY: State University of New York Press.

Smith, T. S. (1989). Ojibwe Persons: Toward a phenomenology of an American Lifeworld. *Journal of Phenomenological Psychology, 20*, 130–144.

Spiegelberg, H. (1982). *The phenomenological movement.* Dordrecht, the Netherlands: Martinus Nijhoff.

Stefanovic, I. L. (1991). Evolving sustainability: A re-thinking of ontological foundations. *Trumpeter, 8*, 194–200.

Stefanovic, I. L. (1992). The experience of place: Housing quality from a phenomenological perspective. *Canadian Journal of Urban Research, 1*, 145–161.

Stefanovic, I. L. (1994). Temporality and architecture: A phenomenological reading of built form. *Journal of Architectural and Planning Research, 11*, 211–225.

Stewart, D., & Mukunis, A. (1990). *Exploring phenomenology: A guide to the field and its literature, second edition.* Athens, Ohio: Ohio University Press.

Sturm, F. (1990). Aesthetics of the Southwest. In N. Markovich, W. Preiser, & F. Sturm (Eds.), *Pueblo style and regional architecture* (pp. 81–92). NY: Van Nostrand Reinhold.

Swentzell, R. (1990). Pueblo space, form, and mythology. In N. Markovich, W. Preiser, & F. Sturm (Eds.), *Pueblo style and regional architecture* (pp. 23–30). NY: Van Nostrand Reinhold.

Thiis-Evensen, T. (1987). *Archetypes in architecture.* Oslo: Scandinavian University Press.

Toombs, S. K. (1992a). The body in multiple sclerosis: A patient's perspective. In D. Leder, (Ed.), *The body in medical thought and practice* (Dordrecht, the Netherlands: Kluwer).

Toombs, S. K. (1992b). *The meaning of iIllness: A phenomenological account of the different perspectives of physician and patient.* Dordrecht, the Netherlands: Kluwer.

Toombs, S. K. (1995a). Sufficient unto the day: A life with multiple sclerosis. In S. K. Toombs, D. Barnard, & R. A. Carson (Eds.), *Chronic illness: From experience to policy.* Bloomington, ID: University Press.

Toombs, S. K. (1995b). The lived experience of disability. *Human studies, 18*, 9–23.

Tuan, Y. (1993). *Passing strange and wonderful: Aesthetics, nature, and culture.* Washington, D. C.: Island Press.

Valle, R. (1998). *Phenomenological inquiry in psychology.* New York: Plenum.

van Manen, M. (1990). *Researching lived experience.* Albany, New York: SUNY Press.

Vesely, D. (1988). On the relevance of phenomenology. In S. Perrella (ed.), Form; being; absence: *Pratt Journal of Architecture, 2*, 54–60. New York: Rizzoli.

Violich, F. (1985). Toward revealing the sense of place: An intuitive "reading" of four Dalmatian towns. In D. Seamon, & R. Mugerauer (Eds.), *Dwelling, place, and environment: Towards a phenomenology of person and world* (pp. 113–136). New York: Columbia University Press.

Violich, F. (1998). *The bridge to Dalmatia: A search for the meaning of place.* Baltimore: Johns Hopkins Press.

von Eckartsberg, R. (1998a). Introducing existential-phenomenological psychology. In R. Valle (Ed.), *Phenomenological inquiry in psychology* (pp. 3–20). New York: Plenum.

von Eckartsberg, R. (1998b). Existential-phenomenological research. In R. Valle (Ed.), *Phenomenological inquiry in psychology* (pp. 21–61). New York: Plenum.

Walkey, R. (1993). A lesson in continuity: The legacy of the builders guild in northern Greece. In D. Seamon, (Ed.), *Dwelling, seeing, and designing: Toward a phenomenological ecology* (pp. 129–157). Albany, New York: SUNY Press.

Weiner, J. F. (1991). *The empty place: Poetry, space, and being among the Foi of Papua, New Guinea.* Bloomington: Indiana University Press.

Wertz, F. J. (1984). Procedures in phenomenological research and the question of validity. In C. Aanstoos (Ed.), *Exploring the lived world: Readings in phenomenological psychology* (pp. 29–48). Carrolton, Georgia: West Georgia College.

Weston, A. (1994). *Back to earth: Tomorrow's environmentalism.* Philadelphia: Temple University Press.

Whone, H. (1990). *Church, monastery, cathedral: An illustrated guide to Christian symbolism.* Longmead, Dorsetshire, Great Britain: Element Books.

Winning, A. (1990). Homesickness. *Phenomenology + Pedagogy, 8*, 245–258.

Winning, A. (1991). The speaking of home. *Phenomenology and Pedagogy, 9*, 172–181.

Wu, K. K. (1993). Pilgrim cathedral. *Architecture and Behavior, 9*, 191–204.

Wu, Z. (1991). The lived experience of being a foreigner. *Phenomenology + Pedagogy, 9*, 267–275.

14

A STORYTELLER'S BELIEFS

Narrative and Existential Research

Herb Childress

Jay Farbstein & Associates, Inc.
San Luis Obispo, CA 93401

1. INTRODUCTION

The task of the contributors to this volume is to discuss the assumptions we bring to our work, and the ways in which these assumptions shape our actions. This chapter is my attempt to describe the assumptions that guide my work as a narrative researcher of human-environment relations.

"Narrative research" is not a term that was used much a decade ago, though it has been practiced for hundreds of years. "Ethnography" used to be the preferred academic term, before ethnography ceased to mean living among people and started to include interviews with relative strangers. And before the concept of social science was developed, ethnography had been preceded by "storytelling."

Whatever the term, narrative research is simply the process of living among people about whom you'd like to know more, watching and listening and talking things over, and then reporting what you saw and heard to some third party. In my case, it is the process of living with American teenagers of small towns and suburbs, trying to understand how they see and how they use the environments that surround them, trying to investigate the questions that the kids bring up rather than the simplistic ones I started with, and then trying to translate their acts and beliefs into language that American adults can comprehend.

It really is that simple.

2. NARRATIVE METHODS

To say that it is simple, of course, is not to say that it is easy. My study of the teenagers of Curtisville, California (Childress, in press) took over a year of living within

Theoretical Perspectives in Environment-Behavior Research, edited by Wapner *et al.*
Kluwer Academic / Plenum Publishers, New York, 2000.

a community and its high school, and another year of writing, reviewing my ideas with my research participants, and writing again. I became an everyday part of Curtisville High School, in part because I believed that the school would be a place of both importance and conflict, and in part because a high school would easily afford a large collection of teenagers to talk with. I took no established role within the high school. I made no attempt to "go undercover" as a student, a pose both ethically questionable and chronologically unlikely. Neither did I pretend to be a teacher, counsellor, aide, administrator, custodian, or coach. I simply walked the halls, stood on the Quad, sat in the back rows of classes and the side of the teachers' lounge, paid attention, and talked to anybody who was interested.

And as I sat, a notebook in my lap, kids would occasionally sit beside me and ask, "What are you *writing* in there all the time?" And I would show them some scribbled notes or a behavior map, and we would have a brief conversation terminated by the beginning of the next period. As the kids gradually learned that I was neither malicious nor disdainful, our conversations grew longer, branched to other topics. They introduced me to their friends, explained to me why someone had said what he'd said or did what she'd done. They teased me to see if I could take a joke, and started including me in their activities.

In short, they reached out to make friends. And only then did I learn.

As both a research actor and a research writer, I make use of the powers of narrative: the power to understand and communicate emotion as well as fact; the power to make unfamiliar lives and circumstances less alien; the power to organize and make coherent without replacing lives with concepts; the power to maintain the depth and ambiguity of what might otherwise seem clean and simplistic. Storytelling, as both a way of thinking and a way of communicating, relies upon interpreting complex characters acting within difficult social, cultural, and physical environments.

Regardless of the initial question that leads me into the field—in this case, the ways that small-town teenagers encounter and think about their environments—I know that I will be attracted to compelling stories: some person or people with intentions who encounter other people with other intentions within a physical and social landscape that shapes their encounter. Their struggles will re-frame my questions and the stories I write in response. And the wonderful liberty of stories invites readers to construct their own meanings based both on my evidence and on their own experience.

3. NARRATIVE ASSUMPTIONS, PERSONAL BELIEFS

In order to fulfil my given task in this chapter, to discuss the assumptions I bring to my work and the ways they shape my actions, I have to take the key word "assumptions" and replace it with "beliefs." From there, I will have something to say.

Assumptions are external and impersonal. They are the constraints we set on a logic problem in order to make it solvable. They are the rules within which we play the game, and can be changed with no threat to the player. Beliefs, on the other hand, are internal and deeply personal. They are the eyes through which we see the world, the master narrative through which we construct meanings. Beliefs are only changed through inner struggle; in making those changes, we redefine ourselves.

Over the past six years of research with suburban and rural teenagers, my life has been changed, by which I mean my beliefs have been changed. The condition, of course, is recursive; as my beliefs have changed, so in turn has my work.

3.1. The Necessary Moral Question

My first and primary belief is that **all personal and social action, including but not limited to research, is inevitably moral.** We cannot make any act in the world without ethical considerations; some people will be enhanced by our work, others left aside. We will choose one topic, and in doing so, not pursue hundreds of others. In my own work, that has led me to research teenagers' settings from the point of view of their teenaged users. This seems straightforward, but is quite rare; far more often, we research youth with the goal of improved administration, of providing kids with more of what we want them to have. But the administrators already have enormous power over teenagers, by virtue of their positions, their social connections, their economic strength. The kids, who have little power even in settings ostensibly designed for them, need someone to help amplify their voices, to explore what they need and want. In this belief, I follow the guidance of Clare Cooper Marcus and Wendy Sarkissian (1986):

"What happens when basic needs have been reasonably accommodated and conflicts still develop? . . . . We believe that such conflicts usually can, and should, be resolved in favor of residents who are most at risk environmentally" (p. 20).

But even the smallest day-to-day aspects of research work have ethical dimensions. Should I stray from the prepared questionnaire when a respondent clearly needs to have someone listen to her story? Should I continue to be seen with one person when that association will limit my access to another social group that I also need to hear from? Should I stay an hour later at school and miss yet another dinner with my wife? I cannot at any moment act without making an ethical choice about goals and priorities, about whose ends will be furthered. If I ignore those questions, the answers will be made through my omission.

This is absolutely not to say that everyone who attends closely to the moral dimensions of their work would make the same decisions that I make. It is simply to say that only through addressing these questions actively and purposefully, by making them a central part of my daily thinking rather than relying passively on someone else's ethical guidelines or human subjects review boards, can I act in good faith.

3.2. Objectivity and Care

A second belief springs directly from the first. **I do not argue that objectivity in research is impossible. However, I am only interested in those questions for which objectivity *is* impossible.** I have a limited time on the Earth, and I cannot honorably spend it in asking questions for which I don't care passionately about the answers. My research participants are individuals with goals and dreams and personalities; I care about their ongoing conditions. I befriend them, and they me.

This personal investment, as well as the ethical belief that drives it, is a root belief of existential thought: that the questions we work to resolve can be divided into *problems*, which stand outside us and can be objectively solved, and *mysteries*, within which we are participants and for which no fixed solution can be obtained (Borowitz, 1965). As Marjorie Grene (1959) put it, "There are steel and stones and mortar, but there is *my* city, which I must hate or love or be indifferent to, live in or leave or come back to" (p. 49). The second city is the one I care about, and the one I must explore.

An example: when investigating the behavior of minors in a juvenile hall, we can explore the correlation between compliance and particular spatial forms of sleeping quarters: linear or podular, single story or split-level. This is an objective question. It stands outside the observer, can be measured, tabulated, agreed upon by all onlookers. But there is a way to watch the same phenomenon as a mystery rather than a problem:

"'The rooms on the bottom are singles,' said Rebecca, the fifteen-year old inmate who was leading our tour of the juvenile hall, 'and the ones on the top are doubles. This wing is for the kids who are pretty good—the RS's, the restricted status kids, are on West Wing.' I asked whether she was in a single or a double. 'I'm upstairs in one of the doubles, but I'm by myself right now. The other girls all got moved, because they're about to fumigate, and I'm leaving today. Do you want to see one of the rooms?' I said sure, and she took us up the stairs. 'This is my room.' Not her *cell*, even though it had a steel door with a 12″-square window on the front. Her *room*."

"She opened the door and went into the room first. There were two paperback books in a little cubbyhole against the far wall, and her bed was made, and there were a pair of canvas slip-on shoes next to the bed with its hard, thin vinyl mattress, and Rebecca—like every girl I'd ever met at the suburban high school I'd studied for a year who had showed me her bedroom during a tour of her house—said, 'It's kind of messy right now.' As though she was opening the wood-panel door to her ranch-house bedroom, with prom pictures on the mirror and a *Cosmopolitan* on the nightstand and socks on the floor, instead of a minimalist eight-by-eight cell with a steel door and dense beige walls and grilled windows and the bedframes bolted to the unpainted concrete floor."

This two-paragraph snippet of my research notes was framed this way because, instead of looking at minors, I looked at Rebecca. Rebecca has desires and aspirations, has a history and a future. By reducing her to a member of an abstract grouping and her behavior to a Likert-scale performance (check one: very obedient, obedient, neutral, disobedient, or very disobedient), I could have treated her as an object of study; but by taking her as a whole person and her acts as meaningful and intentional, I stood in relation to her. To be neutral toward her would have been abominable. Her story has emotional content, which, according to Sartre, means that I care about her future (Sartre, 1976). And by caring, I am morally included in the problem; my work bears a responsibility to her.

3.3. Narrative Holism

This narrative pursuit of knowing is based on a synthetic and holistic conception of human-environment relations. **I believe that people are more than the sums of their traits, that places are more than a physical inventory, and that the complex inter-actions between those people and those places are most profitably described through constructing stories.** As the anthropologist Ruth Behar puts it, I pursue "a mode of knowing that depends on the particular relationship formed by a particular anthro-pologist with a particular set of people in a particular time and place," a mode of knowing that she calls "the most fascinating, bizarre, disturbing, and necessary form of witnessing left to us at the end of the twentieth century" (Behar, 1996, p. 5). My respon-sibility to the people I enter into research with demands that I make witness of their stories, that I present their conditions to a world that normally does not see or hear from them.

This narrative approach, while it is atypical within human-environment research, is much more commonly found in some of the disciplines from which we borrow. In anthropology, Behar and innumerable others like her have followed Clifford Geertz in making the notion of "thick description" central to the definition of the ethnographic task, the holistic baseline from which the field has moved for the past twenty-five years (Geertz, 1973). In material culture studies, Henry Glassie felt compelled to tie tavern songs, hearthside patterns of conversation, home construction materials and techniques, field tools and land management strategies into his understanding of the interwoven people and place of Ballymenone (Glassie, 1982). Similarly, the cultural geographer Nicholas Entrikin describes his narrative study as "seeing things together":

"From the decentered vantage point of the theoretical scientist, place becomes either location or a set of generic relations and thereby loses much of its significance for human action. From the centered viewpoint of the subject, place has meaning only in relation to an individual's or a group's goals and concerns. Place is best viewed from points in between" (Entrikin, 1991, p. 5).

This "point in between" is the shifting and mobile land of the storyteller: neither autobiographer nor constructer of frameworks, but rather (in the word of Mark Kramer, 1995), "host." The literary genre of narrative nonfiction has thoroughly changed our expectations of the "true story." Because of the work of writers such as John McPhee, Joan Didion, and Tracy Kidder, the past thirty years of nonfiction writing has changed from the linear, "just-the-facts" organization of the newspaper journalist to include commentary and digression, to emphasize complexity of character and circumstance, to acknowledge that things are never as simple as they first seem.

3.4. Interdisciplinarity

The responsibility I bear in my work can only exist relative to real people, not to abstractions like "cultural geography" or "environment-behavior research," the two disciplines to which I most belong. **I believe in the individual and his or her circumstances more centrally than in any single field or discipline.** I specialize in the study of human environments because environments seem to help me see the people who live within them. When I study teenagers in their physical and social surroundings—or, more accurately, say, when I study Brady in the high-school theater—I am primarily interested in Brady: her acts, her desires, her goals, her constraints, her emotions. I'm not studying Brady as a member of an abstract grouping, not as one of innumerable teenaged girls or adolescents or college-bound students, but as Brady in her Brady-ness. I don't really care about adolescents or minors or girls; those are mere categories, conceptual clusters that are inanimate and abstract and often irrelevant (Sayer, 1992). They have no existence. Brady not only exists, she has a future. I want her to be joyful, thoughtful, and healthy.

This narrative pursuit of the understanding of one particular person or phenomenon inherently leads to the employment of a number of disciplines. In order to know Brady well, in order to see her "together," I may have to explore her situation through several disciplinary approaches. I may have to learn something about educational policy, about classroom discipline theories, about the cultural origins of the school desk, and the nature of the cul-de-sac in American community planning. I may find it useful to understand Brady's family's orientation to education, their prior mobility, her own career aspirations, or her IQ. For me, however, all of those things are only helpful inasmuch as they help me understand Brady; taken each by itself, they amount to caricatures. It is worth quoting at some length the intellectual Paul Goodman:

"I have been severely criticized as an ignorant man who spreads himself thin on a wide variety of subjects, on sociology and psychology, urbanism and technology, education, literature, esthetics, and ethics. It is true that I don't know much, but it is false that I write about many subjects. I have only one, the human beings I know in their man-made scene. I do not observe that people are in fact subdivided in ways to be conveniently treated by the wide variety of separate disciplines. If you talk separately about their group behavior or their individual behavior, their environment or their characters, their practicality, or their sensibility, you lose what you are talking about. . . . The separate disciplines are the best wisdom we have; I wish I knew them better. But there is a real difficulty with them that we might put as follows: In my opinion, it is impossible to be a good lawyer, teacher, statesman, physician, minister of religion, architect, historian, social worker, or psychologist, without being a good deal of all of them at once; yet obviously—especially today when there is such a wealth of indispensable specialist knowledge—it is impossible to be expert in more than one or two "fields." Again, I do not have an answer; but I prefer to preserve the wholeness of my subject, the people I know, at the cost of being everywhere ignorant or amateurish. I make the choice of what used to be called a Man of Letters, one who relies on the peculiar activity of authorship—a blending of memory, observation, criticism, reasoning, imagination, and reconstruction—in order to treat the objects in the world concretely and centrally" (Goodman, 1962, pp. 4–5).

3.5. Human Agency

As a part of this narrative focus on the individual, I believe in **the power of human agency in the face of structure**. I understand Modernism and Post-Modernism, understand human cognitive development, understand gender-restricted opportunities for American girls; but even within all of those social and biological constraints, strong as they are, there is freedom. It is a difficult freedom, to be sure: the freedom to continually choose and to become what they aspire to, which inevitably includes the commitment to a future that includes the happiness of other people as well as their own. They make their decisions within a social context: "I cannot ignore them, but I am morally committed to them and they to me, because what I do affects them and what they do affects me" (Fontana & Van de Water, 1977, p. 114). The constraints of social, cultural, and environmental structures are interesting and powerful, necessary to our understanding, but the choices made within those constraints are central to the person I want to know.

This again is at the heart of existential beliefs: that "there is no meaning to life except the meaning that a man gives his life by the unfolding of his powers, by living productively. Only constant vigilance, activity, and effort can keep him from failing in the one task that matters—the full development of his powers within the limitations set by the laws of his existence (Erich Fromm, cited in McElroy, 1972). We are what we make ourselves every day.

3.6. The Effects of Research on the Researched

This moral requirement my participants struggle with in their day-to-day acts is exactly the one that I carry as a researcher toward those people I explore: a commitment to our common future. The wholeness of people and events means that those people who work with me will not be able to divide some portion of their lives into an independent self labeled "research subject," a portion that they will be able to disregard during the rest of their lives. I believe that **my research will change my participants**, and that this is a significant and dangerous proposition which is at the same time filled with potential for good. Bronna Romanoff notes that, in research as in therapy, "narratives are agents of change;" the narrator tells the story to an interested listener, who through response helps to co-construct that narrative (Romanoff, 1997). She further suggests that the act of volunteering for interviews or observation, the act of opening one's life to an outsider, indicates a desire for change. It amounts to a statement that the volunteer is interested in expanding his or her narrative, learning new ways to see one's life. She notes that, " 'Informed consent' never includes the possibility of life change, although it probably should."

3.7. The Effects of Research on the Researcher

I also believe that **my research will change me**. I too am interested in expanding my narrative, in learning new ways to see the meanings of the world. If my research and writing doesn't change my beliefs, at least somewhat, then I have committed two sins: I have not opened myself to the people with whom I studied, and I have finished learning. If I leave my work without a significant personal change, then I have accumulated information, but not ideas.

This mutual commitment of researcher and researched to our shared futures and our reciprocal change sounds quite radical within the context of social and behavioral science; after all, most of us have no serious training in crisis counselling or psychotherapy, and perhaps oughtn't to muddle around as though we had. But this commitment and openness is exactly what would be expected of friends, with an equal lack of training. And how can I live with a group of people and try to learn something important about them without at least trying to do them the honor of making friends? How can I treat their lives as important and yet hold them at the same emotional and intellectual distance that I would hold an abstraction like "adolescents?" I do not divide my life into research and non-research acts, any more than I divide my participants' lives into disciplinary fodder or compress them into abstract groups; we all find ourselves in social circumstances, amidst people who matter to us. We pay attention and do our best, making mistakes and trying again.

3.8. The Non-Finality of Research

As a result of this continual and reciprocal change, I believe that **my research will always be incomplete and arguable**, just as my knowledge of another person or situation must always be incomplete and arguable. In the words of Michael Polyani (1969), much of my work will take the form of "tacit knowledge," or "the knowing of unspecifiables, a knowing of more than you can tell" (p. 131). Much of what I learn will be unable to be said, just as much of what I know about Brady—that she is a fierce mix of loyalty and independence, that she will almost certainly never stop learning and becoming, that she is a noble person—could never be objectively verified. And through reading in the next few pages a little of what I know of Brady, you will compare my story with those of people you know, or stories you've read before, and you will create a picture of Brady somewhat or radically different than the picture I hold.

So, in the end, what good is this way of work? After all, the act of "research" is differentiated from other curiosity perhaps only by the intention of reporting back. I intend to tell others what I have learned, even while knowing that language—certainly, my use of the language—will be insufficient. Where does that leave narrative and existential researchers?

Surprisingly, it leaves us in several useful and defensible places. First, it leaves us closer to capturing and conveying the whole person, the whole event, than if we'd never tried. We don't have to be perfect friends to be good friends, and we don't have to be perfect reporters to be good reporters. If it is clear that I care, and can convey my sense of why a person or place is important through a structured series of "sequential, emotional, and even moral experiences that readers undertake" (Kramer, 1995, p. 33), then it becomes easier for the reader to care as well.

Second, it leaves us with a respect for motivation and action and complexity, an understanding of the wholeness of person and circumstance and researcher that allows our writing to be stronger. I have written elsewhere that thinking and writing that substitutes clear frameworks for complex stories is actually a way for the researcher to claim power: power over the situation, power over the reader, power over the participants (Childress, 1998). When an ambiguous situation with difficult characters making decisions is presented synthetically and open-endedly, the reader has the freedom and power to re-interpret and disagree and co-construct. Narrative—as an action of knowing and of writing—is always a balancing act between chaos and taxidermy, between providing insufficient structure to capture and guide the reader and so much structure that the event is dead and the reader has no way to creatively revive it.

Third and finally, the narrative researcher is poised to take advantage of the power of what Carol Weiss calls "conceptual diffusion." Weiss writes that:

"Those social scientists who expect research to be authoritative enough to *determine* policy choices are giving insufficient weight to the many and varied sources from which people drive their understandings and policy preferences. . . . they do not deliberately use research to affect individual decisions. Rather it fills in the background, it supplies the context, from which ideas, concepts, and choices derive" (Weiss, 1980, 385 & 390, emphasis in original).

Research that is memorable, that captures the reader through the presentation of people struggling with significant decisions, stands a strong chance of being brought to the table as part of the context of decision-making. It helps to frame the language, the terms of debate. Fixed concepts must always be examined in the face of the real examples we know; if a researched situation can be displayed so strongly that readers feel that they "know" it, then it will become part of the truth test that they apply to theory and policy.

3.9. Environment-Behavior Research in the Context of the Humanities

My own background in human-environment research was through a school of architecture, a parent discipline itself torn over its identity as a branch of the sciences or of the humanities. And my personal path through that school was equally divided, with courses offered by environmental psychologists and quantitative anthropologists on the one side, by historians and novelists on the other side, and by an uncategorizable group of cultural analysts who, like me, straddled the center.

As my work has progressed and my beliefs have changed, I find myself moving off that center and toward a firmer home in the humanities. The philosopher Richard Rorty has written of the modes of thinking that typify the theorist (which he calls the "ascetic priest," borrowing from Nietzsche) and the novelist. Regarding the theorist, Rorty writes that "Someone dominated by this urge will tell a story only as part of the process of clearing away appearance in order to reveal reality" (Rorty, 1991, p. 6).

Of the novelists, Rorty writes that they are people:

"who enjoyed unweaving the tapestries which the saints and sages had woven. . . . It is comical to think that anyone could transcend the quest for happiness, to think that any theory could be more than a means to happiness, that there is something called Truth which transcends pleasure and pain" (pp. 10, 11).

And Rorty, as he must, chooses:

"I can sum up my sense of the respective importance of Dickens and Heidegger by saying that, if [future generations] were unable to preserve the works of both men, I should much prefer that they preserve Dickens'. . . . the most celebrated and memorable features of his novels is the unsubsumable, uncategorizable idiosyncracy of the characters" (p. 5, 14).

Through reducing lives and actions and intentions to concepts, we reduce the possibility of helping people challenge and re-create their own lives. I believe, finally, that **the power of the humanities to change minds is as important as the power of the sciences to change conditions**. The direction of that change is uncertain, as befits the freedom of the reader to draw his or her own conclusions, but it is the only change that matters to me.

4. A NARRATIVE EXAMPLE

Brady, a junior at Curtisville High School, and her family had volunteered to be among my more intense research participants, and I had stayed by her side constantly—with only the breaks that decency demanded—since I'd met her Thursday morning at school. This excerpt begins on Friday afternoon.

Brady took the bus home after school, while I did Jeff a favor and drove him home. I got to Brady's house at about 3:30; she'd said that the door would be open, but I can't bring myself to just walk into people's houses, so I rang the bell. Brady answered the door and said rapidly, "C'mere, c'mere, c'mere!" leading me toward the kitchen. I followed her, expecting to see her dog or one of her birds doing something silly. Instead, Brady pointed to a folder lying open on the kitchen table. "Read that." I looked at a formally printed sheet that read:

This is to certify that
Brady Louisa Hernandez
has been officially accepted. . . .
and then my eyes scanned diagonally down and right to the big red words,
  "Welcome to Stanford."

She'd applied to Stanford's combined senior-in-high school/freshman-in-college program, without really believing she'd be accepted, but she'd hit the jackpot after all. I looked over at her—one of my favorite kids, all smiles, and braces; as I congratulated her, I reached up to give her a high-five, but she grabbed my hand and pulled me into a hug instead.

Brady wanted to tell her folks, she wanted to run out the door and scream it to the neighbors, but because I was the only one around, I got all of it. She paced non-stop around the kitchen table, telling me all about the program—she had clearly all but memorized the brochures from which she had applied. After about 15 minutes, her mom Ellen got home. Ellen was preoccupied by something, shooed the dog impatiently out the door, but Brady finally slowed her down and led her to the table. "Read this." Ellen did, and it took her about as long as it took me to figure it out. She smiled: "You got in."

"I got in."

"How are you going to pay for it?"

Brady needed more, and crossed the three steps between them and hugged her mom. "I got in, Mom." And then she started to cry, repeating over and over, "I've gotta go, Mom . . . I have to go." Her face was buried in Ellen's shoulder, and she could scarcely breathe. I turned and walked into the back of the kitchen, trying to be out of the way for a few minutes, shuffling around on the vinyl tile.

After their hug, Ellen dialled a couple of different places, trying to reach Brady's father. When she got him, she said, "Hi, Tony. Here, talk to your daughter." Brady took the phone. "Hi, Pop . . . I got accepted to Stanford . . ." Then she laughed. "You and Mom think exactly alike."

We'd been about to go up to the Timber State library anyway so that Brady could work on her current events paper, but the evening now had some scholarship research mixed in. As Brady disappeared into her room to change, Ellen hesitated in the hallway. I said, "You must be having mixed feelings right now."

Ellen looked at me, half smiling, and said, "Yes. Yes, I am. She's . . . well, *she's* ready. I don't know if *we're* ready . . ."

I drove Brady to Timber State. We walked uphill to the library, where Brady went to the periodicals and I rooted around in the scholarship guides in the reference room. I found five good directories, and headed upstairs to meet Brady in the newspaper area. She had three bound volumes of magazines, and had managed to keep her attention on current events rather than next year's college. I looked through the scholarship books; Stanford wanted $27,000 for the coming year, and they guaranteed only $6,000 of that in financial aid. Clearly, Brady had some steep work ahead.

After she took three pages of notes for her paper and reviewed what I'd found in the scholarship guides, Brady and I packed up and left at 5:40. I bought us dinner at La Tapiata before going on to the school to do lighting work for the play, and she shared some carefully considered pieces of her journal (in return, I think, for her having had regular and immediate access to my notebook of writings about her). On our way back to the car after dinner, we walked to her brother Richard's apartment three blocks away. Richard and his roommate Paul were both home, and Brady greeted them both by saying, "Hey, I got in!"

Richard replied, not getting up, "I heard. Good job. How are you going to pay for it?"

I was stunned. A premier university had just told her that she'd made the golden list a year in advance, that she was 16 years old and they wanted her *now* . . . and all three members of her family had offered the same response.

After only a few minutes, we left Richard's to drive to the high school. Brady kept up a continual stream of chatter: "They only accepted fifty people. I was probably like number fifty. My math score on the SAT almost killed me. I hope I don't get there and like barely make it." She also said, between bursts of self-doubt, that she hoped she wouldn't alienate her friends over the next few days by talking non-stop. She'd called her best friend Ida as soon as she'd opened the envelope, and knew that Ida was already sad about Brady's impending departure. I told Brady to imagine a two-minute egg timer, and that she could only talk to any one person about Stanford until the egg timer ran out.

Once we arrived at the theater, we walked up to the lighting booth and had a quick talk with Becki, Brady's trainee on the lights. Ida had the sound board set to go, Becki ran an abbreviated light check, and Brady and I settled into the back row to watch the show (and I to catch up on my notes). Brady's friend Ally came to sit in front of us; Brady told her, "I got accepted to Stanford," to which Ally—whose family had all attended Berkeley—replied, "I'm so sorry for you." She then turned to me and said, "I have a very low opinion of Stanford." Brady was not getting a very satisfactory set of responses to her news.

Once the show began, Brady sat next to me, whispering into her headset in order to coach Becki in the light booth. Just as with the previous night, when I'd spent the show with Brady behind the light board, I was captivated by her absorption in her work. She recited the entire script along with each actor as she thought through the lighting sequence; the flow of narrated Thornton Wilder was periodically broken by hushed comments to Becki over the headset—"You're doing great, just slow your fades down just a little"—and with unheard urging to each of the actors as they moved to their stage cues—"Come on, Scott, come on, up to the light . . . there you go." By the time the show ended, both she and I were exhausted; she'd done everybody's job along with them, even though they hadn't known it.

After congratulating several of the cast after the show, Brady and I were walking to my car when Steve caught up with us. "Do you guys mind if I get a ride home with you?" His place was on our way, so Steve got in the back seat. It turned out that it was his 16th birthday, which we learned because he told us six or seven times. "Today's my sixteenth birthday. Yep, sixteen. Won't be long before I can get my license. And now I'm older than John, so I can give him a hard time. I think my mom's going to have a party when I get home tonight." We wished him a happy birthday, and watched him walk into his family's apartment. As we were driving back out from the parking lot, Brady said, "He's so lonely. That always surprises me when I see someone that lonely."

"Surprises you? Why?"

She smiled. "You keep asking the hardest questions." A long pause. "There are 250 million people in the United States, something like that, and we all speak the same language and we all share the same culture, and it's always surprising when I see someone who can't find people." As we drove down Main Street, we talked about loneliness, about lonely friends, and about how it feels when the people you trust let you down a little . . . and as we stopped in front of her house, she was crying, a tear running down each side of her nose, shining from the streetlight above us. "My parents, I expected that, but not from Richard. I was counting on him . . ." I gave her another hug, longer than the first one seven hours earlier, and then she got out of the car and walked to her front door.

5. CONCLUSIONS

The story of Brady's Friday night is not "about" environments, even though there is much to speak about in terms of environment. The fact, for example, that she was able to talk most openly about her feelings of fear and isolation and betrayal in my car is an environmental fact; we were sitting close together in a quiet space in which no one else could hear or interrupt. It is a case study in the meanings of privacy and personal space . . . or at least, it could be reduced to that if I wanted to use those words.

I have no problems using those words, with using lives as examples of concepts, so long as I also try to convey the lives as being larger and more important than those concepts. No one cares about privacy unless some person is being hindered by its absence or aided through its presence; and my goal is always that my reader care, that he or she enter into a relationship of responsibility and shared future with Brady and other kids like her. Brady's ability to speak within the safety of my car is part of a human story; the ways we analyze and classify and label that story are secondary. Brady is greater and more central to my interest than any set of terms I might use to order her experience.

I do my work among teenagers like Brady because of a strong and interwoven set of beliefs. Through trying to be faithful to those beliefs—beliefs not so much about the right way to do research as about the right way to treat people—I have found that readers, researchers, and research participants are much the same: trying to make sense of an uncertain world, trying to become something greater than we are, and looking for other people who can help. And that is the stuff of which stories are made.

REFERENCES

Behar, R. (1996). *The vulnerable observer*. Boston: Beacon.

Borowitz, E. B. (1965). *A layman's guide to religious existentialism*. Philadelphia: Westminster Press.

Childress, H. (in press). *Landscapes of betrayal, landscapes of joy: Curtisville in the lives of its teenagers*. Albany, NY: SUNY Press.

Childress, H. (1998). Kinder ethnographic writing. *Qualitative Inquiry, 4*(2), 249–264.

Cooper Marcus, C., & Sarkissian, W. (1986). *Housing as if people mattered*. Berkeley: University of California Press.

Entrikin, J. N. (1991). *The betweenness of place: Towards a geography of modernity*. Baltimore: Johns Hopkins University Press.

Fontana, A., & Van de Water, R. (1977). The existential thought of Jean Paul Sartre and Maurice Merleau-Ponty. In J. D. Douglas & J. M. Johnson (Eds.), *Existential sociology* (pp. 101–130). Cambridge: Cambridge University Press.

Geertz, C. (1973). "Thick description: Toward an interpretive theory of culture." in *The interpretation of cultures*. New York: Basic Books.

Glassie, H. (1982). *Passing the time in Ballymenone*. Philadelphia: University of Pennsylvania Press.

Goodman, P. (1962). *Utopian essays and practical proposals*. New York: Random House.

Grene, M. (1959). *Introduction to existentialism*. Chicago: University of Chicago Press.

Kramer, M. (1995). Breakable rules for literary journalists. In N. Sims & M. Kramer (Eds.), *Literary journalism*. New York: Ballantine.

McElroy, D. D. (1972). *Existentialism and modern literature*. Secaucus, NJ: Citadel Press.

Polyani, M. (1969). *Knowing and being* (M. Grene, Ed.). Chicago: University of Chicago Press.

Romanoff, B. (1997, June). *Narrative researchers as agents of change*. Paper presented at the Reclaiming Voice: Qualitative Research in a Post-Modern Age Conference, University of Southern California.

Rorty, R. (1991). Philosophers, novelists, and intercultural comparisons: Heidegger, Kundera, and Dickens. In E. Deutsch (Ed.), *Culture and modernity: East-west philosophic perspectives*. Honolulu: University of Hawaii Press.

Sartre, J. P. (1976). *The emotions: Outline of a theory* (B. Frechtman, Trans.). Secaucus, N. J.: Citadel.

Sayer, A. (1992). *Method in social science* (2nd ed.). London: Routledge.

Weiss, C. H. (1980). Knowledge creep and decision accretion. *Knowledge: Creation, Diffusion, Utilization, 1*(3), 381–404.

15

SEVEN ASSUMPTIONS FOR AN INVESTIGATIVE ENVIRONMENTAL PSYCHOLOGY

David Canter

Centre for Investigative Psychology
Eleanor Rathbone Building
The University of Liverpool
Liverpool L69 7ZA United Kingdom

1. INTRODUCTION

In preparation for the writing of this essay I undertook a daunting task. I dug out many of the environmental psychology papers I have written over the last 25 years and looked at them to see what assumptions underlay those papers. The experience produced a real surprise.

The surprise came from the discovery of the range of topics, as varied as studies of reactions to working in open plan offices (Canter, 1972), furniture arrangements in Japanese apartments (Canter, 1973), human behavior in fires (Canter, 1990), the design of hostels for the homeless (Moore, Canter, Stockley, & Drake, 1995), and the geographical activities of serial rapists (Canter & Larkin, 1993a). There seems to me to be a style to most studies, and to the account I gave of them, that remains remarkably unchanged from my earliest publication in the late 60's in the *Architects' Journal* on "The Need for a Theory of Function in Architecture" (Canter, 1971) to my most recent contribution, this year, to a set of essays edited by Birgit Cold on "Health And Beauty: Enclosure and Structure" (Canter, 1999). Even my current dominant concerns with the actions of criminals have many similarities to earlier studies. All these studies explore how people make sense of their surroundings and act in relation to that understanding. They put the active person at the forefront and attempt to understand how the context of that person's actions provides a basis for conceptualizing their dealings with the world.

But more important than the similarity in the central theoretical perspective is the similarity in *how* those topics are tackled, the way I approach answering the questions they raise. That is the basis of the surprise. For me each of the publications was a

Theoretical Perspectives in Environment-Behavior Research, edited by Wapner *et al.*
Kluwer Academic / Plenum Publishers, New York, 2000.

distinct event. Often it was an account of a particular study that had been carried out in response to a distinct challenge or opportunity. These studies were never part of any predetermined program or previously specified environmental psychology model. Yet an approach emerges from them, a framework that is reflected in common themes and a cumulative clarification of those themes and their relationship to each other. It is as if I was attempting to play jazz improvisations using different motifs, but each improvisation ended up sounding rather similar, although, the tune does evolve and change.

What has been evolving throughout this quarter century of research is a certain type of psychological activity. One that I have called "investigative psychology" because it focuses on examining existing circumstances and naturally occurring patterns of activity in order to solve problems and provide insight. It is thus investigative in a number of senses, but most particularly in the sense that the problems that are tackled have an existence independently of any scientific interest in them. A perspective I first outlined in my inaugural lecture at Surrey University (Canter, 1985a).

They do not reach towards the sort of activity that is investigative in the meaning of "investigative journalism". Nor are they as full-bloodedly "investigative" in the sense that Douglas (1976) advocates in his account of "Investigative Social Research". He sees the truth of social reality as only open to construction from the active participation of teams of people in the processes they are studying. The investigative psychology I am describing is rather more genteel than that. It uses rather more distant forms of contact with its subject matter than the intense immersion in the setting that Douglas insists is so important. It is also different in focusing on the experience and conceptualizations of individuals and the conceptual systems they draw upon to guide their activities. It is much more influenced by the psychological perspective of George Kelly (1955) and his "alternative constructionism", than the field research approaches of Garfinkel (1967) and Goffman (1963) that are a major source of Douglas' methodology. In other words, my quest is to understand the processes going on within individuals that helps to explain their actions in and on the world rather than social processes that are the product of such actions. As is obvious from my writings the other dominant influence beyond George Kelly was the social science methodologist Louis Guttman (Levy, 1996). He took the notion of alternative constructionism into fundamental questions of research process. He thus does share with Douglas the view that scientific truth is constructed rather than discovered, but he showed how research methodologies can assist that construction.

My studies, though, are certainly not solely academic. Some of them are "action research" in the sense that they are carried out with the objective of providing direct input to the decision processes of the people or organizations being studied. Some are even consultancy projects to provide direct guidance to a commissioning client. Others have been carried out with more general scientific objectives in mind. What they all have in common, though, is a search for the meanings and psychological structures that characterize people in their daily situations.

2. AN UNFOLDING LOGIC

To get some understanding of the unfolding logic and therefore the assumptions, theories and methods that are characteristic of my work it will be helpful to have a brief resume of the studies in which I and my colleagues have been involved.

It started with an interest in the nature of aesthetic experience and a desire to study the reactions of artists to their own creative processes. I suppose this grew out of my own dallying with the arts (Canter, 1998) and the mixture of interest in the creative, subjective processes of drama and painting as well as the scientific processes. However, there are rarely opportunities to fund research of such a pure, arcane nature and so I had to compromise by joining a research team in a school of architecture which was looking at the psychological impact of buildings.

This gave rise to a study of open-plan offices, which was formulated as a simple project to look at the effects on workers' performance of being in open-plan offices. The results from this study indicated the active role of people in selecting the sorts of places in which they were prepared to work. So what started as a study of the effects of the environment on people turned out to be a study of the way people influence the environments to which they will be exposed.

Therefore when the research unit I was part of moved on to look at school buildings, I was already primed to be concerned with the meaning that those buildings had for their users and the expectations and patterns of behavior that those meanings brought to play within those buildings. This framework proved most instructive and helped me to develop a rationale for looking at the sources of satisfaction people find in their surroundings. So when I was commissioned to study nurses satisfaction with the design of the hospital wards on which they worked it seemed natural to develop the earlier ideas. This enabled us to show that there was a structure to these satisfactions that may reflect generalized structures to the way we deal with our physical surroundings (Canter & Kenny, 1982a).

In trying to turn these findings into guidelines for design it became apparent, though, that there are many constraints on the ways in which buildings can be created. I discovered that perhaps one of the most underrated of these constraints is the fire regulations. These regulations are not derived from direct behavioral studies of people caught in a building on fire. To fill this gap in behavioral studies I started to examine human actions in buildings on fire. This allowed me to give more emphasis to human actions in buildings then had been possible in the earlier studies, which had focused on verbal accounts, usually questionnaire responses, to the physical environment. These case studies of actual experiences in buildings on fire showed me that even in the most trying circumstances, the actions of people are shaped by an attempt to understand what is going on and to achieve quite mundane objectives.

These predetermined aspects of peoples' reactions are shaped by the organizational context of which they are a part. These studies of fires therefore provided a natural stepping-stone to considering the organizational framework that hinders or possibly facilitates accidents and emergencies. There followed a number of studies of industrial safety. In particular my colleagues and I explored the ways in which the attitudes of the workforce towards safety were reflected in their actions and consequent accidents (Canter & Olearnick, 1986). The success of that work in reducing accidents also served to illustrate the close interplay between peoples' characteristic ways of dealing with the world and their specific actions in possibly unusual circumstances. In other words, "accidents" can be seen to be a natural consequence of particular processes and patterns of ordinary "non-accidental", intentional, activity in a particular setting.

The relationship between prevailing modes of carrying out a job on a day-to-day basis and industrial accidents turns out to be a useful model for other forms of unwanted, counter-productive activities, notably crimes. A fruitful hypothesis is that criminals also illustrate aspects of themselves and their typical ways of dealing with the world when they carry out illegal activities, whether it is arson, theft, murder, or any other crime. This hypothesis is the basis of inferences about the features of unknown offenders and their life styles when they are not committing crimes, that may be derived from how they commit their offences, known as "profiling" (Canter, 1995). Interestingly it is their use of the area in which they live that turns out to offer some of the strongest inferences to link aspects of the crime and features of the criminal.

These studies of criminals are, of course, "investigative" in the strong sense that they contribute directly to police investigations. But they draw on the same approaches and general psychological theories as earlier studies. They are addressed to problem solving in the same way as earlier studies. Therefore, with hindsight, I can now see that most of my earlier studies had an investigative perspective to them. The assumptions that underlay these earlier studies therefore seem to me to provide a general framework for an investigative environmental psychology.

3. ASSUMPTION ONE—RESEARCHERS HAVE STYLES

The similarity of approaches to different sorts of research problems brings me to the conclusion that the most basic assumption I should admit to is that there is a style of working with which I am comfortable. Supervising many student projects over the years has also taught me that many research decisions are guided more by what a researcher is comfortable with than by any inviolate logic. Some people can only think in terms of standardized questionnaires, others open-ended interviews. Some find the world outside the controlled laboratory to be frightening and mystical. Others start their studies by talking to whomever they can and have their likely results formulated before they start collecting any reportable data. My assumption is that there are effective and ineffective ways of working within any style, not that there are right or wrong ways of doing research.

This can be taken as a pragmatic approach, or even one that is opportunistic in the positive sense of taking advantage of the opportunities provided. But it is also a way of providing a more "organic" approach to research questions, that somewhat tongue in cheek I called "holistic" (Canter, 1993b). This approach requires the methodology and results to be a natural part of the problem being studied rather than something that is artificially tacked on only to provide scientific credibility. It does demand a flexible approach to carrying out research, but one that makes sense to the researcher. It is out of this that a research "style" emerges.

The practical point here is that the advantage of having a "style" of research available is that for any question that is presented to the researcher she or he can bring to bear a way of answering that question. The researcher does not struggle to reconstruct the question according to basic principles or to fit it into some predetermined mould. Instead a strategy is brought into play that will produce useful results. To return to our jazz analogy. The good improviser will have a repertoire of techniques for developing the original tune. For some those techniques may, for example, be based upon traditional blues scales, for others the "problem" of elaborating on the melody may be tackled by drawing upon modal forms. The "blues" or "modal" approach will be at the heart of their "style" of improvising.

Within the environmental arena the significance of a research style is particularly pertinent because the origin of much of the research and its development is not entirely under the control of the researcher. Opportunities or funds for research come from organizations external to the researcher. To survive in this applied domain it is often essential to be able to make something worthwhile of what is an offer.

3.1. Learning On the Street Where Research Comes From

The origins of research projects are not often described in academic publications. As a consequence there are some parallels with the way sex was dealt with in Victorian times. Everyone knew it happened but they only found out how and when from private, personal experience. In a similar way we do not usually tell our students about where research comes from or the processes that give rise to its gestation and birth. We expect them to learn about this on the streets.

Learned journals implicitly support the myth that research, and its findings, emerge full-grown from the scientist's loins with no intervention of either God or man. The apparently secret process by which studies are funded, research assistants found, data accessed, and a dozen other hurdles surmounted requires many management and financial skills. If it is field based and/or applied it also requires the support and co-operation of people outside of the research team, as well as direct support from a variety of individuals and organizations who are not necessarily experts in psychology, or scientific studies. Many of these people look to the research activity both to contribute to the growing knowledge and understanding of the science and to have direct benefits either in general policy or in relation to actions that will be carried out as a consequence.

Like jazz performers faced with a new tune, any researchers faced with the challenges of real world research have to draw upon a set of useful habits that will see them through. These habits have to be grounded in effective scientific procedures and ways of thinking. These habits are survival kits that help researchers to cope with the complexities of the tasks that face them. So, in summarizing here the assumptions underlying my studies I suggest that I am really elaborating a research style, not some grand, overarching theory.

3.2. Styles Grow Out of Classical and Romantic Research Traditions

I suggest that just as the major movements in the arts are reflections of classical or romantic traditions so are the styles of scientific activity. The Classical approach is to see everything as a reflection of pure Platonic forms. Beauty is merely a mirror of some ideal that can never be reached. The serenity of Apollo versus the spontaneity of Dionysius. In the history of architecture this gave rise to an analogous but rather different distinction between the romance of form and classical perspective that emphasizes the purity of function. Interestingly, architectural psychology emerged at a stage at which architects were trying to meld form into function, claiming there should be no distinction between the two. However, with the demise of the *Modern Movement* and *International* style, form has emerged again as an entity that has validity independently of function.

In the research tradition the classical is one in which an ideal structure is defined into which the perfect study should fit. From this perspective the everyday is seen as a reflection of pure, ideal forms, whether it be a statue of Venus or carefully crafted factorial designed experiment. The research tradition that imposes a predetermined scientific structure on how studies should be conducted can be seen as classical in many senses of the word.

By contrast, research that sees the everyday as inherently exquisite, that recognizes the need to shape approaches in order to study problems in their own light can be seen as inherently romantic. This may result in a series of controlled experiments each examining a different aspect of the central research question, or as ever more focused exploratory studies investigating a range of perspectives on the problem at hand. Indeed, recently Schneider (1998) has given a rallying call for Romanticism as a basis for a revival in psychology at large.

Most art drifts between the romantic and the classical and I think the same is true for styles of scientific research. Perhaps that dual pull also characterizes much of my activities. An attempt to find and impose structure in the traditions of the Platonic ideal, whilst battling with the need to recognize the richness and complexity of daily life, with all the fascination that that complexity carries. Inevitably scientists lean towards the classical because of the requirements of formality in science. Those that are uncomfortable with this formality may therefore reveal their romantic tendencies through their modes of data collection or the theories they espouse. Yet it has always seemed to me a pity that these two traditions are so often at war with each other. They are complimentary in human thought and together enrich human experience. My own work is inspired by the romantic interest in everyday experience and its rich and enjoyable complexity. But my studies are often shaped by the more classical scientific traditions within which I was educated. So for me speculation is not a comfortable starting point. Theories, as important as they are, are usually the end result of a project not its driving force.

4. ASSUMPTION TWO—DATA SPEAKS THEORIES

There is an important second assumption that emerges from the investigative perspective and the productive tension between classical and romantic traditions. This is that *data is not enough*. Collecting information, for example, on the preferences that people have for one environment compared to another, or giving an account of what people do in one place or another, or recording the number of times a particular location is used, as useful as these pieces of information may be for planning and decision making, they do not add up to an understanding until some explanatory framework is tested against them. We need to know if our observations of activities reveal important aspects of how people typically deal with those settings. We need to see how the preferences expressed enable us to understand the basis of human delight and evaluation.

From their earliest writings environmental psychologists have grappled with these theoretical concerns, separating the discipline off from some of their more descriptive forbears, such as regional geography, architectural history, or urban sociology. We now have a wealth of theoretical formulations to draw upon. From Lynch's writings on the legibility of cities (cf. Banerjee & Southworth, 1990), Altman's (1975) proposals of the centrality of privacy in structuring our transactions with our surroundings through to more recent considerations on the evolutionary mechanisms that drive landscape preferences that the Kaplans have promoted (Kaplan & Kaplan, 1989). I have always assumed that I could make at least some small contribution to this greater understanding. So, my publications have grappled with concepts, models and processes rather than being buried in very extensive data sets or highly complex analyses. I have never believed that all that is needed is good data and that the rest of the scientific process will look after itself.

This is not to say that I see myself as part of that tradition, which regards data as a subservient adjunct to grand theory, dipping into the archives to illustrate points. Quite the reverse, I always have assumed that the data will speak to me and help me to understand the transactions that people have with their surroundings. This is probably more of an act of faith than an assumption. But it was my Ph.D. work that taught me that possibility. I could not make sense of the field experiment findings I had from testing people in large open offices and smaller ones. The effects were there when I tested people in their own offices but they disappeared when I moved them to other offices for testing (Canter, 1972). I was struggling to find an effect of the environment on the respondents but it was the results that made me realize that people were choosing their environments and that was what produced the differences.

5. ASSUMPTION THREE—THEORIES ARE PRACTICAL

However, this interest in building theories (using "building" as a verb and an adjective) should not be taken to imply merely academic concerns. Most of the studies I have carried out have grown out of practical problems, from the earliest studies of office design to recent work on the contribution research laboratories make to industrial innovation.

In more recent years I have tried to explain this search for effective theories as part of practical studies by pointing out the distinction I see between engineering and science. Engineering is concerned with making things work. Science is concerned with understanding how and why things work the way they do. The move from a scientific finding to an engineering application is not inevitable. It requires special development work. Furthermore, there may be engineering discoveries that defy scientific explanation. However, the scientific principles have much further reach and potential over a longer time scale than engineering discoveries. But for me the interest in those theories is their eventual applicability. Some theories and approaches to theory building have more potential for application than others. My work aspires to develop just such applicable theories.

Therefore, my own studies have usually walked the tight rope between theory construction and practical applications. This has broadened the range of institutions that have been prepared to support my activities and in the process has widened the range of influences that have been brought to bear on the studies themselves. This may make the variety of studies look as if they are a magpie collection with no apparent focus. However, it is worth bearing in mind that many of the questions I have sought to answer have been raised by the funding agency before I had any contact with it. Indeed, I have never found it possible to find an organization that would fund a piece of research on the basis of a presentation from a researcher if that organization has not already identified the problem as one with which they need help.

Organizations tend to seek help from people whom they regard as having some special skills to offer. They determine these skills, in the main, on the basis of the previous research of the individual or teams concerned. It, therefore, follows that although the path may not be immediately obvious, applied research will tend to follow a sequence in which studies follow each other in a progression that at least seems logical to the funding bodies. This can be contrasted with theoretical research in which funds can be applied for on the basis of a new idea of what the researcher wants to explore. Although they may be evaluated on their previous successes, when the decision is made as to whether to award the grant they are likely to have more freedom in arguing their case than somebody who is being evaluated by a commercial organization that wants to be sure it will get results it can use. It must be mentioned, though, that all over the world there is a drift away from funding "blue sky" academic research towards funding more "mission oriented" research.

5.1. Theories Should Explain Actions and Experience

The actual substance of the theories that I have attempted to draw from the data in the practical contexts in which I have operated are derived in part from the particular attraction that Environment and Behavior studies have for me. The attraction has never been as a branch of ergonomics, or human factors. Rather, it has been the ways in which studies have allowed an exploration of the basis of delight as well as comfort. More grandiosely, the field explores both those realms that deal with what makes life feasible—protection from the environment, appropriate spaces for action and the like—as well as those matters that make life worth while. This latter is often captured under the heading of aesthetics.

In the early days of the study of people and their physical surroundings the field was called *Architectural Psychology* (Canter, 1970). This enshrined the recognition that there was a determination to connect with the design of buildings and to carry out studies that would help to shape our physical surroundings. Yet ever since those early days there has been a general drift in the field away from a concern with the shaping of designed environments for human benefit. The drift has been towards a focus on human reactions to, and actions on, the physical surroundings, especially in recent years to how people may be degrading those surroundings.

It therefore seems to me that at the heart of environment and behavior studies there is a central conundrum of how to take account of and model the physical surroundings in ways that will capture the qualities of pleasure and delight as well as issues of function and comfort. It also therefore follows that the underlying assumptions that run through the studies in which I have been engaged are assumptions that grapple with this central difficulty of exploring a physical reality whilst dealing with people's inner emotions.

A second consequence of the perspective I have outlined is that because it requires integrated contact with everyday existence as a prerequisite of effective research, it is also naturally relevant to the lives of people other than psychologists. This makes it fundamentally applicable. It may seem strange to see application as a consequence of the theoretical emphasis of the work. Perhaps this is one of the paradoxes of modern psychology? Some of the most exciting developments in theory have come from very direct concern with practical issues.

Of course, the happy consequence of this paradox is that support for research can be found in many locations other than academic institutions. The serious environment and behavior researcher does not stand on ceremony or need to be concerned because he or she is asked to help solve an existing problem. However, applied research does pose special problems of communication. There is a difference between communicating to an academic audience and communicating to people who wish to act on the consequences of those deliberations. This in turn does have consequences for the ways in which the research is carried out. Such constraints may be fruitful in suggesting areas of research, but they are certainly not harmful if they lead researchers to think more clearly about how they are going to communicate the results of their studies.

6. ASSUMPTION FOUR—CONTEXT PROVIDES MEANING

One principle that is a starting point for interpreting the results of many studies is that a person's actions derive their significance from their context. The caresses of a lover are very different in their significance from the lewd fingerings of a rapist, even though the physical behavior itself is identical in both cases. The satisfactions one person feels with his place of work or home may be quite different from somebody else's satisfaction with very similar conditions. Thus, the context that gives significance to our lives is a function both of variations between people and of variations in the settings and how they interact.

This is so fundamental a theoretical assumption that it almost acts as an axiom in my thinking. It is based on the view that our lives are provided with excitement and challenge, or desperation and despair from the transactions between the particular qualities that we bring to a place and the particular qualities of that place. Of course other people in those settings and their actions are often the most significant part of the setting. (I will develop this "social assumption" a little later.) But the point is that the context has a place in space and time that provides the basis for its significance and relevance to the person experiencing it. That person also brings his or her position in social space and time to bear on that interpretation.

A very personal realization of that occurred to me recently when I underwent a minor surgical procedure under a local anaesthetic. Being totally aware of the operating theatre, in what was a potentially threatening and certainly disturbing context, made me acutely aware of the faded floor tiles that I noticed as I climbed onto the operating table and the banal music that was being played in the background. They raised small doubts in my mind about the professionalism of those who were about to operate on me and raised my anxiety. Yet as an environmental psychologist I was aware and also relieved to notice that the surgeon was completely oblivious to these matters as he focused on preparing the tools of his trade.

The powerful significance of the transactions between person and context raise very special problems for any one who would like to understand the processes involved more thoroughly. The fact that generations of psychologists have chosen to ignore the importance of context only serves to illustrate how demanding a challenge it presents.

6.1. Laboratories are Artificial

Taking account of the natural settings in which human experience takes place poses great demands on both theory and methodology. The problem is that human beings are remarkably adaptable. Indeed environmental adaptability is the primary tool in human evolutionary development. But for psychologists it is a capability fraught with dangers as well as delights. People will develop ways of dealing with any setting no matter how artificial. As a consequence the cunning inventions of experimental, laboratory-based psychologists may often bring to the surface skills and propensities which are either non-existent in any other situation. They may at best be superficial or minor capabilities that have no strong function outside of the rarefied milieu of the controlled experiment.

It is helpful to elaborate a little further on this devilish invention which I have called the "Laboratory Experiment". For me it is a situation that is characterized by a large number of artificial constraints. Artificial in the sense that the constraints are devised by psychologists for the purposes of studying problems that are of interest within the realms of psychology, but not necessarily of any significance to anybody else. Constraints in the sense that attempts are made in advance to limit as far as possible all except a very few aspects of the environment that the research participant will experience. These are constraints that limit even more severely the range of responses that the participant will be allowed to generate and that will be measured during the course of the study.

The strength of these controlled experimental studies, it is claimed, is that they allow very clear conclusions to be drawn about the effects that particular aspects of the environment have on particular reactions of the respondent. But this, of course, is precisely their most profound weakness too. They make it extremely difficult to explore and understand how people give shape and meaning to their surroundings and act on it in order to improve their control and subsequent satisfactions. In other words, it is not so much the experimental, scientific precision that makes me uncomfortable with these experiments, but the constipated and restricted models of human beings, their actions and experiences, which are a product of the methods that are used for studying them.

A further problem with the whole concept of the experiment is the way in which it provides a mode of thinking about research that is carried through, unchallenged, into other areas. Field experiments, and the whole vocabulary of "quasi-experimental" design, appear to me to drag inappropriate models into the study of real world phenomena. It seems to me much more appropriate to think of these as quasi-naturalistic studies.

I suppose I have always been most impressed by that stream of psychology that can trace its origins in modern times to William James (1890) and in classical times to Aristotle. This is the view that human beings are best considered in an analogy to Newton's first laws of motion. That is, they are naturally in motion unless acted upon in some way to distort and modify that dynamic. Thus, any framework that looks to a person being triggered, or stimulated to act in a particular way has ignored the basic principle that people already bring with them to the situation far more than they take away. So, besides the methodological implications of this dynamic perspective on human beings, there are also theoretical elaborations that are a natural consequence.

If people are actively in search of control of their surroundings and their satisfactions come from their transactions with their context, then it is reasonable to assume that much of significance to them is available to their conscious awareness. This does not necessarily mean that they will always be aware of the implications and consequences of their actions, but it does accord well with the strong cognitive traditions in psychology and their recognition that even emotional responses are shaped by knowledge and understanding. This certainly means that much of what people say is as valid and necessary to study as what they do and that the two processes of thinking and action cannot be regarded as anything other than reflections of the same system of experience.

This perspective has the further interesting consequence of changing the emphasis with which simulated stimuli are considered. The conventional wisdom among psychologists is that pictures of scenes, or indeed people, can be treated as some sort of surrogate for the actual experience of the place or person. My view is that these representations elicit a particular perspective on the experience, a sort of Brechtian alienation, by which the representation is treated primarily as a symbol rather than as a multi-modular reflection of real experience (cf. Scott & Canter, 1997).

In other contexts Orne (1962) has drawn attention to what he calls the "demand characteristics" of any study. He chose this term to describe the pattern of expectations and implied actions that are built into any laboratory study. But I think this is a more general phenomenon and indeed a fundamental aspect of the way in which we cope with the demands and complexities of daily life. This means that only a sub-set of the possible aspects of experience will be captured in reactions to simulated representations. This view assumes that there is a rich, and possibly continually expanding, repertoire of conceptualizations and associated aspects of experience that can be drawn upon in various ways for particular objectives. Being presented with a photograph and asked for a response to it, elicits a sub-set of that broader range of aspects that are present if the experience is a direct one. Of course, if people are only given the opportunity of expressing a limited, alienated sub-set of the repertoire available, then of course great similarities will be found between the simulation and the actual experience.

6.2. Social is Fundamental

I have sketched out a framework for understanding human actions and experience as drawing upon particular aspects of a cognitively available repertoire that is conceived of as being relevant to the purposes that are salient in any given context. But this is still a strangely heartless model in which repertoires are processed as if in some non-human vacuum. I take it as axiomatic that the purposes and repertoires are defined through contacts with other people. Social processes are thus not superficial additions to fundamental human processes, later chapters in the textbooks of the psyche, but rather they are the fundamental constituents of our world that underlay all purpose and meaning. This is revealed by the fact that the repertoires and purposes draw very directly on contacts with other people and achievements in relation to them. Power and intimacy are dominant purposes for example which are fundamentally social (cf. McAdams, 1988, for a detailed exploration of this).

This theoretical framework seems to me always to have been present in my work, but like the Eskimo carver who is trying to let the fish escape from the stone my experience of psychological research has been to enable these ideas to have some freedom from the morass of possibilities. Indeed, this attempt to summarize my assumptions is yet a further stage in the struggle to give shape to ideas about the ways in which people experience the world

6.3. Purpose Explains

Perhaps an important concept that crept into this attempt to explicate the repertoires of experience is the idea that the repertoires are drawn upon for different purposes. "Purpose" is probably one of the most difficult of notions for psychologists ever to manage, even though it is one of the most readily acceptable ideas in the realms of daily life. Indeed, the legal system is based upon the notion of intention and its relationship to the purpose that the individual had in mind when carrying out an action. Psychologists have always been worried about the logical problem inherent in suggesting that the future, or what follows an action, could somehow be the cause of it. However, this confusion comes about because they have ignored the wonderful quotation from Karl Marx that the difference between the lowliest architects and the most sophisticated is that the former creates a notion of the building in their consciousness before they act on it. So these mental representations of the possibilities are clearly aspects of human action that must be taken into account if those actions are to be understood.

In the evaluations of buildings in which I have been involved the focus on purpose has proven especially helpful (Canter, 1983). A building cannot be evaluated in the abstract. It has to be evaluated in relation to the uses to which its occupants and owners wish to put it. This implies that a building can be good for some uses and not for others. A fairly elementary point, but one almost totally ignored in the building evaluation literature. Most examinations of buildings take a rather Classical perspective seeking an abstract definition of its qualities. Whereas, my approach is more Romantic in the sense of recognizing the here and now qualities of the place for its current uses, but also predicting that the purposes will drift as time and usage produces changes.

7. ASSUMPTION FIVE—STRUCTURES EXPLAIN

The search for the meaning of people's transactions with their surroundings as they experience it in relation to their reasons for being in those locations presents a fairly complex set of interrelated facets. The research task is to distil the many variables that are possibly relevant into the dominant themes that help to explain the processes under study. This is a task that can be accomplished in many different ways, but the one that appeals to me most is one that allows the data to reveal its underlying structure whilst imposing the minimal constraints on that interpretation process.

The crucial idea here is that the meanings that people assign to their actions or that organize their utterances can be established by seeing how those utterances or actions co-occur. It is the patterns of co-occurrence that reveal the underlying meanings of what people say and what people do.

7.1. Means are External, Correlations Internal

There is an important and often poorly understood consequence of the search for meaning in patterns of relationships that questions the fundamental psychological relevance of experimental designs. The laboratory framework is construed as the testing of average differences between groups that experience the treatment condition and those that experience the control condition. The fundamental assumption is made that the participants in the experiment do not interact with the process of experimentation as such. The treatment and control conditions can be regarded as externally imposed by the researcher. They are assumed to be *independent* variables.

The major statistical analysis of this type of research design is to form some summary of the reactions of each of the sub-groups in each of the experimental conditions. These sub-groups are summarized, in effect, by calculating their average response, whether on rating scales or frequencies of behavior. These averages are thus the overall response of a group to the externally structured constraints provided by the experimenter. It follows that it is extremely difficult to understand the meanings that any given individual brings to a situation by comparing the average responses of sub-groups.

By contrast the correlation actually reveals the patterns of co-variance within individuals summarized as general trends across individuals. So if you know there is a correlation between satisfaction with the heating in a room and its lighting you know that the person who likes the heating also likes the lighting and the person who does not does not.

7.2. Research Processes Interacts with Their Products

This brief consideration of the *psychological* differences between measures of central tendency and measures of association makes clear that research processes carry direct and profound implications for the sorts of psychological theory that can reasonably be constructed. The way a study is designed and the sort of data that is collected assume something of the qualities of the people being examined. If a theory is going to be about the active interpretations and purpose oriented actions that people bring to their environmental transactions then the study has to explore processes within the person. These will essentially grow out of correlations between aspects of what each participant in the study says or does.

7.3. Meaning Emerges Out of Structure

The pattern of co-occurrences, associations, and correlations, gives rise to the meanings of what people say or do. For instance, in a study I conducted of prison inmates evaluations of the design of the prisons in which they were incarcerated, it became very clear that the significance of the cell design depended on the regime of the prison (Canter, Ambrose, Brown, Comber, & Hirsch, 1980). The patterns of correlations showed that in prisons that had severe regimes that restricted the movement of prisoners the size and comfort of the cell was crucial to overall satisfaction, in more liberal regimes it was only part of the general pattern. As with many field studies, with hindsight this seems an obvious discovery, but at that stage it was rare for prison architects to consider the ways in which the organization interacts with the design to produce consequences for users.

The methodologies we use for revealing these patterns impose the minimum set of assumptions on the data (explained in detail in Canter, 1985b). Therefore they are honest in representing the underlying material. But the researcher still has to make sense of what the methodology summarizes. This is where the constructive, inventive process of research is recognized.

8. ASSUMPTION SIX—EXPERIENCE IS A CONSTRUCT NOT A GIVEN

The notion of an underlying structure to the experience of the surroundings may give the misleading impression that the framework I operate within is searching for a Classical, Platonic ideal. A search for a pure structure that characterizes reactions to particular places. However, consideration of the purposive orientation and the social dynamic that give significance to places makes it clear that there will be many different ways of describing the experience of a given transaction with the surroundings. In this regard the account that is given of environmental experience has to be seen as a construction that will vary depending on whom is constructing it.

This perspective has especially important practical implications. It helps to explain the inevitable conflicts between the different parties to the building, shaping, managing, and use of places. It thus provides a strong theoretical basis for participative design, not on political grounds but because psychologically that is a productive way to incorporate different perspectives on the experience of places.

8.1. Actions Require Interpretation

The recognition that we are constructing accounts of experience when we are studying them also raises questions about how we approach human actions in places. Central to these questions is the assumption that an action does not inevitably carry the meaning that an observer may naively assign to it. The structural assumption comes to our aid in helping us determining the meaning of actions. The production of patterns of actions due to their regular co-occurrence help in their interpretation. So if one restaurant is typically used by groups of people but another by loners we can glean more of the meanings that these eating places have for their users. Or, more graphically, our findings that the types of actions that occur in rape of a stranger when the offence is committed indoors are rather different from when it is committed outdoors, helps us to understand the perspective and characteristics of serial rapists who consistently offend in one type of location or the other (Canter, 1995).

8.2. Actions Provide Meaning

The processes I have described create a cycle of meaning. The patterns of activity help to give meaning to places that in turn encourage or facilitate similar actions. Out of this process emerges a culturally based structure of place meanings. People draw on this portfolio to give shape to their own aspirations and purposes.

8.3. Preferences Reflect Meaning

One further important consequence of these assumptions is that preferences for particular places are a product of the interrelationships between the meanings and actions that people associate with those places. It may well be possible to capture some of the general qualities that are typical of such places by using the sort of cognitive, evolutionary arguments that Kaplan and Kaplan (1989) have articulated. But the framework of assumptions I have outlined here proposes that the actual cognitive process that people draw upon to form their preferences owe more to current social and cultural patterns of place experience than to innate propensities.

9. ASSUMPTION SEVEN—ALL METHODOLOGIES ASSUME BUT SOME ASSUME MORE THAN OTHERS

To the casual reader it is the methodology that is the most obvious distinguishing feature of my research. The use of Multi-Dimensional Scaling Techniques (MDS) and the interpretation of multi-variate analyses as conceptual structures (Canter, 1982b). These methodologies have been in use for well over 40 years but still appear novel to many psychologists and social scientists. I think this is partly because even applied field researchers are still wedded to experimental models and ways of thinking. They still look back to fashion their research designs on laboratory chemistry as practiced in Victorian times rather than, say, modern astronomy or even archaeology, or, within the social framework, areas of system analysis and structural modeling.

The contructivist approach requires a methodology which is rich in many different ways. What people say and what people do as well as the traces that they leave behind and the records that are collected about their actions must all be drawn upon to provide grit for the psychologist's mill. But this multiplicity of multivariate data sources demands modes of analysis that will not destroy the systems and context that we are trying to study. Even at the risk of clarifying the apparently obvious rather than elaborating the remote and arcane, I have to be confident that what we are finding does reflect a fruitful account of what actually occurs rather than merely playing obeisance to some scientific ritual. As is apparent from many of my publications, I have found the facet approach (Canter, 1985) to research to be the most satisfying way of maintaining scientific rigor without falling into the trap of scientific rhetoric.

This leads to the important assumption that research methodologies, including how the research is organized and the data collected, make fundamental assumptions about the nature of human beings. This is perhaps the most challenging paradox for any psychological research. What we are able to find out about people depends on the way we approach how we will find that out. This self-reflective quality is perhaps the most important assumption of all my work.

REFERENCES

Altman, I. (1975). *The environment and social behaviour*, Monterey: Brooks/Cole.

Banerjee, T., & Southworth, M. (Eds.). (1990). *City sense and city design: Writings and projects of Kevin Lynch*. MIT Press: Cambridge Mass.

Canter, D. (Ed.). (1969). Need for a theory of function in architecture. *Architectural Psychology*, 11–17. (London: RJBA Publications)

Canter, D. (Ed.). (1970). *Architectural Psychology*. London: RIBA Publication.

Canter, D. (1971). Need for a theory of function in architecture. Architect's Journal, 151(289), 38–42.

Canter, D. (1972). Reactions to open-plan offices. *Built Environment, 1*, 465–467.

Canter, D. (1973, September). *A non-reactive study of room usage in modern Japanese apartments*. Presented at Psychological and Built Environment Conference, University of Surrey, published in conference proceedings.

Canter, D., Ambrose, I., Brown, J., Comber, M., & Hirsch, A. (1980). *Prison design and use*. Unpublished Final Report University of Surrey.

Canter, D., & Kenny, C. (1982a). Approaches to environmental evaluation: An introduction. *International Review of Applied Psychology, 31*, 145–151.

Canter, D. (1982b). Facet approach to applied research. *Perceptual and Motor Skills, 55*, 143–154.

Canter, D. (1983). The Purposive Evaluation of Places; A Facet Approach. *Environment and Behaviour, 15*(6).

Canter, D. (1985a). *Applying psychology*. Inaugural Lecture, University of Surrey.

Canter, D. (1985b). *Facet theory approaches to social research*. New York: Springer-Verlag.

Canter, D., & Olearnick, H. (1986). Changing safety culture in a steel works: The zero option, cited in Canter (1996). *Psychology in Action* Dartmouth: Aldershot.

Canter, D. (Ed.). (1990). *Fires and human behaviour*. (2nd ed.). London: David Fulton Publishers.

Canter, D. & Larkin, P. (1993a) The environmental range of serial rapists. *Journal of Environmental Psychology, 13*, 93–99.

Canter, D. (1993b). The holistic, organic researcher: Central issues in clinical research methodology organic researcher. *Psychology in Action*. Dartmouth: Aldershot.

Canter, D. (1995). Psychology of offender profiling. In Bull, R., & Carson, D. (Eds.), *Handbook of psychology in legal contexts*. Chichester: Wiley.

Canter, D. (1995). *Criminal shadows*. London: HarperCollins.

Canter D. (1999). Health and beauty: Enclosure and structure in cold. In ●●, B (Ed.), ●● Avebury: Aldershot.

Douglas, J. D. (1976). *Investigative social research: Individual and team field research.* Beverly Hills: Sage.

Garfinkel, H. (1967). *Studies in ethnomethodology.* Englewood Cliffs: Prentice-Hall.

Goffman, E. (1963). *Behavior in public places.* New York: Free Press.

James, W. (1890). *The principals of psychology.* New York: Henry Holt.

Kaplan, R., & Kaplan, S. (1989). *The experience of nature.* Cambridge, MA: University Press.

Levy, S. (Ed.). (1994*). Louis Guttman on theory and methodology: Selected writings* Dartmouth: Aldershot.

Kelly, G. (1955). *The psychology of personal constructs: Vols. 1 & II.* New York: Norton.

Moore, J., Canter, D., Stockley, D., & Drake, M. (1995). T*he faces of homeless in London.* Dartmouth: Aldershot.

McAdams, D. (1988). *Power, intimacy and the life story: Personological inquiries into identity.* NY: Guilford.

Orme, M. J. (1962). On the social psychology of the psychological experiment; with a particular reference to demand characteristics and their implications. *American Psychologist, 17,* 776–783.

Schneider, K. (1998). *American Psychologist, 53*(3). March.

Scott, M. J., & Canter, D. V. (1997). Picture or place? A multiple sorting study of landscape. *Journal of Environmental Psychology, 17*(4), 263.

16

CROSS-CULTURAL ENVIRONMENT-BEHAVIOR RESEARCH FROM A HOLISTIC, DEVELOPMENTAL, SYSTEMS-ORIENTED PERSPECTIVE

Jack Demick,* Seymour Wapner,** Takiji Yamamoto,***
and Hirofumi Minami****

*Department of Psychology
Suffolk University
Boston, Massachusetts 02114
and Center for Adoption Research and Policy
University of Massachusetts
Worcester, Massachusetts 01605
**Heinz Werner Institute for Developmental Analysis
Clark University
Worcester, Massachusetts 01610-1477
***Research Institute of Human Health Science
10-51-215, Ichibaue-machi
Tsurumi-ku, Yokohama, 230 Japan
****Department of Urban Design
Planning and Disaster Management
Graduate School of Human-Environment Studies
Kyushu University,6-19-1
Hakozaki, Fukuoka, 812 Japan

1. INTRODUCTION

In a recent article in *American Psychologist*, Segall, Lonner, and Berry (1998) have emphatically stated that "... 'culture' and all that it implies with respect to human development, thought, and behavior should be central, not peripheral, in psychological theory and research. To keep culture peripheral, or worse, to avoid it altogether lest it challenge one's own view of reality is myopic and a disservice to psychological inquiry" (p. 1108).

Theoretical Perspectives in Environment-Behavior Research, edited by Wapner *et al.*
Kluwer Academic / Plenum Publishers, New York, 2000.

In line with this, our own holistic, developmental, systems-oriented approach—which utilizes the person-in-environment as the unit of analysis so that environmental context is built into and an essential part of every analysis (see Wapner & Demick, this volume, for a more general description)—has suggested a variety of problems, reviewed below, relevant to the relatively new subfield of cross-cultural psychology.

More specifically, this new subfield has been defined as ". . . the systematic study of relationships between the cultural context of human development and the behaviors that become established in the repertoire of individuals growing up in a particular culture" (Berry, et al., 1997, p. x). Implicit in this definition is the notion that cross-cultural psychology is ". . . a scientific endeavor that shares with more familiar disciplines the use of theories, scientific methodologies, statistical procedures, and data analysis" (Gardiner, Mutter, & Kosmitzki, 1998, p. 3). However, recent critics of this subfield (e.g., Moghaddam & Studer, 1997) have voiced the concerns that most cross-cultural psychologists: (a) fail to conceptualize culture in a holistic manner that encompasses the social, historical, and political contexts and instead treat it as a unidimensional construct that directly determines behavior; and (b) adhere to traditional, positivist frameworks of analysis that lead, for the most part, to cognitively-based laboratory studies. Such traditional approaches, they have argued, discount the roles of power disparities, injustices, and lack of resources that are likely to become evident only through normative (vs. causal) analyses of human functioning. As a result, cross-cultural psychology has been seen not as a critic of mainstream psychology, but rather as an enthusiastic supporter that actually promotes cultural homogeneity rather than diversity.

From our point of view, such criticisms may be unduly harsh. While we agree with Moghaddam and Studer (1997) that there are crucial differences between causal and normative models of human functioning, we do not see this as a function of the politics of contemporary cross-cultural psychologists but rather of the ongoing debate as to whether psychology should remain restricted to the methods of the "natural science" tradition (e.g., laboratory studies seeking to determine general laws of human functioning via quantitative prediction and control of causal relationships) or whether it should augment these with, or substitute for them, "human science" methods (e.g., qualitative phenomenological and/or narrative inquiries) that are more appropriate to the uniqueness of psychology's subject matter (i.e., normative description and understanding of the individual's experience). That psychologists are not necessarily required to buy into one side of this either-or dichotomy is supported by our own research. That is, as we have argued extensively elsewhere (e.g., Wapner & Demick, 1998, 1999), our holistic, developmental, systems-oriented approach to person-in-environment functioning—an elaboration and extension of Werner's (1940/1957) comparative, developmental theory—has characteristically been concerned with describing the relations both among and within the parts (person, environment) that make up the integrated whole (person-in-environment system) as well as with specifying the conditions that make for changes in the organization of these relations. Thus, our approach has been committed to the complementarity of normative explication (description) and causal explanation (conditions under which cause-effect relations occur) rather than being restricted to one or the other.

In turn, this focus impacts our choice of paradigmatic problems (e.g., critical person-in-environment transitions induced by perturbations at all levels of integration) as well as our preferred method of research (i.e., flexibly drawing from both normative/quantitative and causal/qualitative methodologies depending on the level of integration and nature of the problem under examination). This latter notion is consistent with Maslow's (1946) distinction between "means-oriented" and "problem-oriented" research. That is, whereas in the former the method dictates the range of problems that can be studied, the latter gives priority to the phenomenon being studied. Thus, from our perspective, problem-oriented research calls for varying methodologies dependent on the nature of the problem. For example, experimental methods are appropriate for the study of certain aspects of a phenomenon (using examples from our own work, factors such as cognitive style as well as conditions such as planning vs. no planning that affect adaptation to life transitions), while other methods (e.g., narrative) provide access to different aspects of a phenomenon (e.g., experiential changes following such transitions).

Toward demonstrating the usefulness of perspectives that integrate both normative and causal analyses, we now present examples of cross-cultural research generated against the backdrop of our approach (see Wapner & Demick, chapter 2, for a complete listing of our underlying assumptions). Strategically, we will present representative research studies that proceed from less to more complex in terms of our unit of analysis, namely, the person-in-environment system. Then, following a description of each set of studies, we will comment on the way(s) in which: (a) if not readily apparent, the research problem falls under the general rubric of environment-behavior research; and/or (b) one or more of our underlying assumptions has shaped the particular nature of the research.

2. CROSS-CULTURAL ENVIRONMENT-BEHAVIOR RESEARCH FROM OUR APPROACH

2.1. Body and Self Experience: Japan versus United States

2.1.1. Synopsis. Demick, Ishii, and Inoue (1997) extended Wapner and Werner's (1965) longstanding work on the psychology of the body in the context of two cultures. Conceptualizing body experience to develop first within the *sensorimotor* level (body action), then within the *perceptual* level (body perception), and finally within the *conceptual* level (conceptual aspects of body, conceptual aspects of self) of cognitive functioning, we administered a battery of instruments to Japanese and American undergraduates. Findings indicated that the Japanese and Americans exhibited differences in sensorimotor (e.g., Americans moved their bodies through space faster than the Japanese), perceptual (e.g., relative to the Japanese, Americans exhibited greater degrees of body boundary diffuseness and personal space), and conceptual experience (e.g., relative to the Japanese, Americans provided less detail on self-drawings but reported more satisfaction with body/self and higher self-esteem).

2.1.2. Commentary. First, that body and self experience are of significant concern to environmental psychologists as well as to others interested in environment-behavior relations (e.g., architects, planners, sociologists) gains support from the notion that aspects of body and self experience encompass relevant general processes such as spatial relations (e.g., between the individual and others in his or her interpersonal environment), and the impact of physical space on body perception. For example, Wapner, McFarland, and Werner (1963) demonstrated that the physical environmental context affects experience of one's own body. Apparent extension of an outstretched arm was judged to be significantly shorter in a close confined physical context than in an open extended context. Further, the study provides a theoretical framework and classificatory system of heuristic value for the study of body and self experience and, more generally, for the study of environmental experience.

Second, the research on body experience was specifically generated against the backdrop of our assumption that person-in-environment processes are categorized in terms of levels of integration, namely, biological, psychological, and sociocultural. Within the psychological level, processes may be further categorized with respect to sensorimotor, perceptual, and conceptual levels. Such an analysis emphasizes that a more complete and perhaps accurate picture of experience emerges when multiple methods for assessing experience are employed (see: Dandonoli, Demick, & Wapner, 1990; Demick, Hoffman, & Wapner, 1985). For example, had our study been limited only to the assessment of conceptual aspects of body, we might have agreed with Lerner, Iwawaki, Chihara, and Sorell's (1980) finding that, relative to Americans, the Japanese have less favorable views of their bodies' attractiveness. However, assessment of body experience at all levels led us to the opposite conclusion, namely, that the Japanese have a more developmentally advanced sense of body (a more articulated body concept, a less hurried conceptual tempo, and less concern—not distress—with their bodies' attractiveness) than do Americans.

2.2. Values Mothers Hold for Handicapped and Nonhandicapped Preschoolers (Japan, Puerto Rico, Mainland US)

2.2.1. Synopsis. Based on a previous study that focused on the values that mothers held for handicapped and nonhandicapped preschoolers in the United States (Quirk, Sexton, Ciottone, Minami, & Wapner, 1984; Quirk, Ciottone, et al., 1986), this investigation examined the values held by mothers in Japan, Puerto Rico, and the U.S. Mainland for their handicapped and nonhandicapped preschoolers in the contexts of home and school. Specifically, mothers from the three cultures ranked 12 value items related to four value areas: physical; intrapersonal; interpersonal; and sociocultural. Value rankings were different in the three cultures and depended upon the home and school contexts. The presence of a diagnosed orthopedic and/or neurological impairment in the child had no overall effect on the value ratings, but did interact with culture and context to influence mothers' values. For example, among Japanese mothers, cultural awareness was considered differentially more important at home than at school; Puerto Rican mothers attached greater importance to self-confidence for their children at school than at home; U.S. mothers attached greater importance to religion for their children at home relative to school.

2.2.2. Commentary. This investigation employed our assumptions concerning the contextual nature of human functioning and the physical, psychological, and sociocultural aspects of persons and of environments. The study's conceptualization and subsequent findings led us to the generalizations that: (a) the influence of environment has not been fully considered in the traditional study of values (cf. Rokeach, 1973); (b) mothers' values for their children are embedded in the more general value orientations of their cultures (e.g., *amae* or interconnectedness in Japan, *respeto* in Puerto Rico, and *individualism* in Mainland US); and (c) since all groups attached greater importance to physical and sociocultural values for their children at home relative to school and to intrapersonal and interpersonal values at school relative to home, there may be certain universals or functional equivalences in the roles of home and school with respect to child rearing (LeVine, 1977). Thus, the holistic study of values themselves as well as their contextual aspects (e.g., different values for different contexts) appear worthy of attention from environment-behavior researchers.

2.3. Luxury, Necessity, and Amenity in Environmental Design (Japan, US)

2.3.1 Synopsis. This study (Wapner, Quilici-Matteucci, Yamamoto, & Ando, 1990) examined cross-cultural similarities and differences in the experience of necessities (minimum requirements for everyday life), amenities (things, rules, situations, etc. that make for comfort and convenience), and luxuries (things, rules, situations, etc. that provide greater ease, comfort, pleasure, indulgence, or extravagance) in three aspects of the environment (physical, interpersonal, sociocultural) in two sociocultural contexts (Japan, United States). Through the use of questionnaires, it was found that: (a) the two cultures were most similar in terms of basic necessities for the physical environment (e.g., house, clothing, food, water, electricity); (b) the two cultures were most dissimilar with respect to reported necessities, amenities, and luxuries related to the interpersonal and sociocultural aspects of the environment (e.g., while the Japanese emphasized direct interpersonal relationships such as the parent-child relationship, Americans emphasized an abstract approach to interpersonal relationships such as love, sex, and trust); and (c) both groups referenced necessities, amenities, and luxuries unique to the culture (e.g., the Japanese referenced futon, rice, rice cooker, and Karaoke, while Americans identified unpolluted sea, bottled water, boat, and junk food). A follow-up study (Wapner, Quilici-Matteucci, & Cool, 1991) extended this work to four sociocultural contexts (Italy, Japan, Russia, United States) and similarly found: greatest agreement with respect to necessities for the physical environment; and responses unique to each culture.

2.3.2. Commentary. This set of studies was based on the assumption that the environment is comprised of mutually defining physical (e.g., natural and built objects), interpersonal (e.g., father, friend), and sociocultural (e.g., rules and mores of home, community, and other cultural contexts) aspects. The findings further suggest that aspects of environmental design, namely, necessities, amenities, and luxuries, may be construed with respect to these three environmental aspects and not just to the physical aspect of the environment.

2.4. Adaptation to New Environments: Sojourner Behavior and Experience (Japan, US)/Migration (Puerto Rico to Mainland US)

2.4.1. Synopsis. Two separate lines of research deal with the general problem of adaptation to new sociocultural environments, which represents cross cultural research as well as critical person-in-environment transitions (cf. Wapner & Demick, chapter 2, this volume, for a more extension discussion of our work on critical person-in-environment transitions across the life span). First, Wapner, Fujimoto, Imamichi, Inoue, and Toews (1997) conducted a series of studies assessing the experience of Japanese undergraduates on a sojourn in American universities and of American undergraduates on a sojourn in Japanese universities. Findings included the following. Japanese students found the United States to be big, ineffective in transportation, dangerous, inexpensive, and free as compared to Japan. American students found Japan to be small, effective in transportation, safe, regulated, and expensive. Both experienced language barriers and the awareness of being foreigners. Finally, Japanese students in the United States were more satisfied with academics than the American students in Japan.

Second, Lucca-Irizarry, Wapner, and Pacheco (1981), Pacheco, Lucca, and Wapner (1985), and Pacheco, Wapner, and Lucca (1979) studied adolescents migrating from Puerto Rico to the United States and then returning to Puerto Rico. The findings involved both *self* (e.g., adolescents who had traveled more frequently to and from the US mainland presented the most confused sense of identity) and *other* (e.g., those born on the US mainland considered themselves the object of prejudice and discrimination in Puerto Rico more so than native-born return migrants). In a follow-up study, Redondo (1983) compared Puerto Rican and Cuban adolescents' experience of migration to the United States and assessed the usefulness of transitional peer intervention (peer anchor-person) and group counseling for both groups. A more complete synopsis of the findings from our migration studies is found in Wapner and Demick (1998).

2.4.2. Commentary. Collectively, the studies utilized the assumptions of: the person-in-environment as unit of analysis including three aspects of persons and aspects of environments; multiple intentionality; multiple worlds; and adaptation as optimal relations between person and environment. Again, use of such assumptions has clear heuristic potential as well as the potential to uncover significant, previously untapped findings. For example, by conceptualizing adaptation as we have, our migration research has indicated that: the person may conform to the environment (e.g., Puerto Rican migrant adolescents to the US may adopt the dress and customs of mainland US adolescents); the environment may conform to the person (e.g., mainland US schools may provide lessons in both Spanish and English for Puerto Rican adolescents); or, most ideally, the person and the environment may mutually conform to one another (e.g., attempts may be made both at home and at school to help the child experientially integrate the two cultures).

2.5. Cultural Differences in Automobile Safety Belt Use

2.5.1. Synopsis. Several studies are relevant here. First, Wapner, Demick, Inoue, Ishii, and Yamamoto (1986) studied experience and action with respect to automobile safety belt usage in general in Japan and the United States prior to legislation. Questionnaires revealed differences in: (a) general factors preparing individuals for the action/nonaction of using safety belts (e.g., relative to Americans, the Japanese placed higher value on safety belts); (b) precursors triggering the concrete behavior of "buckling up" in the context of the automobile (e.g., for Americans but not for the Japanese, feelings of preoccupation often caused them to forget to buckle up); and (c) action (e.g., the Japanese wore safety belts on the highway more often than Americans) and experience of this action (e.g., relative to Americans, the Japanese felt "virtuous" but not "confident" when wearing safety belts).

Other studies of relevance are those that examine cultural differences in experience and action concerning the introduction of legislation requiring use of automobile safety belts. This research not only provides an example of a critical person-in-environment transition, but also reflects cultural differences. Whereas Japanese strictly adhere to the introduction of such a new law, there is continuous decrease in adherence to the law in Italy and the United States. Indeed, in the United States a significant number of people living in Massachusetts voiced their concern that mandatory safety belt legislation was an invasion of privacy/infringement on human rights and brought it up for a second vote which resulted in repeal of the law and further decrease in safety belt use (see Demick, Inoue, Wapner, Ishii, Minami, Nishiyama, & Yamamoto, 1992; Bertini & Wapner, 1992).

2.5.2. Commentary. First, this set of studies clearly indicates that environmental transitions are not limited to those involving the physical aspect of the environment, but also include those involving the sociocultural aspect of the environment. Second, this set of studies was based primarily on our assumption of transactionalism, namely, that the person-in-environment system is the unit of analysis with transactional (experience and action) and mutually defining aspects of person and environment. This has subsequently led us to the critical problem of how experience and intentionality are translated into action (e.g., Do I, in fact, do what I know I should do or what I want to do?). This problem discussed by Wapner and Demick (chapter 2) shows that our assumptions and our problems operate synergistically, that is, assumptions lead to new problems and vice versa.

3. DISCUSSION AND CONCLUSIONS

Cross-cultural environment-behavior research such as (but not limited to) ours has implications for the problem, theory, and method of interdisciplinary environment-behavior research as well as for the larger field of psychology in the following ways.

1. Problem: A focus such as ours, namely, on the person-in-environment system with mutually defining physical/biological, intrapersonal, and sociocultural aspects of the person and physical, living organisms, and sociocultural aspects of the environment has the potential to broaden the scope of environment-behavior research to include the study of problems that are not limited to the physical aspect of the environment and that are more in line with the complex character of everyday life.
2. Theory: Our work suggests that advances in environment-behavior research as well as in psychological research more generally can be made through the adoption of overarching theories that utilize both normative and causal analyses. That is, the complementarity between normative explication (description) and causal explanation (conditions under which cause-effect relations occur) has the potential to shed light on both cultural similarities *and* differences (here, in environmental processes and transactions) rather than focusing on one or the other as heretofore has been the case (Goodstein & Gielen, 1998).
3. Method: Research such as ours puts the field of environment-behavior research

in tune with recent major developments in the field of psychology and allied disciplines. For example, as Valsiner (1998) has stated, "... human psychological phenomena exist within the *semiosphere*—a sphere of semiotic signs (Lotman, 1992) being constituted and reconstituted by active persons who are involved in processes of acting and reflecting on actions in parallel" (p. 215). Thus, *all* methods traditionally used by both psychologists and other researchers are, in reality, methods of joint or co-construction of information by the researcher and the research participant (cf. Valsiner, 1989, 1997). This is further support for our longstanding assertion that inquiry and knowledge are always biased and that there is no process of "neutral" observation, inquiry, or conclusion in any science (e.g., see Wapner & Demick, 1998, for a summary of this and the related notions of perspectivism, constructivism, and interpretationism).

Finally, such reframing as embodied in the above three areas may also help the larger field of psychology see itself as well as be seen by others as a unified—or in our terminology, a differentiated and integrated—science concerned not only with the study of isolated aspects of human functioning. For example, Kazarian and Evans (1998) have recently called for a "cultural clinical psychology" that parallels our call here for a unified field of cultural environment-behavior studies that can rigorously handle a wide range of problems cutting across various aspects of persons and various aspects of environments in diverse cultural contexts.

REFERENCES

Berry, J. W., Poortinga, Y. H., Pandey, J., Dasen, P. R., Saraswathi, T. S., Segall, M. H., & Kagitçibasi, C. (Eds.). (1997). *Handbook of cross-cultural psychology*. (2nd ed., Vols. 1–3). Needham Heights, MA: Allyn & Bacon.

Bertini, G., & Wapner, S. (1992). *Automobile seat belt usage in Italy prior to and following legislation*. Unpublished study, Clark University, Worcester, MA.

Dandonoli, P., Demick, J., & Wapner, S. (1990). Physical arrangement and age as determinants of environmental representation. *Children's Environments Quarterly*, 7(1), 26–36.

Demick, J., Hoffman, A., & Wapner, S. (1985). Residential context and environmental change as determinants of urban experience. *Children's Environments Quarterly*, 2(3), 44–54.

Demick, J., Ishii, S., & Inoue, W. (1997). Body and self experience: Japan versus USA. In S. Wapner, J. Demick, T. Yamamoto, & T. Takahashi (Eds.), *Handbook of Japan-United States environment-behavior research: Toward a transactional approach* (pp. 83–99). New York: Plenum.

Demick, J., Inoue, W., Wapner, S., Ishii, S., Minami, H., Nishiyama, S., & Yamamoto, T. (1992). Cultural differences in impact of governmental legislation: Automobile safety belt use. *Journal of Cross-cultural Psychology*, 23(4), 468–487.

Gallistel, C. R. (1980). *The organization of action: A new synthesis*. Hillsdale, NJ: Erlbaum.

Gardiner, H. W., Mutter, J. D., & Kosmitzki, C. (1998). *Lives across cultures: Cross-cultural human development*. Boston: Allyn & Bacon.

Goodstein, R., & Gielen, U. P. (1998). Some conceptual similarities and differences between cross-cultural and multicultural psychology. *International Psychologist*, 38(2), 42–43.

Kazarian, S. S., & Evans, D. R. (Eds.). (1998). *Cultural clinical psychology: Theory, research, and practice*. New York: Oxford University Press.

Lerner, R. M., Iwawaki, S., Chihara, T., & Sorell, G. T. (1980). Self-concept, self-esteem, and body attitudes among Japanese male and female adolescents. *Child Development*, 51, 847–855.

LeVine, R. A. (1977). Child-rearing as cultural adoption. In P. H. Leiderman, S. R. Tulkin, & A. Rosenfeld (Eds.), *Culture and infancy: Variations in the human experience*. New York: Academic Press.

Lucca-Irizarry, N., Wapner, S., & Pacheco, A. M. (1981). Adolescent return migration to Puerto Rico: Self-identity and bilingualism. *Agenda: A Journal of Hispanic Issues, 11*, 15–17, 33.

Maslow, A. H. (1946). Problem-centering vs. means-centering in science. *Philosophy of Science, 13*, 326–331.

Moghaddam, F. M., & Studer, C. (1997). Cross-cultural psychology: The frustrated gadfly's promises, potentialities, and failures. In D. Fox & I. Prilleltensky (Eds.), *Critical psychology: An introduction* (pp. 185–201). London: Sage.

Pacheco, A. M., Lucca, N., & Wapner, S. (1985). The assessment of interpersonal relations among Puerto Rican migrant adolescents. In R. Riaz-Guerrero (Ed.), *Cross-cultural and national studies in social psychology* (pp. 169–175). Amsterdam, The Netherlands: Elsevier Science.

Pacheco, A. M., Wapner, S., & Lucca, N. (1979). Migration as a critical person-in-environment transition: An organismic-developmental interpretation. *Revista de Ciencias Sociales (Social Sciences Journal), 21*, 123–157.

Quirk, M., Ciottone, R., Minami, H., Wapner, S., Yamamoto, T., Ishii, S., Lucca-Irizarry, N., & Pacheco, A. (1986). Values mothers hold for handicapped and nonhandicapped preschool children in Japan, Puerto Rico, and the United States mainland. *International Journal of Psychology, 21*, 463–485.

Quirk, M., Sexton, M., Ciottone, R., Minami, H., & Wapner, S. (1984). Values mothers hold for handicapped and nonhandicapped preschoolers. *Merrill-Palmer Quarterly, 30*, 403–418.

Redondo, J. P. (1983). *Migration as a critical transition: A comparison of the experience of Puerto Rican and Cuban adolescents.* Doctoral dissertation, Clark University, Worcester, MA.

Rokeach, M. (1973). *The nature of human values.* New York: Free Press.

Segall, M. H., Lonner, W. J., & Berry, J. W. (1998). Cross-cultural psychology as a scholarly discipline: On the flowering of culture in behavioral research. *American Psychologist, 52*(10), 1101–1108.

Turvey, M. T. (1977). Preliminaries to a theory of action with reference to vision. In R. Shaw & J. Bransford (Eds.), *Perceiving, acting, and knowing.* Hillsdale, NJ: Erlbaum.

Valsiner, J. (1989). *Human development and culture.* Lexington, MA: Heath.

Valsiner, J. (1997). *Culture and the development of children's action* (2nd ed.). New York: Wiley.

Valsiner, J. (1998). The development of the concept of development: Historical and epistemological perspectives. In W. Damon (Series Ed.) & R. M. Lerner (Vol. Ed.), *Handbook of child psychology: Vol. 1. Theoretical models of human development* (5th ed., pp. 189–232).

Wapner, S., & Demick, J. (1998). Developmental analysis: A holistic, developmental, systems-oriented perspective. In W. Damon (Series Ed.) & R. M. Lerner (Vol. Ed.), *Handbook of child psychology: Vol. 1. Theoretical models of human development* (5th ed., pp. 761–805). New York: Wiley.

Wapner, S., & Demick, J. (1999). Developmental theory and clinical child psychology: A holistic, developmental, systems-oriented approach. In W. K. Silverman & T. H. Ollendick (Eds.), *Developmental issues in the clinical treatment of children and adolescents* (pp. 3–30). Boston: Allyn & Bacon.

Wapner, S., Demick, J., Inoue, W., Ishii, S., & Yamamoto, T. (1986). Relations between experience and action: Automobile seat belt usage in Japan and the United States. In W. H. Ittelson, M. Asai, & M. Carr (Eds.), *Proceedings of the second USA-Japan seminar on environment and behavior* (pp. 279–295). Tucson, AZ: University of Arizona Press.

Wapner, S., Fujimoto, J., Imamichi, T., Inoue, Y., & Toews, K. (1997). Sojourn in a new culture: Japanese students in American universities and American students in Japanese universities. In S. Wapner, J. Demick, T. Yamamoto, & T. Takahashi (Eds.) *Handbook of Japan-United States Environment-Behavior Research: Toward a Transactional Approach* (pp. 283–312) New York: Plenum.

Wapner, S., McFarland, J. H., & Werner, H. (1963). Effects of visual spatial context on perception of one's own body. *British Journal of Psychology, 54*, 41–49.

Wapner, S., Quilici-Matteucci, F., & Cool, K. (1991). Cross-cultural comparisons of necessities, amenities and luxuries in the physical, interpersonal and sociocultural aspects of the environment. In T. Niit, M. Raudsepp, & K. Liik (Eds.), *Environment and social development: Proceedings of the east-west colloquium in environmental psychology* (pp. 47–56). Tallinn, Estonia: Tallinn Pedagogical Institute.

Wapner, S., Quilici-Matteucci, Yamamoto, T., & Ando, T. (1990, July). Cross-cultural comparison of the concept of necessity, amenity and luxury. In Y. Yoshitake, R. B. Bechtel, T. Takahashi, & M. Asai (Eds.), *Current issues in environment-behavior research: Proceedings of the third Japan-United States seminar* (pp. 21–32), Kyoto, Japan. Tokyo: University of Tokyo.

Wapner, S., & Werner, H. (1965). An experimental approach to body perception from the organismic-developmental point of view. In S. Wapner & H. Werner (Eds.), *The body percept* (pp. 9–25). New York: Random House.

Werner, H. (1940/1957). *Comparative psychology of mental development.* New York: International Universities Press. (Originally published in German, 1926 and in English, 1940).

17

THE GEOGRAPHY OF HOSPITALS

A Developing Approach to the Architectural Planning of Hospitals

Yasushi Nagasawa

Department of Architecture
Graduate School of Engineering
The University of Tokyo 7-3-1 Hongo
Bunkyo-ku, Tokyo 113 Japan

1. INTRODUCTION

A review of hospital planning over the years reveals that the underlying assumptions of planners, designers, and researchers are best seen as constituting a process of evolution. Such assumptions have evolved in conjunction with, although granted at sometimes not a completely simultaneous pace, the social and philosophical theories/paradigms/revolutions of the respective many periods. Both the building types and methods of treatment thought appropriate for persons who were sick were tied closely to the changing perceptions of what exactly sickness and health were and what were their causal factors.

The following pages begin with an historical overview of both the building types and environments (a word that was obviously not in use during the first stages) that have been used to house the sick, and the varying attitudes toward sickness and disease. (Nagasawa, 1995) This overview brings us in the end to the present time and the situation in which we currently find ourselves in hospital planning.

2. EVOLVING HISTORY OF ASSUMPTIONS AND PROBLEMS

2.1. Early History: Pre-Florence Nightingale

In ancient times the primary place of treating, curing, or caring for those who were ill was the home. However, in some cases this was neither possible nor was it thought acceptable. The former includes cases of people who didn't have homes in

Theoretical Perspectives in Environment-Behavior Research, edited by Wapner *et al.*
Kluwer Academic / Plenum Publishers, New York, 2000.

which to be kept or family members who could care for them. Latter cases of in acceptance sometimes occurred as a result of prevailing superstitions, psychological fear, or religious dogma. It is easy to imagine that where sickness was thought to be retribution for acts resulting in a "fall from grace" or that disease was the "due reward" for those who were somehow "unclean", the conditions under which the sick were able to be maintained were limited. At the time it was rare for new buildings to be designed as hospitals (Thompson & Goldin, 1975). Most often, other building types/institutions were appropriated and converted. Both the types of buildings that were appropriated (churches, convents, palaces, and prisons) and the names of the post-conversion facilities (alms house, pest house, sick house, and lunatic asylum) clearly reveal to us today the very different understanding of sickness and its place in society at that time. From this, some light is shed on what would have been the assumptions and problems of providing care at the time.

2.2. Florence Nightingale and Early Decentralization

The first building type which was designed as a hospital, the function of which was clearly described, is what we now call the Nightingale Hospital. Curiously enough, Florence Nightingale (1820–1910), the first person who defined the function of hospital buildings was neither a physician nor an architect but was a nurse. In her writings on the subject we find, "It may seem a strange principle to enunciate as very first requirement in a hospital that it should do the sick no harm." (Nightingale, 1863) To some today this might seem to be a case of stating the obvious but what is of importance here is that from such a statement one can easily assume that the above mentioned converted accommodations for the sick at that time provided only an extremely poor physical environment for the sick. Based on her experience at Scutari, a hospital converted from Turkish barracks, Nightingale proved that the mortality rate of hospital patients could be reduced through the provision of a better physical environment than was then common; i.e. she proposed increased penetration of sunlight, greater circulation of fresh air, and more appropriate room temperatures. She also tested and verified that the proximity of various wards was a key factor in the spread of disease and the rate of recovery or death. In the end, she proposed a pavilion type hospital consisting of 2-story high "Nightingale Wards" and she specified high ceiling heights, a spacious area around each bed, and the layout of wards with greater spacing on an ample site. This decentralized configuration of wards/buildings was disseminated throughout the world as a typical hospital at the end of 19th century.

2.3. The Birth of Modern Medical Technology

The end of the 19th century and beginning of the 20th century saw continued remarkable discoveries from the scientific fields (e.g. anesthesia/sterilization techniques, bacteria, X-rays, and antibiotics, etc.) that were crucial to changing our understanding of the causes of sickness and disease and/or their treatment and which further fueled the development of increasingly sophisticated western medical technologies. Once again assumptions were challenged and in the end altered (Thorwald, 1956).

2.4. Economic Imperatives and the Centralization of Hospital Function

The exponential development of technology continues in the twentieth century and is increasingly affecting our lives. One area in which advanced technology is seen as being fully utilized is modern hospital buildings. In order to improve the economical efficiency of expensive medical equipment and to make full use of scarce professional human resources such as physicians, nurses, radiologists and clinical laboratory technicians, new functional units such as radiology departments, path labs, operating theaters, CSSD, dispensing and pharmaceutical departments, central catering and medical records departments started to appear in hospitals in the early part of this century. This has meant a move toward centralization of hospital departments and functions.

At the same time, the development of building-technologies has enabled us to work or stay in completely artificial environments for many hours every day. Such artificial environments were easily accepted into the new centralized departments of modern hospitals. Without sufficient consideration, artificial environment technologies were also applied to areas used as healing places for patients. Instead of providing natural sunlight, fresh open air circulation and optimum room temperatures as recommended by F. Nightingale, artificial illumination, mechanical ventilation, and electrical heating/cooling systems were adopted. Additionally, the centralization of hospital functions created various complicated movements of people and materials in hospitals. Hospital staff move among various departments during their working hours. In-patients may have to walk long distances from their wards each day. Out-patients, especially in Japan, have to visit various diagnostic or therapeutic departments and wait in each for hours.

Thus, hospital designers began to search for more compact building shapes in order to reduce costs by reducing external wall-to-floor ratios. The aim was to shorten walking distances between relevant departments and to find more economical solutions to material handling. (Freisen, 1975) The result is that modern hospitals do not look like homes so much as factories. The compact building shapes, however, suited situations such as in Japan where ample site areas were not available.

In response to an increasing demand for scientific data to support the planning decisions of modern centralized hospitals, quite a number of studies on hospital buildings, mainly based on the method of Post Occupancy Evaluation (POE), have been carried out (Nuffield, 1955; Yoshitake, 1964; Thompson & Goldin, 1975). Most of the research is "problem-oriented" and "quantitative analysis based". These efforts resulted in the successful dissemination of practical data for improving the efficiency of medical, nursing, and administrative functions in modern hospital buildings.

In the end however, or at least at this point in time, based on experience gained during several decades of designing functional hospitals in the twentieth century, there currently exists some doubt about the contemporary design trend of compact modern hospitals in terms of running costs, difficulties in expanding and changing after operation and occupation, and vulnerability against natural disasters, including fires and earthquakes, as well as human-originated disasters such as cross infections and explosions.

3. CURRENT ASSUMPTIONS AND METHODOLOGIES

3.1. General Assumptions of Hospital Geography Studies

The idea for this study came from a simple question: Why do people have only inhuman and negative images of modern high-tech hospital buildings in spite of the extensive efforts made by hospital architects, planners, and others in pursuit of better hospital function?

It occurred to me and I have subsequently assumed that researchers in the field of hospital architectural planning perhaps overlooked something among the very important issues of hospital concepts while we have been preceding to fulfill the requirements of rapidly growing medical technologies in hospital services. I realize we also should have been searching wider to provide better "healing environments" for patients in hospitals. In other words, we have not been concept-oriented researchers as much as system-oriented researchers.

Hospital geography studies have evolved from Weeks's (1986) first proposal of the "Geography of Hospitals" in which he posited that a hospital building is not a single building but rather a complex of various buildings similar to the idea of different buildings in villages or towns in which each building has not only its own function but also its own distinguishable appearance. A church looks like a church, the city hall like a city hall, and so on. As the village grows or changes, buildings are added or removed, and appearances can change. The concept of growth and change is critical to the concept. Likewise, it follows that hospitals should be designed that facilitate continued growth and change long after occupation and operation.

It is our belief that the rather limited initial range of the geography of hospitals concept can be expanded by connecting it to other fields of study, e.g. studies on cognition and wayfinding. Therefore, our hospital geography studies are developing the idea further by introducing the geographic approach to hospital planning in new ways in order to provide better healing environments, in terms of layout and operations, for patients who are forced to become accustomed to previously unexperienced complicated hospital facilities.

3.2. Methodologies

With the target of more humane hospital environments in mind a series of surveys, focusing on both out-patients and inpatients, was carried out. The method of Post Occupancy Evaluation (POE) has been considerably applied to studies on planning of various types of buildings, for example hospitals, schools, libraries, and housing complexes in Japan, for the past 50 years. This rather traditional method was extensively employed in these surveys. Also, methodologies of other academic fields such as behavioral sciences, environmental and cognitive psychology including space syntax, wayfinding, and space cognition, were utilized, aimed at obtaining a more panoramic perspective of the physical and psychological situations of patients in a hospital environment (Lang, 1987; Passini, 1984; Ralph, 1976).

We acknowledge the inherent weaknesses in aspects of some approaches and recognize the limitations of attempting a scientific approach based on the analysis of quantifiable data that can only be judged by the brain. Something must be said of the value of experience and belief as guides. Another recently growing limitation in studies of hospital environments in Japan involves the changing notions of privacy and more is said about this later.

Our approach intentionally involves a multiplicity of studies because our experience supports the notion that individual studies do not necessarily lead to findings that are useful. That is to say, of course individual studies may show isolated problems and ones gets an indication of the bigger picture from them but as a group there is greater synthesis. In the case of the studies presented here, small "gaps" (discussed in following sections on case studies and respective assumptions and conclusions) between patients' expectations and the actual physical environment and operational systems within hospitals were discovered in individual studies. However, they did not take on the considerable significance that they now have until it was found that such gaps existed in the findings of most of the studies.

4. CASE STUDIES: SURVEYS ON SPACE COGNITION

4.1. Surveys on Space Cognition of Out-patients

4.1.1. Assumptions and Methodologies. In out-patient departments in Japanese hospitals there are two major problems from the point of view of a patient. Firstly, there is the problem of finding one's way in a complicated layout involving many facilities. Secondly, there is the problem of time; it is not uncommon for patients to spend many hours just waiting. These conditions are not new and we knew of them going into the studies. We expected to address the problems directly. For example, in the former case we assumed the solution to be a good layout of each department with the clear positioning of reception areas and a clearly ordered system of guidance signs. In the latter case, much has been done on the part of administration through the introduction of better appointment systems and from an architectural point of view we could expect to improve things by providing elements that offer better psychological relief during waiting hours.

Since we additionally assumed these two problems to be major obstacles in the pursuit of better healing environments, eight surveys (OP-1-8) on space cognition of out-patients were conducted (Nagasawa, 1993).

4.1.2. Survey OP-1 (1989/90). Position, posture and eye direction of patients, family attendants and staff were recorded at 15–30 minute intervals by direct observation supplemented by videotape recordings and described on floor plans, including furniture shape and position, of an entrance hall and several waiting spaces in two general hospitals. In the main waiting area, patients tended to stay where they could see the reception or sub-waiting areas. In sub-waiting areas, patients tended to stay in front of designated Consulting/Examination (C/E) rooms and usually looked at the doors of C/E rooms. Generally speaking, patients want to stay closer to C/E rooms despite the fact that the physical condition of sub-waiting areas is poorer than main-waiting areas. Vacant chairs in waiting areas were not 100% occupied even in cases where the number of waiting people exceeded the number of chairs.

4.1.3. Survey OP-2 (1989/90). 8 and 15 patients respectively were followed from the entrance and from the reception area to C/E rooms of out-patient departments (OPDs) in two general hospitals and their behavior, place, and time were described on survey sheets. Each case provided quite valuable information on the patient situations in OPDs.

4.1.4. Survey OP-3 (1989/90). Enquiries to staff at the main reception area and several sub-reception areas of OPDs were recorded in two general hospitals and classified into two groups, i.e. as requiring "information only" or "information as solution" to a more pressing need. The former includes "Where should I go next ?", "What procedure is needed?", and "Should I expect to wait here?". The latter includes "Can I get my ID card back?", "Can't I see the physician yet?", and "Can I leave the hospital now?". Most patient enquiries were about procedural information and the first part of the process of visiting an OPD was the most difficult to be understood.

4.1.5. Survey OP-4 (1989/90). Staff were asked, during a couple of weeks, to record the contents, place, time, and date of patient enquiries received while outside of their designated divisions. 62 and 58 cases were reported respectively, in two general hospitals. Most enquiries occurred in elevator halls and at entrances to long corridors. Enquiries were mostly about the place patients were asked to go, and included cases where the destination wasn't clear or the destination was clear but patients did not know where it was located.

4.1.6. Survey OP-5 (1989/90). Procedures for visiting OPDs were collected at several hospitals and there existed slight differences. As well, subjects were asked to visit an imaginary hospital and predict procedures. In comparisons of actual and imaginary procedures remarkable differences are seen in reception areas and at accounting counters. In imaginary procedures people tend not to differentiate among the four different receptions areas of accounting, paying, prescription, and drug delivery. More than 75% of subjects used the terms "entering department" and "paying at cashier". More than 50% used "waiting for examination" and "being called to C/E rooms" and more than 25% used "showing insurance card", "filling out consultation application form", "showing ID card", "following staff direction", "receiving patient records", and "waiting for consultation". Overall, people had quite stereotyped images of OPD visiting procedures, which seem quite similar to "scripts" (Schank & Aberson, 1977).

4.1.7. Survey OP-6 (1989/90). Analysis was made on the location of radiology, pathology, and physiology laboratories in 47 general hospitals, completed between 1964 and 1984, equipped with more than 200 beds. Only 12 cases were identified as having the three departments located in one vicinity; 17 cases are identified as having the radiology department located on the same floor but far from the others.

4.1.8. Survey OP-7 (1992). Out-patients were asked to indicate where they were on a floor plan and staff were asked, during one week, to record the contents, place, time, and date of patient enquiries received while outside of designated divisions, in three hospitals. Of those who were able to identify their location, there were two different types of people, i.e. element-type and route-type people. Element-type people identified their location according to a particular physical element, e.g. elevator, stair case, round shaped waiting space. Route-type people, identified their location according to a pathway followed. The frequency of enquiries increased proportionally with the size of buildings/departments. The more deeply patients got into a building, the more frequently they used route-type indicators and inquired of hospital staff. Particular places such as a lobby facing outdoors or a loft with high a ceiling are easily used by element-type people.

4.1.9. Survey OP-8 (1992). Space syntax analysis (Hillier & Hanson, 1984) was conducted as well as a count of the number of people encountered during rounds in the OPDs of three hospitals. One "step" is defined as one straight-line axis and the number of steps were counted from OPD reception areas to C/E rooms. In the first hospital there was a long step-one axis and the reception counter of specialty departments faced to a step-two axis with the exception of the radiology department which faced to a step-four axis. In the second hospital, specialty receptions faced to a step-two axis. In the third hospital, the step-axes were complicated in spite of the small building size. The number of people encountered decreased with an increase in the number of step-axes.

4.2. Conclusions

At the beginning of this section we introduced two problems that most patients face in Japanese hospitals (wayfinding and waiting) and the fact that we thought solutions might be straightforward. However, what we found was that the problems may be more complex than realized and require further consideration.

For example, in the case of wayfinding, despite many clearly posted direction boards or signage posts provided by hospitals, patients still lose their way because of interpretation problems; symbols that have meaning for hospital staff may not be understood by patients. Specifically, in OP-4 we found that many enquiries were observed/recorded on the way from C/E rooms to laboratories or radiology departments and again this partly stems from the difference in the naming of rooms/departments within hospitals, e.g. patients will usually not imagine the physiology laboratory as a place of cardiography examinations. Signage continues to be a source of previously identified "gaps".

Again we emphasize that a reconsidered more responsive approach to the layout of each department's facilities and a more "sympathetic" i.e. more in line with patient expectations, signage system would be beneficial. For example, having x-ray, pathology, and physiology labs on the same floor would be a marked improvement. From the results of OP-8 it would also be recommended that corridor systems be simplified.

As for the issue/environment of waiting patients, during survey hours, frequent voice announcements were recorded, i.e. every 1.9 minutes on average and 1.4 minutes at maximum. Patients tried to listen to every announcement, wondering if they were relevant to them. This kind of information can be called "controlled information", officially provided by the hospital.

Video recording in Survey OP-1 and the results of Survey OP-2 showed that, even though patients often napped on chairs, read magazines/newspapers, and watched TV, most of their waiting time was spent collecting information on their waiting situation, in various ways, e.g. looking into the reception and sub-waiting areas, inquiring at the reception desk or to passing hospital staff, asking nearby patients, listening to other patients' conversations, looking at watches and clocks on the wall, looking around and at patients who stood after being called. In some cases it appeared they preferred standing rather than occupying vacant seating in order to better obtain information. This kind of information, which is obtained by patients themselves, can be called "uncontrolled information".

Patients expect confirmation of their situation at every step and even though they have to wait for a long time they expect hospital staff to realize that they have arrived at the reception area, their records have been transported and distributed to each physician, they are in the waiting area, and their waiting time is becoming shorter. They were very anxious whether or not they are called while away to use WC facilities.

The degree to which waiting patients rely on information that they get from non-hospital-provided sources was previously underestimated and greater consideration of this would result in a more psychologically secure environment for patients.

4.2. Surveys on Space Cognition of Inpatients

4.2.1. Assumptions and Methodologies. In talking about the life of patients in the hospitals of her day, Florence Nightingale said, "As long as they are hospital inmates, they feel as hospital inmates, they think as hospital inmates, they act as hospital inmates, not as people recovering.". (Nightingale, 1863) In fact we find that patients do behave "like patients". They harbor fears of staff looking at them in an unfavorable light and they are uncomfortable with the rigidness of a typical inpatient's daily routine.

At this point it would be appropriate to consider the context of inpatient studies. We have plenty of data accumulated from nurse response studies. However, a continuing difficulty lies in the absence of information gained from the point of view of the inpatients themselves; we have had little direct indication of their psychological stance. This stems in great part from the general "taboo" of interacting with inpatients directly. Such interaction is usually reserved for hospital staff only and outside intervention is guarded against. As mentioned when discussing methodologies (3.2.), growing privacy orientations continue to further reduce chances to interact with inpatients directly and this restricts current methodologies (it can take years to get rights to enter the private domain of inpatient lifestyles). However it was determined critical to make more direct contact and permission was gained both from hospital authorities and inpatients to conduct the following inpatient studies. Since passing out questionnaires often doesn't result in information that is as useful, we opted for direct interviews and rather than ask, for example, "how do you feel about such and such" (distorts the study) we would ask more directly provable questions such as "do you know where the phone is' (this reveals more).

Although through these two difficult studies we managed to tap into an until now not often taken advantage of resource, we acknowledge certain limitations. Interviews with a limited number of subjects will always result in the question of how well the selected subjects represent the opinion of the majority, particularly when it can be anticipated that only certain types of inpatients volunteer for such studies. We may be seeing only the "tip of the iceberg" of a larger inner world so to speak.

Two surveys (IP-1-2) on space cognition of inpatients were conducted. (Nagasawa, 1996)

4.2.2. Survey IP-1 (1992). Inpatients were asked to indicate "In which bedroom are you accommodated?", "In which bedroom are your acquaintances accommodated?", and "Where do you get to know other patients?". All the interviews were tape-recorded and answers were described on the floor plan of three Japanese hospitals. Location identifying elements were classified into two groups as elements located on the main route, e.g. elevator, staircase, entrance, nurses' station, linen storage; or as elements which are physically and psychologically impressive, e.g. outside staircase which can be seen through the windows, WCs, wagon carts in the corridor, and bedrooms in which friends are accommodated. The former was more frequently utilized to identify patients' own bedrooms. "Nurses" stations' was recorded more frequently than expected and this implies staff activities in nursing units are considerably impressive for inpatients in identifying location. In addition, patients tended to describe their range of walking around within the hospital as smaller than it actually is. Strong consciousness was observed of patients "movement from one ward/bedroom to another". Communication between patients started when they helped each other or moved to another bedroom, and also in WCs, washing rooms, and in corridors.

4.2.3. Survey IP-2 (1993). Periodic observations and interviews with five patients were carried out for four weeks after their admittance to a Japanese general hospital in 1993. All interviewees had no previous experience of admittance to the hospital. The main contents of the interviews were daily activities, human relationships, the range of accustomed space in the hospital, and variety of objects used by them. Although varying in sequence from patient to patient, five types of inpatient behavior and cognition were regularly found to occur. They are: a. receiving information about space use from hospital staff and other patients, and verifying this information through activities; b. acting according to expected patient behavior; c. using spaces in a manner other than intended; d. visiting places within the hospital where the patient is unknown; and e. developing an identity for various places.

4.2.4. Conclusions. The discovery from the findings of the out-patient (OP) studies of the different ways in which patients made use of the two kinds of information, i.e. controlled and uncontrolled, was a new one for our research. Subsequently this affected assumptions being made while developing inpatient (IP) studies and the existence of these two kinds of information was also assumed to be a factor in inpatient cognition.

In fact, Survey IP-2 showed a similarity in new patients' admittances to a ward. They started by receiving controlled information and gradually obtained uncontrolled information. Controlled information includes orientation rounds conducted by nurses and items found in a guide book for inpatients, e.g. prescribed wake up and meal times, emergency evacuation procedures, etc. Uncontrolled information in most cases includes that which is discovered by inpatients during the course of their self-determined "experience tours" of hospital facilities that constitute their personal hospital experience.

Over a period of several weeks the studies monitored the patterns of inpatients' behavior and cognition of their environment according to the duration of their stay in a hospital. Both behavior in spaces and cognition of the environment varied from the programmed space use intended by hospital planners. As a result, the following suggestions in hospital design are made:

- expect use of space outside of the program
- emphasize spatial legibility of the whole building
- recognize the importance of windows as a means to prevent isolation
- consider patients' negative perception of nurses' stations, as places around which privacy is unavailable, in planning their location
- include spaces for patients where observation by hospital staff is not apparent
- design spaces where patients can gather and communicate

5. DISCUSSION AND CONCLUSIONS

At end of the 20th century we find we have highly technologically developed pristine modern hospitals that people don't want to visit: they still don't suit most people's common notions of health or health supporting environments.

As discussed, our studies reveal that there exists an extensive "gap" between patients' expectations and the actual physical environment and operational systems within hospitals and our conclusion is that this is a major cause of the current inhumane/negative images of hospital environments. Although the degree of gap is not absolutely quantifiable, due to study method limitations, the fact that it is extensive is clear and would not have been possible without having made the methodological decision of conducting multiple small studies in one area. The results of the individual studies (isolated small gaps) didn't demand the same attention that the greater synthesis, the pattern that emerged from the series of studies as a whole, did.

Such gaps, or any finding that indicates a separation between desired and actual conditions, act as important indicators to further research. They are in fact the first step in the process of changing assumptions. In the case of our studies, for example, we newly discovered that patients judged their various situations based on attention to controlled information officially provided by hospital administration, supplemented by uncontrolled information.that patients obtained by their own efforts. The importance of distinguishing between the two types and the conclusion/recommendation that improved hospital planning account for the greater facilitation or provision of opportunities for both are a result.

Even a cursory review of the historical overview presented reveals how assumptions are continually influenced by the impact of technological development on both general lifestyle and the hospital environment; this as a continuing condition. Although a clear model of relationship has not been finalized, i.e. we don't know exactly the degree to which technology leads and people follow or people's ideas lead and technological changes in the environment follow, we know the relationship needs to be fully addressed.

The situation of health and medical architecture is physically, socially, and culturally different in every country and problems of hospital design in software and hardware are complicated with political, economical, and social factors. Accordingly, it is difficult to propose any clear-cut solution for the near-term, and we must continue to study and test alternatives. However, this much we do know. The speed with which technology changes affect our lives and the far-reaching extent to which it impacts all aspects of our lives can be looked at positively in the following sense. We easily imagine that change, even great change, is possible. This allows us to further assume that, given a concerted effort by researchers who possess a clear goal, new ways, better solutions can be readily open to us.

Strategically, we repeat the recent conclusion that while fulfilling the requirements of medical technologies in hospital services we have perhaps overlooked issues important to providing better "healing environments" for patients in hospitals and have not been concept-oriented researchers as much as system-oriented researchers.

This insight, allows to address anew the techniques and knowledge handed down from Florence Nightingale and her directive that the ultimate hospital environment is the one that most resembles a home environment. In this endeavor, it is quite possible and advantageous to try to find better healing environments for future generations based on the geographical aspects which were discussed in this paper.

REFERENCES

Freisen, G. (1975). *Concept of hospital planning*. World Hospitals, Winter. London: International Hospital Federation.

Hillier, W., & Hanson, J. (1984). *Social logic of space*. Cambridge: Cambridge University Press.

Lang, J. T. (1987). *Creating architectural theory*. New York: Van Nostrand Reinhold Company Inc.

Nagasawa, Y. (1993). Hospital out-patient department based on patients' behavior and cognition: Studies on hospital geography 1. *Journal of Arch. Plan. Eng., AIJ-452*, 75–84.

Nagasawa, Y. (1995). *Hospital architecture as geographical environment* (pp. 143–145). Hospital Management International, London.

Nagasawa, Y. (1996). A study of hospital environment based on changes of patients' consciousness and behavior over time lapse: studies on hospital geography 2. *Journal of Arch. Plan. Eng., AIJ-483*, 121–128.

Nightingale, F. (1863). *Notes on hospitals* (3rd. ed). London: Longman, Green, Longman, Roberts, and Green.

Nuffield Provincial Hospitals Trust. (1955). *Studies in the function and design of hospitals*. Oxford University Press.

Passini, R. (1984). *Wayfinding in architecture*. New York: Van Nostrand Reinhold Company Inc.

Ralph, E. (1976). *Place and placelessness*. London: Pion.

Schank, R. C., & Abelson, R. P. (1977). *Scripts, plans, goals, and understanding*. New York: Erlbaum.

Thompson, J. D., & Goldin, G. (1975). *The hospital: A social and architectural history*. New Haven and London: Yale University Press.

Thorwald, J. (1956). *Das jahrhundert der chirurgen*. Stuttgart: Steingruben Verlag.

Weeks, J. (1986). *Geography of hospitals* (p. 13). World Hospitals, September, London: International Hospital Federation.

Yoshitake, Y. (1964). *Studies in architectural planning*. Tokyo: Kajima Institute Publishing.

18

SYMPATHETIC METHODS IN ENVIRONMENTAL DESIGN AND EDUCATION

Takashi Takahashi

Graduate School of Science and Technology
Niigata University 2-8050 Igarashi
Niigata 950-21 Japan

1. INTRODUCTION

This paper discusses the adoption of a new interpretation of the idea of "globalization". Unlike previous interpretations of the word, which have focused on picking up only information common to all countries, in this paper it will mean to also pick up locally derived information including peculiarities, and to share it as common knowledge. In this sense, we can say that the new globalization is nurtured by localization.

As well, the idea of a "sympathetic method" in design processes, a concept which would help us move beyond stereotypical architectural education, from a "problem-solving" process to a "problem-finding" process, is introduced. Some case studies of this method are presented which will make evident the importance of providing conditions in which students can better sympathize with the many issues related to a particular project.

Finally, the relationship between these two ideas, that "globalization", by widening our shared knowledge base and providing more chances of discourse with others of diverse values, will encourage the "sympathetic method" and transform or awaken students' sympathy, will be discussed.

2. THEORETICAL ASSUMPTIONS

2.1. Problem-Finding Process in Design

The main assumptions of modern architectural design were, and still are, based on architectural determinism or environmental determinism. Determinism includes, I

Theoretical Perspectives in Environment-Behavior Research, edited by Wapner *et al.*
Kluwer Academic / Plenum Publishers, New York, 2000.

think, two aspects of Person-Environment relations. One is a positive relation which means that the environment supports persons or behaviors. For example, if a mother cradles her baby in her arms, a good relationship develops between them and the baby can develop with psychological stability. The other is a negative effect relation which means that the environment forces a specific behavior or moulds behaviors contrary to natural inclination. For example, the layout of rooms in a house forces occupants to adapt a specific behavior or lifestyle and exclude other behaviors.

Now however in the field of Environment-Behavior studies, deterministic theories are undergoing some criticism. New concepts of Person-Environment relations have appeared. For example, Trait, Interactional, Organismic, and Transactional approaches have been proposed by Altman and Rogoff (1987). Another approach is Canter's classification of theories (1985) into Strong determinism, Weak determinism, Interactionism, Weak transactionalism, and Strong transactionalism.

We can see a definite shift from a deterministic approach to a transactional approach. In the latter theories, environment and person are considered as a united whole, and the united whole produces serial events over a period of time. Various factors are not independent, but are aspects of events.

The transactional approach can be seen in the field of biology. Dubos (1965) wrote that every mechanical part of the human body reacts to the environment in the same way as other living things. But a living person responds to the environment, and response means that the individual can discover new aspects of oneself in that process. This biological model is an example of a model which is not explained by cause and effect relations.

This new response of Person-Environment relations has the following characteristics.

(a) Person-Environment relations vary according to the degree of sympathy which a person has with the situation or the environment.
(b) We are very sensitive to a P-E relationship which we are experiencing for the first time.
(c) The new experience soon result in negative or positive attitude within us.

Such an attitude gives rise to a problem-finding process in design instead of a problem-solving process, or engineering approach in design.

We must enhance our own experience of Environment-Behavior relations. This is very similar to fieldwork or field research in the cultural-anthropological area. We also need to continue to observe and record the E-B situation (using POE studies—Post-Occupancy Evaluation). So we should view our daily life as a serial design process, in which we must constantly observe the changing process of E-B relationships.

Such considerations lead us to the new perspective in which we are all designers of Environment-Behavior relationships. Given this approach, what special contributions can architects or designers make?

2.2. Architectural Model of Environmental-Behavior Situation

The transactional approach treats environment and behavior not as separate elements, but as holistic entities which constitute aspects of an environmental situation.

The approach describes in reality Environment-Behavior situations inside and outside of buildings. Although it appears to be a very sophisticated concept, it is very difficult to design a research program according to the concept. How do we observe the events, and how do we analyze the results.?

In architectural design, we operate the physical environment by setting up buildings elements. On the other hand our behavior is influenced by the physical settings. Moreover when we behave in a physical setting which we have not experienced before, we can find a new environment-behavior situation. Considering the above, I assume an Environment-Behavior relation model in architectural design as follows.

(a) Fixed-stable relation
(b) Environment deterministic relation
(c) Situation-problem finding relation

Fixed-stable relationships can been seen in vernacular architecture in which stable Environment-Behavior relations have been established over many years. In the relations between a person and daily-tools, like furniture, the same situation can be found.

The second deterministic relationship has been recognized as an explicit principle in design when the modern movement appeared. Of course too rigid a relation between environment and behavior should be avoided but we continue to need this principle in design.

The last relation of situation-problem finding is created in the process of daily life. As for architectural design, the start of using a new building which is usually the end of architectural design and construction, can be considered as the start of designing the Environment-Behavior situation. POE is an example of this type of an analysis. Moreover knowledge on those types of relations in each culture will be verified by globalization.

In this context users or students of architecture have a similar role in formulating Environment-Behavior situation. Actual modes of analysis and research problems will be discussed following description of case studies in architectural education.

3. BACKGROUND

Internationalization and globalization continues and common interests in architectural/environmental design methods and education are expanding. At the same time, we see the emergence of urban/architectural planning and design issues particular to regions and cultures. Unlike previous interpretations of the word globalization, which have focused on picking up only information common to all countries, in this paper it will mean to also pick up locally derived information including peculiarities, and to share it as common knowledge; to assemble it into a common global knowledge base.

For instance, in Japan, architectural education in universities started with the arrival of British architect Josiah Conder. Thus the education was strongly affected by British ways in the early years. However this early form of globalization was superseded by local factors after the Great Kanto Earthquake in 1923, when wood and brick masonry construction proved susceptible in earthquakes. The necessity of developing more fire and earthquake resistant ways of construction was focused on, lead to growth in research and education in structural engineering. As a result, unlike European and American universities, departments of architecture in Japan came to include the three fields of structural engineering, environmental engineering, and architectural design and planning.and they are contained within engineering departments.

In the current curriculum of the Department of Architecture at the University of Tokyo we find: Students of architecture need learning in a wide range fields but at the same time, it is desirable for them to choose study fields according to their disposition and capabilities. In light of this, compulsory subjects are kept to a minimum, while students are encouraged to choose subjects of their own will and to attend lectures of other departments or faculties. The aim of this policy is to develop a variety of human resources. Also, special subjects are taught by outside experts as part of the curriculum.

The mornings are allocated to lectures, while the afternoons to design lessons, exercises, experiments, etc. Lectures can be roughly divided into structural and planning subjects. Structural subjects mainly deal with buildings as "structures" and include such courses as Building Structures, Building Materials, and Building Construction. Meanwhile, planning subjects primarily consider buildings as "space" and include such courses as Architectural Planning (for analyzing relations between buildings and human behavior), Environmental Engineering, Mechanical Equipment for Buildings, Architectural Design, and History of Architecture. Along with academic advancement, subjects of architecture tend to be divided into an increasing number of specialities. On the undergraduate level however, all students are expected to receive a comprehensive education regardless of their specialties. Main lectures are supplemented with exercise or experiment courses and in addition, there are architectural design and drawing courses as part of the core of the whole curriculum. Architectural design and drawing courses are regarded as the most important of all subjects at the Department of Architecture in that those courses will enable students to acquire comprehensive judging capabilities and creation techniques required in architecture. All of the teaching staff are in charge of those courses, in which students are given a theme every one or two months to be made into drawings, models, etc., based on individual or joint design. Each of such works is reviewed by several instructors at a time.

As stated in the curriculum, design lessons (studios) are considered an occasion to integrate the three fields. In fact however, structural design and mechanical equipment planning are rarely approached; structural and environmental engineering are researched and taught separately and are not integrated with design and planning.

As well, another problem which deserves greater attention lies in the aims of design studios. For example, when I was a student, the first project was to "design a one-story, reinforced concrete house, for a nuclear family of four, on a flat and rectangular site in a residential area in Tokyo." I wondered how anybody could design a house with only those conditions and what the educators were expecting students to create. I had no idea and in the end helplessly imitated and combined the designs and of houses I saw in magazines.

In this studio, since nothing about the quality of the environment of the house was expected to be mentioned, we had to establish the goals by ourselves but being the first project, I was not aware of it. I did however recognize instinctively that I could not design anything from only the given conditions. The given theme did not awaken my motivation to design; in other words, I could not feel any sympathy for the project. I remember that proportion and size, and the appearance of eaves hanging out from the concrete frame were the points of discussion about my design at the presentation. The standpoint of criticism was dependent on the professors' interests and again I could not feel any sympathy for the criticisms.

In an era of growing concern in sustainability of the environment, architectural education, especially design education, should be transformed to include the following:

(a) Architecture should be considered as a constituent of the environment in a broader sense.

(b) In the process of designing, students' interests should be led to how buildings are built, sustained, and demolished, and also be made to grasp a realization of the cycles in the environment, including nature.

(c) Students should be made aware that design has meaning not only in constructing physical environments or artifacts but also in supporting person-environment relations.

(d) Following (c) above, programs should be developed so that students can recognize that the completion of the building, which used to be the end of the design process, is a transition point to a new person-environment relations, in other words the beginning of environment sustaining design.

4. GLOBALIZATION ENCOURAGES SYMPATHY

When we regard that "globalization does not mean homogenization by one region dominating over others but interactions between particular societies and broader situations" (Kang, S., 1997), occasions such as this symposium, in which the interaction can be recognized and considered, are important. Environmental education must offer chances of recognition and practice of such interactions as well.

From a cultural point of view, it can be said that "men are cultural beings that always stand at the meeting point of various cultures". This gives rise to the next problem. A famous Japanese actress who works in Tokyo and Paris, says she constantly feels cultural pressure in both cities. She can find relief from this pressure only in international flights—places with no nationality. Globalization does remove some borders but on the other hand it also raises awareness of the cultural bubble surrounding oneself. A "culturally blank" space, such as an international flight where people can escape from such pressure, may become increasingly necessary. We need to find education subjects that can clarify this kind of situation.

In this time of globalization and localization, architectural education must satisfy the following.

(a) It is advantageous to have persons from as many cultures as possible both within the student body and faculty as a micro-level nesting in the macro-trend of a multi-cultural world.

(b) Opportunities to experience and realize diversity must be provided. Discussions with foreign students on cultural similarity and difference offer many chances.

(c) It is not a matter of "anything goes" as in current trends in architectural design after Modernism and Post-modernism, and perhaps we should reconsider the discipline of "abstraction" characteristic of early modernism. This may provide an answer to "statelessness".

5. MOVING BEYOND THE STEREOTYPICAL IN ARCHITECTURAL EDUCATION

How can we move beyond a stereotypical design process, in which we comprehend the relations between man and environment as fixed patterns and allot forms in stereotyped patterns of building types or in life style, and how do we establish a design process that can propose a transitional framework of the relation between man and environment?

In the process of design education, everyone involved in a particular project should learn from each other through discussion and other acts, and it is desirable that they develop greater sympathy for the person-environment relations aspects of the particular project as they go through the process.

If that happens then, design education can change from problem-solving studies, more typically found in the field of engineering, to "problem-finding" studies. As will be seen in the following examples of projects given, it can be pointed out that sympathy in the person-environment system plays a significant role in motivation and is key to generating images of environments from experiences in everyday life.

5.1. Designing Multi-Family Housing in Nezu (Old Downtown Neighborhood in Tokyo)

In this project we attempted to consider collective living and the design of multi-family housing in an old downtown neighborhood near the University of Tokyo where commercial and residential areas are mixed. It is usual to think that a stranger cannot experience a place like the inhabitants who have been living there for a long time. However, even in such areas where active communication and a strong local community exists, people do not know the complete details of each others lives as evidenced in this quote from a relevant interview. "We refrain from inquiring into each others' private lives". It is sometimes possible that researchers investigating an entire area by in-depth interview and observation may grasp an overall image of that area better than its inhabitants. In this project, based on this hypothesis, participants did year-long field research, established a database, and then, based on the database, each participant presented a design proposal.

Because of the complicated and often inaccessible factors of land possession and economic problems as related to redevelopment, we could not make proposals sufficiently practical to persuade inhabitants. However, participants showed a sympathy to conditions of the person-environment systems of the site, and made diverse proposals that might not have been possible from a more typical process of problem-solving.

5.2. Designing a House Based on "Behavior Settings"

The next example was a seven week project for third year students. Normally in a problem-solving type design process, single "rooms" correspondent to particular activities are adopted as basic units in design, and drawings are generated from them. In this project, freedom from this architectural structure and form that restrict design in the former problem-solving processes was pursued by considering "behavior settings".

This project was characteristic in its attempt to achieve more greatly enriched architectural images before making definitive design drawings. Architect Ito Toyoo, who assigned the project, said "Presupposing the relationship between our currently extended sense of physical self and architecture, what would an architectural skin be? In thinking about this question and keeping human behavior as the central issue, I have always wanted to reconsider architecture from the standpoint of how one should cover settings for behavior. For example, we first required the students to configure settings for behavior that they each considered important, using only furniture. On a flat floor, small groupings of furniture were formed for various ways of gathering and various postures, such as for eating, sleeping, working, and relaxing, were arranged on a flat floor. Since furniture reflects people's silhouettes, the settings of furniture were sufficient to imagine how people gather and behave." (Suzuki, 1997)

The theme was urban housing with the site to be determined by students to accommodate a minimum family structure of one couple and two children. Students were required to:

1. Present a panel showing visually at least five proposals of an ideal lifestyle
2. Make a model of the settings, using only minimum elements
3. Provide shelter/covering for the settings to make a room
4. Add structure
5. Rearrange into architectural form and present a model and panels

The process resulted in unique proposals from those students who consciously, or unconsciously, had sympathy with the educators' aims.

5.3. "Camping": A Studio Project

The Great Hanshin Earthquake, that occurred in January of 1996, had a great effect on Japanese society. In the field of architecture and environmental design, not only issues of structural stability of man-made structures but also support for inhabitants who met with disaster were raised. In light of this situation, the studio project, "Camping" was prepared for fourth year undergraduate students. After the earthquake, many forms of temporary housing (much of which is still occupied) emerged; people lived in parks, in tents, etc.; "desperate wisdom" was applied variously to the problem of housing. In this project, students were to design a simple shelter which one could carry and build easily in disaster situations. In this case as well, it was assumed that both the students' and educators' awareness of the problems of earthquake disaster and sympathy for the suffering of the people had greatly affected the design process.

6. ENHANCING SENSITIVITY IN EDUCATION

As a prerequisite to environmental education, an enhanced sensitivity is required to the quality of person-environment relations in daily life. However the question is, is it possible to obtain this sensitivity through education? The controversy of "nature" versus "nurture" as regards human ability can also be applied to sensitivity. However, as in the diverse discussions on person-environment interaction/transaction, a harmonious process of both nature and nurture, that is transitional over time, must be considered.

Here I would like to make two points regarding the transitional process. Firstly, sensitivity is not taught directly but is infectiously passed from one to another. It seems that sympathy enables this. Secondly, that antipathy to another's sensitivity also occurs in the same way; in both processes, sleeping sensitivity can be aroused by the influence of others. We should set up situations in education where this influence is enhanced.

Globalization, in general, gives us a chance to develop sensitivity for the quality of the environment through sympathy or antipathy that occurs as a result of bringing into contact culturally diverse elements and relationships, whether considered to be of higher or lower quality, which thereby present new person-environment relations/ systems. A similar globalization of educational methods will play a critical role in enhancing the possibility of such transfers of sensitivity.

REFERENCES

Altman, I., & Rogoff, B. (1987). World views in psychology: Trait, interactional, organismic, and transactional perspectives. In Stokols, D. and Altman, I. (Eds.), *Handbook of Environmental Psychology* (Chap. 1, 7–40). Somerset, NJ: John Wiley & Sons.

Canter, D. (1985). *Applying psychology*. Augural Lecture at the University of Surrey.

Dubos, R. (1965). *Man adapting*. New Haven, CT: Yale University Press.

Kang, Sang-jung (1997, July). Beyond orientalism and globalization. *The International House of Japan, 8*(2), 11–14.

Suzuki, T. (1997). *Designing architecture from "behavior settings", designing person-environment system* (edited by the Architectural Institute of Japan) (pp. 260–261). Shoukoku-sha.

19

CULTURAL ASSUMPTIONS UNDERLYING CONCEPT-FORMATION AND THEORY BUILDING IN ENVIRONMENT-BEHAVIOR RESEARCH

Urban Planning and Life-World Design

Hirofumi Minami* and Takiji Yamamoto**

*Department of Urban Design
Planning and Disaster Management
Graduate School of Human-Environment Studies
Kyushu University
6-19-1, Hakozaki, Fukuoka, 812 Japan
**Research Institute of Human Health Science
10-51-215 Ichibaue-machi, Tsurumi-ku
Yokohama, 230 Japan

1. INTRODUCTION

More than a decade ago, Hagino, Mochizuki, and Yamamoto (1987), in their handbook chapter of "Environmental psychology in Japan," posted a list of tasks for "future environmental psychologists in Japan." They include such tasks and questions as the following: How can we ease the conflict between rapid industrialization and traditional Japanese culture? How do we design habitats to better house our elderly and keep them in touch with other people? How do we design environments for our elderly where they can continue to do useful and confidence-building work? How do we design educational and relaxing playgrounds in a very small area? The list goes on. While looking over the list we cannot but feel some cynicism. During the intervening years there has been a wide range of construction of public and private facilities, a variety of city-renewal projects, and much "resort" making all over the country. Yet, the living situation for the elderly citizens and educational environments for the young people in this country do not seem better since then; there are numerous signs telling of crises

Theoretical Perspectives in Environment-Behavior Research, edited by Wapner *et al.*
Kluwer Academic / Plenum Publishers, New York, 2000.

among these populations which are at least partially due to environmental qualities surrounding their everyday life.

There are two possibilities accounting for the lack of provision of good environments. One is that sufficient knowledge and guidance by Environment-Behavior (E-B hereafter) studies and professionals, which are already available, were not properly implemented in the actual construction process. This is a view shared by some founders of the field in the United States (see Sommer, 1997). Considering the vastness of the construction and building projects in a country like Japan, the number of and relative social power assigned to workers in E-B research are far from sufficient to be effective in making visible differences. There is, however, another possibility: the models and implicit assumptions underlying current E-B research, which are essentially cultural products of the Euro-American intellectual climate, do not fit some of the socio-cultural backgrounds of the Japanese people. While recognizing both possibilities as equally plausible, an effort is made in the present paper to further elaborate on the second possibility.

Such reflections on the cultural assumptions underlying current E-B research is particularly significant and ripe considering the historical context of our age when basic paradigms of society and science are questioned. There is increasing awareness of the limitations of modern rationalism, which are based on 1) a mechanistic view of nature, 2) the identification of human agents as a rational being, 3) the concept of society as a collection of individuals and 4) a monotheistic or universalist world view (Yamamoto, 1992). One of the strongest manifestations of these paradigms of modernity is seen in the practice of urban planning, which has remodeled the constellation of most of Japanese cities after the second world war.

The present paper focuses on the nature of urban planning, especially on the guiding principles that transformed traditional Japanese town into modern urban structures. In-depth, longitudinal ethnographic data on a city renewal project (Minami & Tanaka, 1995; Minami, 1997) was used as an exemplar case to examine working assumptions underlying modern urban planning. Analyses of naive concepts or naive psychology (Heider, 1958) in the discourse of urban planning and environmental design in general, both by professional and lay people, were also conducted. In accordance with basic premises of cultural psychology, which posits that cultures (or culturally constituted realities) and psyches (or reality constituting psyches) make each other up (Stigler, Shweder, & Herdt, 1990), we focus on collective representations as clues to hidden assumptions in both academic endeavors and praxis of the environmental design field.

2. EXAMINATION OF CULTURAL ASSUMPTIONS UNDERLYING E-B RESEARCH AND PRACTICE OF URBAN PLANNING

2.1. Vitalism vs. Rationalism in Urban Planning

One of the most common phrases used in the discourse of Japanese urban planning is "*machi-zukuri* (town making)" and "*machi-okoshi* (town awakening)." Upon reflection such phrases appear rather odd since the practice of *machi-zukuri* is applied not to new towns which have to be built but to towns which already exist and in most cases have a long and considerable history. The "making" of towns in this case implies re-modeling and strengthening its identity through the active leadership of the government and the active involvement by the citizens. A metaphor of the "awakening" of a town might better capture the cultural ethos of urban planning in Japan. A synonym, *kasseika* (vitalization) is also used in conjunction with *machi-zukuri* projects.

Naive psychology underlying these concepts is an idea that a town is a living entity, which has to be awakened or re-vitalized time after time so as to reach its maximum functioning and use for its inhabitants. Here we see a version of vitalism in conceiving the working of towns. As a means to vitalize and re-vitalize towns, such cultural practices as seasonal festivals have traditionally been used. This is reflected in present-day practices of *machi-okoshi*, in which social events featuring historical and cultural heritages of the town are accommodated for the appeal to public attention. Here the emphasis is more on temporal activation of the lives of citizens than on permanent and structural changes.

Such a conception of town planning casts a contrast with modern theories of urban planning, which are western in origin. A text by Le Corbusier (1929), translated into English as "The city of tomorrow and its planning," provides a seminal expression of the basic philosophy of modern urban planning. Le Corbusier (1929) begins his ideas of urban planning in a metaphor of two ways: "the pack-donkey's way and man's way," in which he describes "Man walks in a straight line because he has a goal and knows where he is going; he has made up his mind to reach some particular place and goes straight to it. The pack-donkey meanders along, mediates a little in his scattered-brain and distracted fashion, he zigzags in order to avoid the larger stones, or to ease the climb, or to gain a little shade; he takes the line of least resistance."

Most continental cities were built in the Pack-Donkey's Way, according to Le Corbusier, where "the city has grown year by year, its plan being dictated by the existing roads leading to it," and "the result is an ingenious series of adaptations made during many centuries." The plan in this case is "a fine one of a curvilinear type" (Fig. 1).

While "the winding road is the pack-donkey's way, the straight road is man's way." The "rectilinear" plans (Fig. 2), therefore, are the victory and "a work of the mind" in Corbusier's philosophy since "a modern city lives by the straight line, inevitably; for the construction of buildings, sewers and tunnels, highways, pavements," and "the circulation of traffic demands the straight line." Geometrical order, among other things, becomes a guiding principle in creating modern cities.

In the ethnographic study of an urban renewal project of a town in Hiroshima city, which preserved pre-modern qualities of a densely housed downtown community (Fig. 3), we observed decreased social encounters after the remodeling of the town into grid patterns and the enlargement of roads to ease the traffic (Minami, 1997). The renewed town (Fig. 4) fits perfectly with Le Corbusier's plan of urban structure, i.e., a well designed grid pattern and hierarchical ordering in road system. This new town structure, however, is accompanied by decreased communal life or "vital" aspects in the context of Japanese city life. This leads to the next topic.

2.2. Socio-fugal vs. Socio-petal Biases

Tacit in the concepts of crowding and personal space (Sommer, 1967), both of which are landmark topics in much of the E-B studies of the seventies, is the idea that people have a basic need for space. The topic of crowding was chosen as the topic which

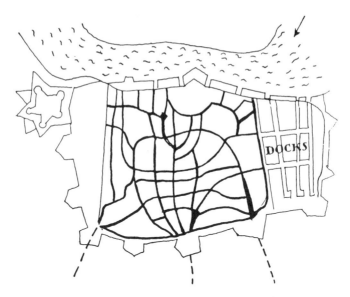

Figure 1. The plan of a city based on the "Pack-Donkey's Way": Antwerp in the seventeenth century (Le Corbusier, 1929).

most captures problematic urban conditions where enough space is lacking for individual residents. The notion of personal space and buffer-zones further elaborated theoretically the ways in which individuals maintain physical distance from others in regulating social encounters. In looking at human transactions, space and distance are taken as "the figure in the ground" (in Gestalt Psychology terms) of relatedness. Here we see cultural biases of "individualism" (Sampson, 1989; Triandis, Bontempo, Villareal, Asai, & Lucca, 1988), or they can also be characterized as "socio-fugal" biases using Osmond's (1957) well-known concepts.

The after-effects of urban renewal in the aforementioned study (Minami, 1997) were decreased social encounters due to such environmental factors as increased spacing between houses, separation from neighbors by walls, locked doors with interphones, and enlarged roads mostly for the convenience of the car traffic. These are factors of socio-fugal dynamics, which accompany modern urban planning.

The traditional Japanese town nurtured communal life through the sharing of many intermediate regions like streets and common facilities for everyday life. The boundaries between houses and individual rooms were not clear-cut. Here the connection and closeness in social life space are taken as the figure in the ground of divisions of individual space. They may be characterized as "socio-petal" biases based on "collectivism" in the social climate of Japanese culture.

For vital transactions of people it is necessary that there be a certain degree of concentration of population. Depopulation of an area (*kaso-ka*) is taken seriously as a symptom of social decline. Density and crowding in this case are not vicious conditions, but an asset of city life.

Another key concept in modern urban planning, "zoning," might be interpreted in a new way using socio-petal and socio-fugal biases as underlying principles of environmental design. In the practice of zoning, differentiation of functions and separation of zones, which are planned to concentrate on particular functions, are achieved: the

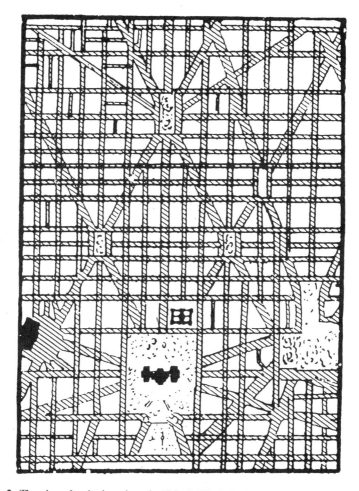

Figure 2. The plan of a city based on the "Man's Way": Washington (Le Corbusier, 1929).

residential, commercial, and recreational zones are thus created. The traditional downtown area in Japan, however, is characterized by its mixture of functions and interlacing of commercial, recreational, and residential space in a densely packed housing structure. The concept of zoning emphasizes more on socio-fugal aspects, whereas the traditional Japanese downtown is socio-petal in nature.

2.3. The Domain of "The Public"

In deliberation on the Japanese space concept and its manifestation in the form of address systems in cities, Funahashi (1996) contrasted the district-node type in Japanese cities with the path-landmark type (street-based system), common in most planned cities in the west. Ever since Lynch's (1960) seminal work on "The image of the city," legibility afforded by a well-structured visual image of the city is taken as a basic condition of a good city. Such a city is easy to grasp conceptually and provides better accessibility to a wider range of visitors and residents: it is for the service of the public in

Figure 3. Road pattern of D-town (in Hiroshima city) in the pre-renewal state.

general. The Japanese cities, according to Funahashi (1996), with their maze-like road systems and lack of logical structure in address codes, are created so as to do service to long-time residents: local logic understandable only by the inner-circle are dominant here.

Despite the apparent confusion and difficulties experienced by outsiders, the plans of the Japanese cities are designed to provide temporal and contextual qualities, with the gradual unfolding of scenes as a person goes deeper and gets acclimated with the inner culture. Roads in such a plan are not purely public domain but intermediate or buffer-space between the outside and inside (Funahashi, 1996). A Japanese architect, Kurokawa (1983), also epitomized the "philosophy of roads" as a guiding principle of traditional Japanese architecture in which intermediate or semi-private regions and the fusion of functions are actualized. Maki (1992), another influential architect and theoretician, proposed a concept of "*oku* (inwardness)" as uniquely Japanese spatial scheme and deep structure of collective conceptions of dwelling among Japanese people. Such a spatial concept is in contrast to the "philosophy of plaza (agora)"

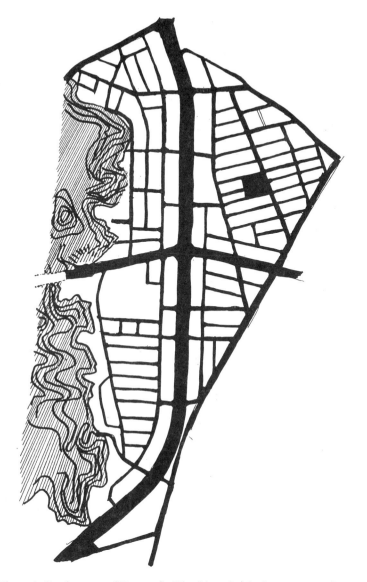

Figure 4. Road pattern of D-town (in Hiroshima city) in the post-renewal state.

and concept of centrality in the west, in which public domain is clearly identified and separated from private domains.

The inefficiency of urban planning in Japanese society is often attributed to the lack of public consciousness by which rational regulations and constraints of private interests are attained through public debate and political negotiations. The exclusion of intermediate space and its replacement by "public" space, however, may not result in enriched urban environments in the cultural context of Japanese cities. Thus, the sterility of many "new towns," which are created by principles of modern urban planning, and the logic behind the separation between public and private regions may be understood as due to mis-match between the design model and the socio-cultural context. Further conceptual and empirical investigation into the "public domain" is needed to examine the cultural-appropriateness of certain E-B concepts and design models.

3. DISCUSSION AND CONCLUSION

3.1. Pack-Donkey's Way and Man's Way as Two Cultural Norms

The central theme and task given to each author in this volume is to clarify one's own assumptions in his or her research activity and practice. Using Gadamer's (1976) expression, such a task is philosophical in its most authentic sense, that is, to "give an account of all his prejudices and all his self-evident assumptions." This laborious task is important since "his 'Sitz im Leben' is determined by this requirement by his own unique act."

While examining cultural assumptions underlying E-B research and urban planning practices in the previous section, the contrast was made between traditional Japanese dwellings together with their naive concepts, and ideas introduced by modern urban planning and Environment-Behavior research, most of which are western in origin. This contrast based on nationality and region, however, may not be appropriate. These should rather be viewed as different sets of cultural norms, which can manifest in various localities and personalities.

Using Le Corbusier's (1929) impressive metaphor these norms can be epitomized in his two ways: the pack-donkey's way and man's way. Apparently, Le Corbusier and his followers prefer man's way, because it is more rational, efficient, in control, and makes better sense. The present authors, while recognizing the necessity to use rational and linear-type planning as means to solve natural and social problems embedded in urban life, pose the question of what if the other way might have a complementary role in making not "man's" but "human's" or a "humane" environment.

Instead of using an animalistic metaphor, we may better term the other way as "life-world design (LWD)," in contrast to urban planning. LWD considers cities or human dwellings in general more in organismic and contextual terms rather than mechanistic terms (Pepper,1942; Altman & Rogoff, 1987). The city is a living entity (organism) and thus, is born, grows and ages. A significant part of it consists of life-world aspects, i.e., "culturally transmitted and linguistically organized stocks of interpretive pattern" (Harbermas, 1981). As such, a city as life-world is something to be understood: communicative action rather than operative control becomes salient mode of transaction. Some characterization, albeit very tentative, of the two modes of urban and town design is provided in Table 1.

Table 1. Two Forms of Cultural Assumptions Underlying Urban and Town Design

	"Man's Way" (Urban Planning)	"Pack-Donkey's Way" (Life-world Design)
Logical Type	System	Life-world
World Hypothesis	Mechanism	Organicism & contextualism
Guiding Principle	Rationality	Vitality & historicity
Operative Mode	Control	Communication
Value Orientation	Individuality	Communality
Social Organization	Socio-fugal	Socio-petal
Action Mode	Planning	Participation

In accordance with characteristics of LWD, the principle means of investigation are more interpretive, using ethnographic and qualitative methodologies (Minami, 1997). The research is also incorporated with practice with its participatory paradigm. Participatory action research (Meteo, 1994), thus, is seen as an ideal research strategy from this perspective.

3.2. Transactionalism in Its Multiple Manifestations

While our reflections on the cultural assumptions presumably working in a current E-B research and urban planning took the form of a dichotomy, such as "vitalism vs. rationalism," "socio-petal vs. socio-fugal biases," and "system vs. life-world," our intention does not lie in strengthening dichotomous thinking in this field. As always the case, the point is in the relative emphasis or selective attention directed toward either pole of the continuum in the awareness of people in different socio-cultural milieu.

Some aspects of vitalism, for example, are evident in the organicist's world assumptions (Altman & Rogoff, 1987), in the organismic-developmental perspective, and its recent modifications (Wapner, 1987). The needs for more density in central city areas are also proposed by keen observers of city life, such as in the typical American city of New York (White, 1988). The renovation and re-activation of downtown areas in most large cities in the U.S. have been central concerns of city developers. The activation of communal life through seasonal festivals and the temporal unfolding of scenes is the very quality that Altman and Rogoff (1987) have conceptualized in their account of "transactionalism." Current debates on "new urbanism" also cast new light on the present issue (Brown, Burton, & Sweaney, 1998).

We believe that the awareness of such counter-effects from the implementation of some of the basic E-B concepts and principles of urban planning is heightened by the contextual and culture-specific investigation of person-environment transactions. The basic premise guiding such an investigation is that there are multiple worlds which have multiple goals of optimal person-environment fit (Wapner, 1987) and multiple means in the cultural stock of spatial organization and social regulative devices. Our task is to find ways of conceiving generalizable (not universal) principles out of local contexts of human dwelling in its multiple manifestations.

REFERENCES

Altman, I., & Rogoff, B. (1987). World views in psychology: Trait, interactional, organismic, and transactional perspectives. In D. Stokols & I. Altman (Eds.), *Handbook of environmental psychology* (pp. 7–40). New York: Wiley.

Brown, B. B., Burton, J. R., & Sweaney, A. L. (1998). Neighbors, households, and front porches: New urbanist community tool or mere nostalgia? *Environment & Behavior, 30*(5), 579–600.

Funahashi, K. (1996). Living in towns: legibility and the structure of cities. In Nakajima, T. & Onho, R. (Ed.), *Psychology of dwelling*. Tokyo: Shokoku-sha Publishing Co.

Gadamer, H-G. (1976). *Philosophical hermeneutics*. (Translated and edited by D. E. Linge), Berkeley: University of California Press.

Hagino, G., Mochizuki, M., & Yamamoto, T. (1987). Environmental psychology in Japan. In D. Stokols, & I. Altman (Eds.), *Handbook of environmental psychology* (pp. 1155–1170). New York: Wiley.

Harbermas, J. (1981). *The theory of communicative action*. (Vol. I. II). Boston: Beacon Press.

Heider, F. (1958). *The psychology of interpersonal relations*. Hillsdale, NJ: Lawrence Erlbaum Associates, Inc.

Kurokawa, K. (1983). *Michi no Kenchiku: Chukan Ryoiki e (The architecture of roads: Toward intermediate region)*. Tokyo: Kinokunia Publishing Co.

Le Corbusier (1929). *The city of tomorrow and its planning*. New York: Payson & Clarke Ltd.

Lynch, K. (1960). *The image of the city*. Boston: The MIT Press.

Maki, F. (1997). *KIOKU NO KEISHOU (Images of memory)*. Tokyo: Chikuma Shobou.

Meteo, M. (1994). Participatory action research within community: The interface of ordinary and scientific knowledge. Keynote address at the 23rd International Congress of Applied Psychology, Madrid, July 17 to 23, 1994.

Minami, H. (1997). Urban renewal and the elderly: An ethnographic approach. In S. Wapner, J. Demick, T. Yamamoto, & T. Takahashi (Eds.), *Handbook of Japan-united States environment-behavior research: Toward a transactional approach*. (pp. 133–148). New York: Plenum Press.

Minami, H., & Tanaka, K. (1995). Social and environmental psychology: Transaction between physical space and group-dynamic processes. *Environment & Behavior, 27,* 43–55.

Osmond, H. (1957). Function as a basis of psychiatric ward design. *Mental Hospital, 8,* 23–29.

Pepper, S. C. (1942). *World hypothesis: A study in evidence*. Berkeley: University of California Press.

Sampson, E. E. (1989). The challenge of social change for psychology: Globalization and psychology's theory of the person. *American Psychologists, 44*(6), 914–921.

Sommer, R. (1967). *Personal space: The behavioral basis of design*. Englewood Cliffs, NJ: Prentice-Hall, Inc.

Sommer, R. (1997). Benchmarks in environmental psychology. *Journal of Environmental Education, 17*(1), 1–10.

Stigler, J. W., Schweder, R. A., & Herdt, G. (1990). *Cultural psychology: Essays on comparative human development*. Cambridge: Cambridge University Press.

Triandis, H. C., Bontempo, R., Villareal, M. J., Asai, M., & Lucca, I. (1988). Individualism and collectivism: Cross-cultural perspective on self-group relationships. *Journal of Personality and Social Psychology, 54*(2), 323–338.

Wapner, S. (1987). A holistic, developmental, systems-oriented environmental psychology: Some beginnings. In D. Stokols & I. Altman (Eds.), *Handbook of environmental psychology* (pp. 1433–1465). New York: Wiley.

White, W. H. (1988). *City: Rediscovering the center*. New York: Doubleday.

Yamamoto, T. (1992). *Toward a creation of new paradigm of life culture*. Tokyo: Fuji Social Education Center.

20

RESIDENTIAL CROWDING IN THE CONTEXT OF INNER CITY POVERTY

Gary W. Evans[1] and Susan Saegert[2]

[1] Department of Design & Environmental Analysis
Cornell University
Ithaca, NY 14853-4401
[2] Graduate Center CUNY
Environmental Psychology
New York, NY 10036

1. INTRODUCTION

The accumulation of stressful experiences leads to increased risk for adverse outcomes as the adaptive capabilities of the organism are increasingly challenged. If this proposition is correct, then an important omission in environmental stress research has been the examination of crowding, noise, pollution, heat, poor housing quality, and other physical stressors in isolation of the naturalistic contexts in which they normally occur (Bronfenbrenner, 1979; Cohen, Evans, Stokols, & Krantz, 1986; Petrinovich, 1979). This chapter examines residential crowding from an ecological perspective. Crowding is one among several, adverse environmental conditions more likely to occur for poor households and among ethnic minority families. The concentration of low income households in the inner city brings with it heightened exposure to environmental stressors. Housing in the inner city is often both relatively expensive and small. Those with limited financial resources must compromise on the size as well as the quality of their housing. Moreover, low income, minority families often must accept housing in less desirable neighborhoods characterized by poor physical conditions such as noise, pollution, physical decay, as well as cope with inadequate public services. These constraints on residential choice are magnified for ethnic minorities by housing discrimination or the need to be proximate to others who share a common language and culture.

The confluence of many poor families in distressed neighborhoods further exacerbates the stressful aspects of the environment by contributing to stressful social

Theoretical Perspectives in Environment-Behavior Research, edited by Wapner *et al.*
Kluwer Academic / Plenum Publishers, New York, 2000.

conditions ranging from high crime rates to large numbers of single parent households whose resources for supervising children and keeping up their own households are stretched thin. These community level stressors put pressure on the household as a unit and on individual members. The results of community stressors plus the strains on households and individuals living in poverty can lead to more problematic social interactions within the household and poorer mental and physical health among its members. These poverty-related characteristics of households and individuals are in themselves also stressful.

Various psychosocial stressors including financial pressure, unemployment, family turmoil, family separation and loss, death and serious illness, and exposure to violence, characterize both the community and the daily lives of low income, inner city families (Garbarino, Durbrow, Kosteleny, & Porter, 1992; McLoyd, Jayaratne, Cebello, & Boquet, 1994; McLoyd, 1990; 1998; Richters & Martinez, 1993; Rutter et al., 1974; Wilson, 1987; Work, Cowen, Parker, & Wyman, 1990). The immediate neighborhoods of low income, urban children are often marked by characteristics that contribute to additional demands upon families. These neighborhood qualities include high mobility, low occupancy rates, abandoned buildings and grounds, greater numbers of commercial and industrial facilities, and inadequate municipal services and amenities (police, fire, schools, parks, and other recreational facilities) (Mc Loyd, 1998; Taylor, Repetti, & Seeman, 1997; Wandersman & Nation, 1998). Low income, minority inner city families are also more likely to live amidst poor environmental conditions including crowding and substandard housing; to be exposed to elevated noise levels from various sources (vehicular traffic, airports, factories, neighbors); to have less ability to regulate the immediate climate of their residences (temperature and humidity); and to suffer elevated exposure to noxious pollutants and allergens including lead, smog, particulates, and dust mites (Evans, in press; Taylor et al., 1997).

Poor, ethnic minority children living in inner city neighborhoods experience substantially more environmental and psychosocial stressors related to poverty in comparison to the more frequently studied white suburban samples (Attar, Guerra, & Tolan, 1994; Brown & Cowen, 1988; Dubow, Tisak, Casuey, Hryshko, & Reid, 1991; Gad & Johnson, 1980). Children in lower SES homes are exposed to more chaotic and unpredictable environments than their more economically advantaged counterparts (Matheny, Wachs, Ludwig, & Phillips, 1995). This confluence of psychosocial and physical stressors in the immediate and surrounding environments of inner city families may compound the effects of environmental stressors like residential crowding on children's well being. Yet research on environmental stressors typically examines their effects on children and families with methodological controls for socioeconomic status. If we are correct in hypothesizing that the adverse impacts of environmental stressors are magnified when they intersect with the myriad of physical and psychosocial stressors accompanying inner city poverty, then current estimates of the potential psychological consequences of exposure to environmental stressors may substantially underestimate their true impact.

This chapter is organized into four major sections. We begin by explaining why crowding in the context of the multiple stressors accompanying inner city poverty is likely to have more negative effects on children and families than when crowding is examined in isolation of its ecological context. We then review prior research on crowding, poverty-related stressors, cumulative risk factors in human development, and multiple stressor exposure in adults. As a science, the field of environment and behavior has the opportunity, indeed the obligation, to empirically examine basic assumptions, whenever feasible, that lie at the foundations of major theories, models, and frameworks. Thus the third section of the chapter provides some preliminary data on the multiplicative effects of residential crowding and inner city stressors on developmental outcomes. We investigate a small sample of ethnic minority children who reside in public housing located in the inner city. Finally, we summarize several important conceptual and methodological assumptions embedded in research on environmental stress and psychological health.

2. CROWDING IN THE CONTEXT OF POVERTY

There are several reasons why crowding in the context of inner city stressors is expected to produce more dysfunctional outcomes among families and children than has been reported heretofore. First, stressors are most likely to interact when they are chronic and intractable (Lepore & Evans, 1996). Families in poverty are unfortunately deeply rooted in a plethora of uncontrollable, chronic demands. Residential crowding shares these characteristics.

Second, in addition to their chronicity and intractability, many of the stressors accompanying poverty, including high residential crowding, occur at higher intensities and across a wider range than typically found in the often-used samples of college undergraduates or community studies of middle class residents. An assumption underlying environmental stressor research specifically, and the field of environment and behavior more generally, is that the extent of presumed variance in environmental conditions is adequate to detect effects. Range restriction truncates estimates of association (Ghiselli, Campbell, & Zedeck, 1981). As an example of the potential role of poverty to influence the extent of stressor exposure, we uncovered a range of household density from 11 people per room to 2 people per room among an urban sample in India (Evans, Palsane, Lepore, & Martin, 1989). This range is more than double any reported statistics on residential density among North American community studies, where nearly all of the empirical research on crowding has been conducted. Its also worth noting that the correlates of residential density uncovered in the India study were substantially greater than those typically reported in the North American literature.

Third, stressors that draw upon the same coping resources because of parallel demands are more likely to produce multiplicative, adaptive demands on the human organism (Lepore & Evans, 1996). In the second section of this chapter we review findings suggesting several parallel outcomes of crowding and inner city stressors on families and children.

A fourth reason for synergistic effects of crowding and the psychosocial and environmental stressors accompanying inner city poverty may arise from shared proximal processes. Proximal processes are the underlying engines of human development. They consist of: "reciprocal interaction between an active, evolving biopsychological human organism and the persons, objects, and symbols in its immediate environment. To be effective the interaction must occur on a fairly regular basis over extended periods of time with progressively more complex interactions of person and environment" (Bronfenbrenner & Morris, 1998, p. 996).

One critical proximal process for early childhood development appears to be responsive parenting offered in a warm, supportive manner (Bronfenbrenner, 1979; Bronfenbrenner & Morris, 1998). Below we also provide evidence that both residential crowding and poverty, respectively, coincide with less responsive, harsher parent-child interactions.

2.1. Parallel Outcomes and Shared Proximal Processes: Crowding and Poverty

A primary reason why residential crowding and inner city poverty may have multiplicative, adverse effects is because they exhibit parallel developmental outcomes. Furthermore, crowding and some of the stressors accompanying inner city poverty may share a common, underlying proximal process of unresponsive, harsh parent-child interaction.

2.2. Crowding

Psychophysiological stress is associated with crowding. Elevated blood pressure is correlated with crowded living conditions (D'Atri, 1975; Evans, Lepore, Shejwal, & Palsane, 1998; Paulus, McGain, & Cox, 1978). Blood pressure and indices of sympathetic arousal are increased by laboratory exposure to crowding (Aiello, Epstein, & Karlin, 1975; Aiello, Nicosia, & Thompson, 1979; Evans, 1979). Congested commuting conditions and crowded shopping experiences are associated with elevated neuroendocrine hormone activity (Evans & Carrere, 1991; Heshka & Pylypuk, 1975; Lundberg, 1976).

Many studies have explored linkages between crowding and psychological health. Interior, residential crowding has been associated with strained social relationships (McCarthy & Saegert, 1979; Evans et al., 1989; Evans & Lepore, 1993; Lepore et al., 1991), and higher levels of psychological distress (Evans et al., 1989; Gove & Hughes, 1983; Lepore et al., 1991).

There is also evidence linking crowding to cognitive development. Residential crowding is associated with behavioral problems in school (Booth & Johnson, 1975; Evans et al., 1998; Maxwell, 1996; Saegert, 1982) and with reading deficits (Murray, 1974; Saegert, 1982; Wedge & Pitzing, 1970). Residential crowding has also been linked to delayed, infant cognitive development (Wachs & Gruen, 1982).

Crowding has also been correlated with motivational problems, possibly indicative of learned helplessness. Reduced persistence on challenging puzzles follows experimental exposures to crowding (Evans, 1979; Sherrod, 1974). Multiple indices of helplessness have also been associated with longterm residential crowding (Baum, Aiello, & Calesnick, 1978; Baum, Gatchel, Aiello, & Thompson, 1981; Evans et al., 1998; Rodin, 1976).

Developmental outcomes of residential crowding may be mediated by proximal processes related to parent-child interaction. Residential crowding challenges the adaptive resources of parents which may, in turn, influence their interactions with their children. Crowding in the home may erode family social support resources. Higher density is related to more family conflict and less parental supervision of children (Booth & Edwards, 1976; Gove & Hughes, 1983; Mitchell, 1971); with feelings of anger and irritation among young children (Saegert, 1982); and with diminished parental responsiveness (Bradley & Caldwell, 1984; Evans, Maxwell, & Hart, 1999; Evans et al., 1998;

Wachs, 1989). Residential crowding also erodes social support among adult family members (Evans et al., 1989; Evans & Lepore, 1993; Lepore et al., 1991).

2.3. Inner City Stressors

Psychophysiological stress has been relatively neglected as an outcome of the multiple stressors experienced by inner city residents. However the stress literature does suggest that cardiovascular processes would be negatively affected by many of them. Children respond to angry encounters among adults with heightened cardiovascular activation (El-Sheik, Cummings, & Goetsch, 1989). Cardiovascular reactivity to laboratory challenges (e.g., mental arithmetic) is elevated among young children facing ongoing, background stressors such as family disruption (Matthews, Gump, Block, & Allen, 1997). These reactivity studies have not examined inner city stressors per se, instead examining relations between degree of chronic stressor exposure and reactivity. Children's cardiovascular reactivity may be sensitive to family social climate. Boys from families with more authoritarian parenting and less socially supportive families, two characteristics related to high stress, low income families, have elevated cardiovascular reactivity (Woodall & Matthews, 1989).

Studies relating psychological health to inner city conditions are more common. Inner city, black children have greater psychological distress relative to more affluent white and black children (Children's Defense Fund, 1991; Edelman, 1985). Family income is negatively associated with behavioral disturbance (Attar et al., 1994; Duncan, Brooks-Gunn, & Klevanov, 1994). Children perceiving greater family economic hardships have more distress and anxiety plus lower self esteem (Mc Loyd et al., 1994). Lower SES preschool and elementary school children manifest more behavioral problems and aggressive behaviors at school than middle income children (Dodge, Pettit, & Bates, 1994; Guerra, Huesmann, Tolan, Van Acker, & Eron, 1995). Similar trends have been noted among adolescents (Conger, Conger, Elder, Lorenz, Simons, & Whitbeck, 1994).

Outcomes of exposure to violence among young children include anxiety and fear, re-experiences of violent episodes, disrupted sleeping and eating, and interpersonal relationship difficulties (Garbarino et al., 1992; Martinez & Richters, 1993; Osofsky, 1995; Richters, 1993; Richters & Martinez, 1993). Experiences of violence appear to be related as well to aggressive behavior in children and increased probability of violent behavior as they grow older (Garbarino et al., 1992; Sampson, 1992).

Lower levels of competence whether measured by academic achievement or behavioral skills, indicate that the multiple disadvantages experienced disproportionately by minority inner city children are associated with lower levels of competence (Gibbs, 1984; Mc Loyd, 1990). Studies often show lower levels of academic achievement and lower social competency as assessed by teachers and peers for poorer, more disadvantaged children but tend to focus more on potential protective factors to account for variance in outcomes otherwise attributed to disadvantage (Garmezy, Masten, & Tellegen, 1984; Masten et al., 1987; Wyman, Cowen, Work, & Parker, 1991). One exception is a study by Duncan et al. (1994) reporting that IQ scores for 5 year olds were more affected by income than by living in a female headed household.

The term learned helplessness is infrequently used in the literature on inner city stressors. However conceptually related phenomena are discussed, particularly in the violence research (Osofsky, 1995). One impact of early exposure to violence has been termed loss of meaning. In addition children living in violent communities suffer from "severe constrictions in activities, exploration, and thinking, for fear of re-experiencing the traumatic event." (Garbarino et al., 1992; p. 92).

Residents of low income neighborhoods are constantly assaulted by physical evidence of neighborhood neglect and decay, including dilapidated housing, abandoned buildings, vandalism and graffiti, and litter. Moreover the occurrence of illicit social activities (e.g., drug trafficking, gang activity) may compound a collective sense of anomie and hopelessness. This combination of physical and social elements in many low income neighborhoods leads to excessive fear of crime and feelings of despair (Wandersman & Nation, 1998).

Some studies of poverty have examined proximal processes that may help us to understand how the child's daily experiences of poverty could lead to negative, developmental outcomes. More authoritarian parenting and lack of responsiveness have been associated with unemployment, single parenthood, and neighborhoods low in social capital (Kaslow, Deering, & Rascusin, 1984; Lewinsohn, Roberts, Seeley, Rohde, Gotlib, and Hops, 1994; Mc Loyd, 1990; 1994; Tulkin, 1977). Income is significantly related to family cohesion and conflict in rural, black families (Brody, Stonemant, & Flor, 1996). Learned helplessness and proximal processes may be inter-related as well. Lower SES mothers report less self efficacy with respect to their abilities to influence their children's development (Tulkin, 1977).

There is more direct evidence of nurturant parenting as a proximal process that accounts for some of the harmful effects of poverty upon children. Economic hardship that causes paternal rejection of daughters sometimes accounts for girls' emotional adjustment problems (Elder, Nguyen, & Caspi, 1985). The degree of nurturant parenting by mothers and fathers is an important pathway linking poverty to adolescents' emotional well being (Conger et al., 1992; Conger, Ge, Elder, Lorenz, & Simons, 1994; Mc Loyd et al., 1994). Younger children also appear sensitive to more coercive, less nurturant parenting practices associated with poverty (Dodge et al., 1994).

2.4. Summary

One line of reasoning that leads to the prediction of multiplicative effects of crowding and inner city stressors on children's psychological well being is shared outcome. Crowding and the community and household stressors associated with life in inner city neighborhoods adversely affect psychological distress and cognitive competency. Crowding elevates psychophysiological stress and is associated with motivational deficits related to learned helplessness. The picture is less clear with respect to inner city stressors and either psychophysiological stress or motivation, although suggestive trends are evident.

Although crowding, inner city stressors, and their predicted multiplicative effects may directly impact children, it is worth considering the potential for synergism because of shared, proximal processes that underlie the interactive effects. As indicated above, exposure to both residential crowding and inner city stressors are linked to more restrictive, authoritarian parenting practices as well as less nurturant, responsive parent-child interaction. Thus not only do residential crowding and inner city stressor exposure share several common developmental outcomes, they also appear to have in common a potent, underlying proximal process–parental responsiveness to children. This leads to the prediction that the hypothesized, multiplicative effects of poverty and crowding on physiological, socioemotional, and cognitive developmental outcomes will be mediated by parent-child interaction.

3. CUMULATIVE RISK AND MULTIPLE STRESSORS

The other major reason we believe inner city stressors may exacerbate the harmful effects of chronic crowding on children and families is because of prior research documenting the cumulative impacts of multiple risk factors in children as well as other research verifying the potential for physical and psychosocial stressors to interact.

3.1. Cumulative Risk

According to research on cumulative risk factors among children, as the total number of stressors or risk factors increases, so do negative outcomes. Risks typically include marital discord, low social status, overcrowding, paternal criminality, maternal psychiatric status, and admission into care of local authorities (Barocas, Seiffer, & Sameroff, 1993; Richters & Martinez, 1993; Rutter, 1983; Sameroff, Seiffer, Baldwin, & Baldwin, 1993; Sameroff, Seifer, Barocas, Zax, & Greenspan, 1987). This literature reveals: 0–1 risks typically show little impact; whereas with exposure from 2 to 4 or 5 risks adverse effects accumulate, approximately linearly. Effects from additional risk exposures appear to asymptote. At a certain point children may become so depleted of coping resources that further risk accumulation causes no further damage.

A handful of studies have examined the potentiating role of poverty or associated stressful conditions on responses to other stressors. Although this literature is very small, trends suggest elevated vulnerability to stressors as a function of conditions associated with poverty. For example, the harmful effects of multiple, maternal hospital admissions on young children's psychological health are significantly greater among children living in families with greater interpersonal conflict (Rutter, 1983). Attar et al. (1994) showed that children from low versus moderate quality neighborhoods (defined in terms of income, welfare, housing, and violent crime) who also had experienced more life events were more aggressive concurrently and one year later.

3.2. Multiple Stressors

There are a few studies indicating multiplicative, interactive effects of psychosocial stressors with the physical environment, including crowding. One approach has examined the interaction of noise and workload on stress. Under acute noise, psychophysiological responses rapidly habituate, however this pattern can be interrupted by increased workload demands (Carter & Beh, 1989; Conrad, 1973; Ising, Dienel, Gunther, & Markert, 1980; Mosskov & Ettema, 1977). Some analogous data are available in the field. Cardiovascular morbidity is not clearly linked to occupational noise exposure (Evans, in press). However among workers with higher cognitive workload, this relation is manifested (Welch, 1979). Noise in concert with demanding tasks has significantly greater health impacts than noise exposure alone.

Another paradigm has examined the interactive effects of physical and psychosocial stressors. Evans, Allen, Tafalla, and O'Meara (1996) conducted two studies on noise and psychosocial stressors. In one study, noise induced significantly larger elevations in blood pressure and decreases in task motivation when it followed a short period of acute stress (giving a speech) relative to a no stress, control condition. A second study replicated these interactive noise and psychosocial stressor findings but using a naturalistic stressor (final examinations). Fleming, Baum, Davidson, Rectanus, and McArdle (1987) found that cardiovascular reactivity to a challenging mental task was exacerbated by chronic neighborhood crowding.

Additional evidence concordant with our ideas about the impacts of psychosocial stressors associated with poverty and environmental stressors emanates from studies of psychosocial stressors at work and in the community. Cesana and colleagues (1982) found that the adverse impacts of noise on neuroendocrine functions were elevated by shift work. Investigators have also uncovered positive interactions between job stress and noise on blood pressure (Cottington, Matthews, Talbottt, & Kuller, 1983; Lercher, Hortnagl, & Kofler, 1993). Job stress also amplified the impacts of toxic fumes on physical symptoms among laborers in a rubber plant (House, McMichael, Wells, Kaplan, & Landerman, 1979).

Turning to the community, Caspi and colleagues (1987) found that poor neighborhood quality exacerbated the relation between stressful daily events (e.g. argument with spouse) and both concurrent daily negative mood as well as mood on the subsequent day. The influence of daily events on negative mood was double for those living in poor versus good neighborhood conditions. In two studies of air pollution and mental health, Evans, Jacobs, Dooley, and Catalano (1987) found that smog had little impact on the greater Los Angeles population but was significantly increased psychological distress among the subset of Los Angeles residents who had also recently experienced one or more major stressful life event.

One article warrants more in depth description within the context of this chapter. Lepore, Evans, and Palsane (1991) examined residential density and some of the chronic strains associated with poverty. In one study the positive relation between density and psychological distress was amplified among male heads of households in India who had also experienced social hassles in their home. Respondents indicated how often in the past two months they had experienced social hassles in their home as indicated, for example, by quarrels, heavy family responsibilities, or overnight guests. The interactive effects of residential crowding and social hassles in the home were replicated in a prospective, longitudinal study of male and female, American college students living in off campus apartments. These two crowding studies, although with adults and restricted to self-report measures, are in accord with our central thesis. The impacts of crowding on well being may be greater among low income families than among middle income families.

3.3. Summary

Two types of paradigms have been used to study the inter-relations among physical and psychosocial stressors with human well being. These paradigms were not derived to examine the central issue of this chapter, the contextual implications of poverty for research on environmental stressors like crowding. Nonetheless the child cumulative risk studies and the multiple stressor studies are conceptually analogous to our aim since they investigated whether exposures to multiple social and environmental demands could elevate dysfunction. Although both the cumulative, development risk and multiple stressor literatures, respectively are small, several trends are in accord with our hypothesis that the negative effects of chronic residential crowding on families and children may be amplified by the myriad of psychosocial stressors experienced by low income, inner city families.

4. CROWDING AND INNER CITY STRESS: A PRELIMINARY STUDY

We designed a preliminary study to investigate the basic, underlying theoretical premise presented in this chapter. This theoretical premise consists of three propositions. One, families living in poverty will manifest stronger relations between environmental stressors, like crowding, and well being relative to similar relations uncovered among middle and high income populations. Two, researchers interested in environmental stressors and well being have underestimated their impact on children and families. This has happened because investigators have isolated the effects of specific environmental stressors and examined their singular impact with statistical controls for socioeconomic and demographic conditions. Three, the accumulation of exposure to residential crowding and the multiple stressors accompanying inner city poverty will influence common proximal processes related to parent-child interaction. Thus we would expect the interactive effects of crowding and inner city stressors to be mediated by parenting practices.

The present preliminary investigation examines propositions two and three directly. To truly address proposition one, a sample of low and middle or high income children would be necessary. Herein we examine a small sample of ethnic minority, inner city families, living in public housing in East Harlem. We focus our investigation on the potential multiplicative consequences of household crowding and multiple, inner city stressors on psychological health, psychophysiological stress, and intellectual competency. We also investigate the role of parenting practices in this system. As depicted in Fig. 1, we expect inner city stressors and residential density to interact, with the effects of density on stress amplified among families exposed to more inner city stressors. We also hypothesize that this multiplicative effect of density and inner city stressors on developmental outcomes will be mediated by parenting practices.

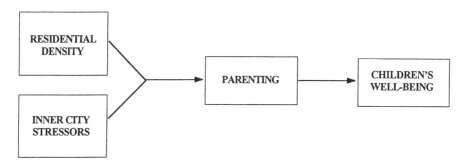

Figure 1. Schematic diagram of the underlying hypothesis. The interactive relations between residential density and inner city stressors with children's well being are mediated by the proximal process of parent-child relationships.

4.1. Method

4.1.1. Participants. The sample was recruited by word of mouth and flyers in several large, public housing project in East Harlem, a predominantly low income, African American and Hispanic neighborhood in New York City. The participants were 19 girls and 21 boys in grades three and four ($M = 9.8$ years of age) who were African American (66%) or Latino (34%). The majority of the households were headed by single women (68%) with 68% having a high school education or less. The median income level of households was $12,250; 76% were unemployed; and 68% received some type of public assistance.

4.1.2. Procedures. Data were collected in two home visits by trained graduate students. The data were collected independently in one-on-one interviews with the child and his/her mother, respectively. Only one child per household was a participant in the study.

Density was determined by dividing the number of rooms by the number of people residing in the home. Anyone staying overnight three or more nights per week was considered a resident. Because the sample were all from the same public housing development, determination of room counts was straightforward since every apartment had one kitchen, one living room, and a variable number of bathrooms and bedrooms.

As an index of exposure to inner city stressors, we used the Life Events and Circumstances Checklist (LEC) (Work, Cowen, Parker, & Wyman, 1990; Wyman, Cowen, Work, & Parker, 1991). This 32 item checklist describes diverse stressful experiences endemic to poverty. It includes acute events (e.g., "a close family member died") but primarily consists of chronic circumstances (e.g., "your child has witnessed serious family arguments"). The parent reports, "yes" or "no", whether events or circumstances have occurred in the child's life. The scale has undergone extensive psychometric development and appears to well represent the array of common stressors that occur among low income, inner city populations in the United States.

There are various ways one could attempt to model exposure to inner city stressors associated with poverty in order to study its multiplicative effects with crowding on children's well being. We examined inner city stressor exposure in several ways. First, we counted the total number of stressors experienced by the child. Although this count has some direct, main effects on several outcome variables, it did not interact with density to predict developmental outcomes. The total count of stressful life experiences from the LEC obscures the number of different areas or domains in which stress has been experienced. Thus in a second approach we examined the different domains of stressors embedded in the LEC. The LEC has five factors which include family turmoil, poverty, exposure to violence, illness/death, and family separation (Work et al., 1990; Wyman et al., 1991). The number of factors (0–5) children are exposed to would seem like the most straightforward metric for indexing multiple stressor exposure associated with poverty. To our surprise, however, inspection of the data revealed little variance in such an index. Perhaps indicative of the extremely challenging conditions of living in public housing in East Harlem, nearly all children (83%) had been exposed to four or all five of the inner city stressor domains. All of the children in our sample were poor; all had witnessed violence; 93% had experienced family separation; and 90% had experienced death or serious illness in a close relative.

Therefore we decided to focus our analyses on family turmoil as a measure of inner city stressor exposure that might interact with residential density. It was the only inner city stressor domain with a reasonable amount of variance. Recall that another reason inner city stressors are posited to interact with residential crowding to exacerbate negative developmental outcomes is because of the greater intensity and range in these stressors experienced by inner city families. While our assumption about intensity was clearly borne out by the data, we were wrong with respect to our expectations about a broad range of exposure to various domains of inner city stressors. With the exception of family turmoil, inner city African American and Latino families living in public housing in East Harlem appear to have uniformly high levels of inner city stressor exposure.

Conceptually the stressor domain of family turmoil also seems highly salient to residential density. As noted above a major reason why stressors may have multiplicative impact is because they operate via one or more shared, proximal processes to influence development. Crowding provokes interpersonal conflict and interrupts ongoing family life. The Family Turmoil subscale of the LEC consists of seven questions (yes/no) about issues such as the child experiencing frequent residential relocations, exposure to upsetting family arguments, and having a close family member with an alcohol or drug problems. A simple count of yes responses (0–7) is employed as the index.

Four measures were used to index developmental outcomes and one to evaluate the hypothesized proximal process of parenting. To index psychological distress among children we employed the Children's Behavior Questionnaire (CBQ). This 26 item scale asks parents to rate on a three point continuum (0 = doesn't apply, 1 = applies somewhat, 2 = certainly applies) a list of common childhood symptoms indicative of behavioral conduct disorders (e.g., bullies other children, has stolen things on one or more occasions) as well as symptoms of depression and anxiety (e.g., often worries, worried about many things; often appears miserable, unhappy, tearful or distressed). The CBQ has undergone extensive psychometric development including measures of test-retest reliability, inter-rater reliability, and was found to be internally consistent for the present sample (a = 0.83). The CBQ discriminates between normal and psychiatric outpatients and correlates well with clinical diagnosis by child psychiatrists. See Rutter, Tizzard, and Whitmore (1970) as well as Boyle and Jones (1985) for further details.

Psychophysiological stress was indexed by neuroendocrine measures. Overnight, 12 hour urine collection was completed and kept refrigerated with a preservative. The volume of urine was measured, small samples extracted, and deep frozen (–70 C). The catecholamine samples were also pH adjusted to reduce oxidation. High performance liquid chromatography with electrochemical detection (Riggin & Kissinger, 1977) was used to estimate the amount of epinephrine and norepinephrine. Creatinine was also assayed to help control for differences in body mass and incomplete urine voiding. Urinary cortisol was also assayed but only marginally related (although in the predicted direction) to the interaction of density and family turmoil, so is not included herein. Urinary neuroendocrine levels are valid indicators of chronic stress (Baum & Grunberg, 1995; Grunberg & Singer, 1990; Lundberg, 1985).

Intellectual competence was assessed with the Scholastic Competence subscale of Harter's Competency scale (Harter, 1982). Children indicate in a forced choice format which of two, bipolar behavioral descriptions are really true or sort of true of them. Sample items on the Scholastic Competency subscale include: "Some kids are really good at their school work." vs. "Other kids worry about whether they can do the school work assigned to them." "Some kids feel they are just as smart as other kids their age." vs. "Other kids aren't so sure and wonder if they are as smart." The subscale is reliable and exhibits a stable and coherent factor structure across several, heterogeneous sociodemographic samples (Harter, 1982).

A scale to measure parent-child interactions was developed from two, previously used scales. The first scale came from Conger, Ge, Elder, Lorenz, & Simons (1994) and was designed to assess parental hostility. The mother indicated how often hostile behaviors such as being angry, yelling at, criticizing, etc. occurred. The second scale is part of the National Survey of Families and Households and indexes of more overt manifestations of punitiveness including slapping, hitting, spanking, etc.) (Hashima & Amato, 1994). Both scales have proven reliable in prior research and differentiate parental styles between low and middle income parents. Both scales also predict negative developmental outcomes. We combined the two scales into one reliable index consisting of 12 items (1 = never − 4 very often) (a = 77).

4.2. Results and Discussion

Because of length restrictions and our interest herein on the multiplicative effects of residential crowding and family turmoil, we focus our analyses and discussion on the interaction of these two variables plus examine the role of parenting as a hypothetical mediator of the interaction (see Fig. 1). Multiple regression is used throughout to test for the interactive and mediational relations posited in Fig. 1. All of the predictor variables and the mediator variable are centered (x-M) to reduce multicollinearity (Aiken & West, 1991). The multiplicative term (Density Centered X Family Turmoil Centered) is entered into the regression equation after the main effects for each of the centered variables (density, family turmoil). A multiplicative effect is indicated by a significant increment in ΔR^2 for the interaction term. The regression plots are plotted across density levels at the mean and at one standard deviation above and below the moderator (family turmoil).

As can be seen in Fig. 2, family turmoil and density interact to predict children's psychological distress ($\Delta R^2 = 0.114$, $F(1,36) = 6.75$, $p < 0.014$). The relation between density and psychological distress is stronger as a function of the extent of family turmoil. Children in families with low levels of turmoil are largely unaffected by density. Similar patterns emerge for epinephrine ($\Delta R^2 = 0.147$, $F(1,29) = 5.02$, $p < 0.03$ (Fig. 3), norepinephrine ($\Delta R^2 = 0.267$, $F(1,29) = 11.17$, $p < 0.002$ (Fig. 4), and scholastic competence ($\Delta R^2 = 0.093$, $F(1,36) = 4.75$, $p < 0.036$ (Fig. 5). The degrees of freedom are smaller for the psychophysiological indices because we were unable to collect urine specimens from all children. As can be seen, the impacts of residential density appear greatest at high levels of family turmoil. At low levels of turmoil, density has little or no relationship to psychophysiological stress. Children's perceptions of their scholastic competence are also a joint function of residential density and family turmoil. Children living in crowded residences rate themselves as less academically competent relative to other children their own age. This trend is substantially elevated in households with high family turmoil.

Looking at these outcome measures together, a clear pattern is evident. Low income, minority children living in public housing are especially vulnerable to the negative effects of chronic residential crowding if they live in households with more family turmoil. It is also worth noting that given the small sample size and the low statistical power of interaction terms (Aiken & West, 1991), these multiplicative effects are quite strong. We believe the reason for this is because many of the dysfunctional outcomes associated with residential density on children occur because of negative alterations in

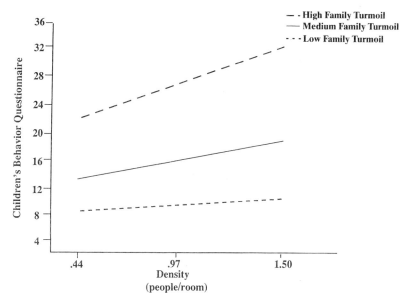

Figure 2. Relations between residential density and maternal ratings of children's psychological distress (Rutter Children's Behavior Questionnaire) as a function of family turmoil.

the day to day, social interactions of families in the home. The interruptive, intrusive nature of too much unwanted social interaction in crowded households is exacerbated among families that also suffer from other sources of disruption and instability. Chronic, uncontrollable stressors that place similar adaptive demands on the organism are prime candidates for multiplicative effects (Lepore & Evans, 1996). The level of chaos in fam-

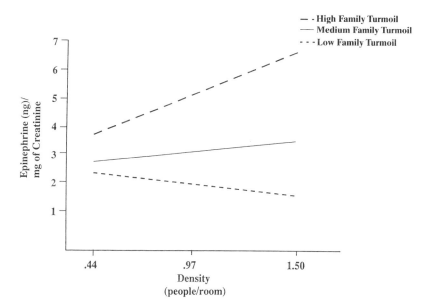

Figure 3. Relations between residential density and children's psychophysiological stress (overnight urinary epinephrine) as a function of family turmoil.

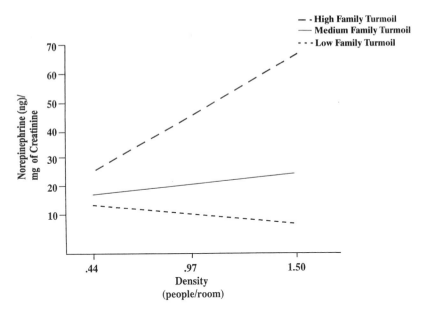

Figure 4. Relations between residential density and children's psychophysiological stress (overnight urinary norepinephrine) as a function of family turmoil.

ilies with greater turmoil and elevated crowding appears to be magnified by the confluence of these two stressful conditions relative to families that have experienced either family turmoil or crowding alone.

The extremely high levels of inner city stressor exposures (i.e., violence, financial pressure, family separation, death and serious illness, family turmoil) in our sample further substantiate our argument that inner city, ethnic minority families are

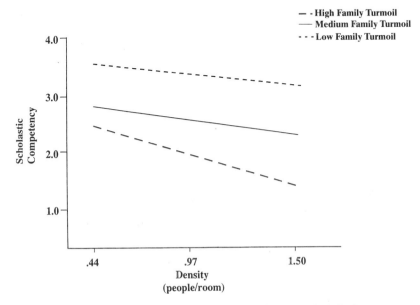

Figure 5. Relations between residential density and children's self reports of scholastic competency (Harter Competency Scale) as a function of family turmoil.

confronted with a broad array of extremely demanding, aversive social conditions that place enormous demands on family and individual coping resources. Not surprisingly we also see unusually high levels of psychological distress and very low perceived academic competency among the children in this sample relative to normative data for the CBQ (Rutter et al., 1970) and the Scholastic Competency Scale (Harter, 1982). Insufficient information are available on norms for overnight urinary stress hormones to draw similar conclusions.

A second reason the uniformly, high levels of stressor exposure is important is because this precluded our ability to directly test our model of multiple, inner stressor exposure and crowding (see Fig. 1). Restricted variance in predictor variables renders extremely low statistical power (Ghiselli et al., 1981). It is worth reiterating that the one inner city stressor with sufficient variance, family turmoil, did reflect the hypothetical model, interacting with residential crowding to amplify developmental outcomes.

Crowding and many of the psychosocial stressors accompanying poverty may share certain proximal processes related to parent-child interaction. In particular both crowding and poverty have been associated with less responsive, more authoritarian parenting styles. Thus we were also interested in modelling whether the interactive effects of residential crowding and inner city stressors on development could be at least partially explained by the underlying, proximal process of parent-child interaction. Therefore parental punitiveness was hypothesized to mediate the significant, interaction of residential density and family turmoil on children's chronic stress (see Fig. 1).

The analytic technique to investigate this mediational pathway builds upon the above analyses using multiple regression. After the main effects for density and family turmoil are in the regression equation, the parenting term (centered on its mean) is forced into the regression equation prior to the interaction term. If the significant interactive effects of density and family turmoil are mediated by parenting, the previously significant multiplicative term (i.e., significant ΔR^2 term for the interaction) will no longer be significant (see Baron & Kenny, 1986; Evans & Lepore, 1997 for more details).

When parenting is forced into each of the above equations, evidence is found for mediation for punitive parenting style on psychological distress, perceived scholastic competence and epinephrine but not norepinephrine. The ΔR^2 for the interaction term is no longer significant for the Children's Behavioral Questionnaire ($\Delta R^2 = 0.031$; $F(1,35) = 2.38$), for epinephrine ($\Delta R^2 = 0.082$; $F(1,28) = 3.18$), or for scholastic competency ($\Delta R^2 = 0.043$; $F(1,35) = 2.30$). For norepinephrine, however, parenting does not shrink the increment in R^2, $\Delta R^2 = 0.272$, $F(1,28) = 11.14$, $p < 0.002$.

These results support the hypothesis that parental punitiveness conveys some of the adverse effects of the interaction of residential crowding and family turmoil on children's well being. Psychological distress, perceived scholastic competency, and some physiological stress indicators fit this trend. We do not know why norepinephrine did not follow this pattern.

The relations between residential crowding, a physical parameter of the ecological context, and developmental outcomes, are modified by a psychosocial, contextual factor associated with inner city poverty, family turmoil. Low income, ethnic minority children living in crowded public housing with high levels of family turmoil are especially vulnerable to the negative effects of crowding. This is manifested across multiple methods including maternal ratings of psychological health, children's self-reports of intellectual competency, and neuroendocrine measures of stress hormone activity.

Moreover these adverse, symbiotic relations appear to be partially explained by elevated levels of parental punitiveness among these families.

Our findings underscore the importance of considering the potential developmental impacts of environmental stressors like crowding within their natural, ecological context (Bronfenbrenner, 1979; Petrinovich, 1979). Proximal processes of parent-child support and responsiveness are negatively affected by household crowding. The extent of negative impact appears to be contingent upon the occurrence of family turmoil, a risk factor significantly elevated by poverty. The direction and level of proximal processes within the immediate, microenvironment are sensitive to contextual factors such as socioeconomic status (Bronfenbrenner & Morris, 1998). Many studies have documented, both in the laboratory and the field, that crowding is harmful to children (Evans, in press; Saegert, 1982; Wachs & Gruen, 1982). When crowding occurs in concert with the multitude of community and household stressors accompanying inner city poverty, the impacts on developmental dysfunction are substantially greater than found in prior crowding studies.

5. ASSUMPTIONS IN ENVIRONMENTAL STRESS RESEARCH

In this chapter we have briefly alluded to several assumptions, both in the environmental stress area and in the field of environment and behavior more generally. We briefly re-iterate those assumptions here and point out an overarching, meta-assumption that pervades environmental stress research. The study of environmental stressors like crowding, noise, pollution, or temperature, typically isolates the stressor in order to draw causal inferences about possible, negative outcomes. This isolation of the environmental stimulus typically includes methodological and/or statistical techniques that control for other potentially stressful physical conditions (e.g., other environmental stressors) and psychosocial conditions (e.g., socioeconomic status). Although this traditional, experimental approach to research has obvious merits in terms of internal validity, it may lead to misspecification of the relations between environmental stressors and health. Methodological or statistical isolation of singular physical or psychosocial stressors may significantly underestimate their true impact on families and children. This may be particularly true in ecological contexts such as poverty where children and families face a confluence of psychosocial and environmental stressors. The effects of multiple physical and psychosocial stressors on human health and well being may not be adequately characterized by existing research which has typically examined these stressors apart from their natural, ecological context (Bronfenbrenner, 1979; Cohen et al., 1986; Petrinovich, 1979).

There are also some potentially important analytic implications of the ecological perspective. The common practice of statistically controlling for income or other indicators of socioeconomic status with covariance may be incorrect. Use of statistical controls assumes that the control variable (i.e., covariate factor) does not interact with the independent variable (Cohen & Cohen, 1983). The cumulative risk research challenges another underlying statistical assumption present in nearly all environment and behavior research, including the environmental stress literature—namely linear relationships. As the reader may recall there is typically a linear function of number of risk factors experienced and dysfunction among children. However after four or five risks are experienced, additional risk exposure leads to slight or no additional pathology (Barocas et al., 1993; Rutter, 1983; Richters & Martinez, 1993; Sameroff et al., 1987; 1993). If multiple stressor exposure leads to an asymptotic, nonlinear function with developmental outcomes, use of the general linear model will underestimate the extent of association (Cohen & Cohen, 1983). More care is called for in visually inspecting scatterplots and regression plots.

A final assumption we have discussed is the operation of proximal processes as the underlying mechanisms that drive human development (Bronfenbrenner & Morris, 1998). Too much research in environment and behavior has focused on descriptive associations of physical and person variables as they predict human behavior. The field needs to move to a deeper level of thought in order to analyze why and how the physical environment can affect human behavior (Evans & Lepore, 1997).

Use of the stress perspective, as any other scientific paradigm, carries inherent biases and limitations. A meta-assumption underlying this chapter is that certain characteristics of the physical and social environment, both singly and when embedded in ecological context, lead to dysfunction. The choice of environmental and psychosocial variables, outcome measures, and proximal processes, are all shaped by an overarching model of pathology—what characteristics of the child's surroundings can lead to dysfunction?

If we re-state the issue as one of examining environmental conditions and contexts that promote and sustain human competence, rather than dysfunction, the choice of environmental factors, contextual factors, and developmental outcomes changes. Perhaps of even greater importance, we can then also begin to conceptualize and measure proximal processes within a constructive context (Bronfenbrenner & Morris, 1998; Tulkin, 1977). What physical and social environmental characteristics might have the capability to foster the development of competence and character (Kaplan, 1995; Wandersman & Nation, 1998)? Answers to such a question would enable us to develop more effective strategies to curtail the growing chaos in the lives of children, youth, and families.

The choice of paradigm also directs policy and programmatic interventions. To the extent that the experience of poor children living in poor communities goes over the tolerable limit for healthy development, it is extremely important to reduce the burden of stress borne by these children and their parents. It is an open question as to whether interventions of a more positive nature are required to level the playing field for poor inner city children, or whether simply eliminating the multiple stresses associated with living in poverty would be sufficient. After reviewing the literature on the developmental consequences of poverty for children, McLoyd (1998) concluded that the surest way to eliminate negative consequences would be to reduce the poverty children experience. However, she pointed out, the United States has consistently preferred programs that provide services rather than adequate income to poor households. The service approach could benefit from better knowledge about contexts and processes that support child development. However, well conceived programs for children and families in poverty often founder when they encounter the realities of engaging parents and children experiencing the multiple stressors associated with poverty. Shore of increasing the income of households in poverty, the material and social stressors poor families experience must also be reduced.

ACKNOWLEDGMENTS

We are especially grateful to the families of East Harlem, New York City who were willing to share information and ideas with us about the environment of poverty and its impacts on them. We appreciate the assistance of Gretchen Susi, Kira Krenichyn, Raymond Coddrington, and Colin Thomas-Jensen in data collection. Urie Bronfenbrenner kindly provided critical feedback on an earlier draft of this chapter. Support for this work came from the John D. and Catherine T. Mac Arthur Foundation Network on Socioeconomic Status and Health, the Bronfenbrenner Life Course Institute at Cornell University, the National Institute of Child Health and Human Development, 1 F33 HD08473-01, The Graduate Center, CUNY, and the Edna McConnell Clark Foundation.

REFERENCES

Aiello, J. R., Epstein, Y., & Karlin, R. (1975). Effects of crowding on electrodermal activity. *Sociological Symposium, 14*, 42–57.

Aiello, J. R., Nicosia, G., & Thompson, D. (1979). Physiological, social, and behavioral consequences of crowding on children and adolescents. *Child Development, 50*, 195–202.

Aiken, L., & West, S. (1991). *Multiple regression.* Los Angeles: Sage.

Attar, B., Guerra, N., & Tolan, P. (1994). Neighborhood disadvantage, stressful life events and adjustments in urban elementary school children. *Journal of Child Clinical Psychology, 23*, 391–400.

Barocas, R., Seiffer, R., & Sameroff, A. (1993). Defining risk: Multiple dimensions of psychological vulnerability. *American Journal of Community Psychology, 13*, 433–447.

Baron, R., & Kenny, D. (1986). The moderator-mediator variable distinction in social psychological research: Conceptual, strategic, and statistical considerations. *Journal of Personality and Social Psychology, 51*, 1173–1182.

Baum, A., Aiello, J. R., & Calesnick, L. (1978). Crowding and personal control: Social density and the development of learned helplessness. *Journal of Personality and Social Psychology, 36*, 1000–1011.

Baum, A., Gatchel, R., Aiello, J. R., & Thompson, D. (1981). Cognitive mediation of environmental stress. In J. Harvey (Ed.), *Cognition, social behavior and the environment* (pp. 513–533). Hillsdale, NJ: Erlbaum.

Baum, A., & Grunberg, N. (1995). Measurement of stress hormones. In S. Cohen, R. C. Kessler, & L. Gordon (Eds.), *Measuring stress* (pp. 175–192). NY: Oxford.

Booth, A., & Johnson, D. (1975). Crowding and family relations. *American Sociological Review, 41*, 308–321.

Boyle, M., & Jones, S. (1985). Selecting measures of emotional and behavioral disorders of childhood for use in general populations. *Journal of Clinical Psychology and Psychiatry, 26*, 137–159.

Bradley, R., & Caldwell, B. (1984). The HOME inventory and family demographics. *Developmental Psychology, 20*, 315–320.

Brody, G., Stonemant, Z., & Flor, D. (1996). Parental religiosity, family processes, and youth competence in rural, two parent African American families. *Developmental Psychology, 32*, 696–706.

Bronfenbrenner, U. (1979). *The ecology of human development: Experiments by nature and design.* Cambridge, MA: Harvard University Press.

Bronfenbrenner, U., & Morris, P. (1998). The ecology of developmental process. In W. Damon, & R. Lerner (Eds.). *Handbook of child development, vol. 1* (pp. 992–1028). New York: Wiley.

Brown, L., & Cowen, E. (1988). Children's judgments of event upsettingness and personal experiencing of stressful events. *American Journal of Community Psychology, 16*, 123–135.

Carter, N., & Beh, H. (1989). The effect of intermittent noise on cardiovascular functioning during vigilance task performance. *Psychophysiology, 26*, 548–559.

Caspi, A., Bolger, N., & Eckenrode, J. (1987). Linking person and context in the daily stress process. *Journal of Personality and Social Psychology, 52*, 184–195.

Cesana, G, Ferrario, M., Curti, R., Zanettini, R., Grieco, A., Sega, R., Palermo, A., Mara, G., Libretti, A., & Alergri, S. (1982). Work-stress and urinary catecholamine excretion in shift workers exposed to noise. *La Medicina del Lavoro, 2*, 99–109.

Children's Defense Fund. (1985). *Black and white children in America: Key Facts.* Washington, DC: Children's Defense Fund.

Cohen, J., & Cohen, P. (1983). *Applied multiple regression/correlation analysis for the behavioral sciences, 2nd. ed.* Hillsdale, NJ: Erlbaum.

Cohen, S., Evans, G. W., Stokols, D., & Krantz, D. S. (1986). *Behavior, health, and environmental stress.* New York: Plenum.

Conger, R. D., Conger, K., Elder, G., Lorenz, F., Simons, R., & Whitbeck, L. (1992). A family process model of economic hardship and adjustment of early adolescent boys. *Child Development, 63,* 526–541.

Conger, R. D., Ge., X., Elder, G., Lorenz, F., & Simons, R. (1994). Economic stress, coercive family process, and developmental problems of adolescents. *Child Development, 65,* 541–561.

Conrad, D. (1973). The effects of intermittent noise on human serial decoding performance and physiological response. *Ergonomics, 16,* 739–747.

Cottington, E., Matthews, K., Talbott, E., & Kuller, R. (1983). Occupational stress and diastolic blood pressure in a blue collar population: The Pittsburgh noise-hypertension project. *Annual Meeting for the Society of Epidemiology.* Winnipeg, Manitoba.

D'Atri, D. (1975). Psychophysiological responses to crowding. *Environment and Behavior, 7,* 237–251.

Dodge, K., Pettit, G., & Bates, J. (1994). Socialization mediators of the relation between socioeconomic status and child conduct problems. *Child Development, 54,* 649–665.

Dubow, E., Tisak, J., Casuey, D., Hryshko, A., & Reid, G. (1991). A two-year longitudinal study of stressful life events, social support, and social problem solving skills: Contributions to children's behavioral and academic adjustment. *Child Development, 62,* 583–599.

Duncan, G., Brooks-Gunn, J., & Klebanov, P. (1994). Economic deprivation and early childhood development. *Child Development, 65,* 296–318.

Edleman, M. (1985). The sea is so wide and my boat is so small. In H. Mc Adoo, J. Mc Adoo (Eds.), *Handbook of health psychology* (pp. 456–489). Hillsdale, NJ: Erlbaum.

El-Sheikh, M., Cummings, E., & Goetsch, V. (1989). Coping with adults angry behavior: Behavioral, physiological, and verbal responses in preschoolers. *Developmental Psychology, 25,* 490–498.

Elder, G., Nguyen, T., & Caspi, A. (1985). Linking family hardship to children's lives. *Child Development, 56,* 361–375.

Evans, G. W. (1979). Behavioral and physiological consequences of crowding. *Journal of Applied Social Psychology, 9,* 27–46.

Evans, G. W. (in press). Environmental stress and health. In A. Baum, T. Revenson, & J. E. Singer (Eds.), *Handbook of health psychology.* Mahweh, NJ: Erlbaum.

Evans, G. W., Allen, K., Tafalla, R., & O'Meara, T. (1996). Multiple stressors: Performance, psychophysiologic, and affective responses. *Journal of Environmental Psychology, 16,* 147–154.

Evans, G. W., & Carrere, S. (1991). Traffic congestion, perceived control, and psychophysiological stress among urban bus drivers. *Journal of Applied Psychology, 76,* 658–663.

Evans, G. W., Jacobs, S. V., Dooley, D., & Catalano, R. (1987). Stressful life events and chronic stress. *American Journal of Community Psychology, 15,* 125–134.

Evans, G. W., & Lepore, S. J. (1993). Household crowding and social support: A quasi-experimental analysis. *Journal of Personality and Social Psychology, 65,* 308–316.

Evans, G. W., & Lepore, S. J. (1997). Moderating and mediating processes in environment-behavior research. In G. T. Moore, & R. W. Marans (Eds.), *Advances in environment, behavior, and design, vol. 4* (pp. 255–285). NY: Plenum.

Evans, G. W., Lepore, S. J., Shejwal, B., & Palsane, M. N. (1998). Chronic residential crowding and children's well being: An ecological perspective. *Child Development, 69,* 1514–1523.

Evans, G. W., Maxwell, L. E., & Hart, B. (1999). Parental language and verbal responsiveness to children in crowded homes. *Developmental Psychology, 35,* 1020–1023.

Evans, G. W., Palsane, M. N., Lepore, S. J., & Martin, J. (1989). Residential density and psychological health: The mediating effects of social support. *Journal of Personality and Social Psychology, 57,* 994–999.

Fleming, I., Baum, A., Davidson, L., Rectanus, E., & Mc Ardle, S. (1987). Chronic stress as a factor in physiologic reactivity to challenge. *Health Psychology, 6,* 221–237.

Gad, M., & Johnson, J. (1980). Correlates of adolescent life stress as related to race, SES, and levels of perceived support. *Journal of Clinical Child Psychology, 9,* 13–16.

Garbarino, J., Durbrow, W., Kosteleny, K., & Porter, C. (1992). *Children in danger: Coping with the consequences of community violence.* San Francisco: Jossey-Bass.

Garmezy, N., Masten, A., & Tellegen, A. (1984). The study of stress and competence in children: A building block for developmental psychopathology. *Child Development, 55,* 97–111.

Ghiselli, E., Campbell, J., & Zedeck, S. (1981). *Measurement theory for the behavioral sciences.* San Francisco: Freeman.

Gibbs, J. (1984). Black adolescents and youth: An endangered species. *American Journal of Orthopsychiatry, 54,* 6–21.

Gove, W., & Hughes, M. (1983). *Overcrowding in the household.* New York: Academic.

Grunberg, N., & Singer, J. E. (1990). Biochemical measurement. In J. Cacciopo, & L. Tassinary (Eds.), *Principles of psychophysiology* (pp. 149–176). NY: Cambridge.

Guerra, N., Huesmann, L., Tolan, P., Van Acker, R., & Eron, L. (1995). Stressful events and individual beliefs as correlates of economic disadvantage and aggression among urban children. *Journal of Consulting and Clinical Psychology, 53*, 518–528.

Harter, S. (1982). The perceived competence scale for children. *Child Development, 53*, 87–97.

Hashima, P., & Amato, P. (1994). Poverty, social support, and parental behavior. *Child Development, 65*, 394–403.

Heshka, S., & Pylypuk, A. (1975). *Human crowding and adrenocortical activity.* Paper presented at the Canadian Psychological Association. Montreal.

House, J., Mc Michael, A., Wells, J., Kaplan, B., & Landerman, L. (1979). Occupational stress and health among factory workers. *Journal of Health and Social Behavior, 20*, 139–160.

Huston, A., Mc Loyd, V., & Coll, C. (1994). Children and poverty: Issues in contemporary research. *Child Development, 65*, 275–282.

Ising, H., Dienel, D., Gunther, T., & Markert, B. (1980). Health effects of traffic noise. *International Archives of Occupational and Environmental Health, 47*, 179–190.

Kaplan, S. (1995). The restorative benefits of nature: Toward an integrative framework. *Journal of Environmental Psychology, 15*, 169–182.

Kaslow, N., Deering, C., & Rascusin, G. (1994). Depressed children and their families. *Clinical Psychology Review, 14*, 39–59.

Lepore, S. J., & Evans, G. W. (1996). Coping with multiple stressors in the environment. In M. Zeidner, & N. Endler (Eds.). *Handbook of coping* (pp. 350–377). New York: Wiley.

Lepore, S. J., Evans, G. W., & Palsane, M. N. (1991). Social hassles and psychological health in the context of crowding. *Journal of Health and Social Behavior, 32*, 357–367.

Lepore, S. J., Evans, G. W., & Schneider, M. (1991). The dynamic role of social support in the link between chronic stress and psychological distress. *Journal of Personality and Social Psychology, 61*, 899–909.

Lercher, P., Hortnagl, J., & Kofler, W. (1993). Work noise annoyance and blood pressure: Combined effects with stressful working conditions. *International Archives of Occupational and Environmental Health, 65*, 23–28.

Lewisohn, P., Roberts, R., Seeley, J., Rohde, P., Gotlib, I., & Hops, H. (1994). Adolescent psychopathology II. Psychological risk factors for depression. *Journal of Abnormal Psychology, 103*, 302–313.

Lundberg, U. (1976). Urban commuting: Crowdedness and catecholamine excretion. *Journal of Human Stress, 2*, 26–34.

Lundberg, U. (1985). Catecholamines. In A. Steptoe (Ed.), *Assessment of sympathetic nervous function in human stress research* (pp. 26–41). London: Ciba Foundation.

Martinez, P., & Richters, J. (1993). The NIMH community violence project II. Children's distress symptoms associated with violence exposure. *Psychiatry, 56*, 22–35.

Masten, A., Garmezy, N., Tellegen, A., Pelligrini, D., Larkin, K., & Larsen, A. (1987). Competence and stress in school children: The moderating effects of individual and family qualities. *Journal of Child Psychology and Psychiatry, 29*, 745–764.

Matheny, A., Wachs, T. D., Ludwig, J., & Phillips, K. (1995). Bringing order out of chaos: Psychometric characteristics of the confusion, hubbub, and order scale. *Journal of Applied Developmental Psychology, 16*, 429–444.

Matthews, K. A., Gump, B., Block, D., & Allen, M. (1997). Does background stress heighten or dampen children's cardiovascular responses to acute stress? *Psychosomatic Medicine, 59*, 488–496.

Maxwell, L. E. (1996). Multiple effects of home and day care crowding. *Environment and Behavior, 28*, 494–511.

Mc Carthy, D., & Saegert, S. (1979). Residential density, social overload and social withdrawal. *Human Ecology, 16*, 253–271.

Mc Loyd, V. C. (1990). The impact of economic hardship on black families and children: Psychological distress, parenting, and socioemotional development. *Child Development, 61*, 311–346.

Mc Loyd, V. C. (1998). Socioeconomic disadvantage and child development. *American Psychologist, 53*, 185–204.

Mc Loyd, V. C., Jayaratne, T., Cebello, R., & Boquet, J. (1994). Unemployment and work interruption among African American single mothers: Effects on parenting and adolescent socioemotional functioning. *Child Development, 65*, 562–589.

Mitchell, R. (1981). Some social implications of high density housing. *American Sociological Review, 36*, 18–29.

Mosskov, J., & Ettema, J. (1977). Extra-auditory effects of short term exposure to aircraft and traffic noise. *International Archives of Occupational and Environmental Health, 40*, 165–173.

Murray, R. (1974). The influence of crowding on children's behavior. In D. Canter, & T. Lee (Eds.), *Psychology and the built environment* (pp. 112–117). Chichester, UK: Wiley.

Osofsky, J. (1995). The effects of exposure to violence on young children. *American Psychologist, 50*, 782–788.

Paulus, P. B., Mc Gain, G., & Cox, V. (1978). Death rates, psychiatric commitments, blood pressure, and perceived crowding as a function of institutional crowding. *Environmental Psychology and Nonverbal Behavior, 3*, 107–116.

Petrinovich, L. (1979). Probablistic functionalism. *American Psychologist, 34*, 373–390.

Richters, J. (1993). Toward a developmental perspective on conduct disorder. *Developmental Psychopathology, 5*, 1–4.

Richters, J., & Martinez, P. (1993). Violent communities, family choices, and children's chances: An algorithm for improving the odds. *Developmental Psychopathology, 5,* 609–627.

Riggin, R., & Kissinger, P. (1977). Determination of catecholamines in urine by reverse phase, liquid chromatography with electrochemical detection. *Analytic Chemistry, 49*, 2109–2111.

Rodin, J. (1976). Crowding, perceived choice and response to controllable and uncontrollable outcomes. *Journal of Experimental Social Psychology, 12*, 564–578.

Rutter, M. (1983). Stress, coping, and development. In N. Garmezy, & M. Rutter (Eds.). *Stress, coping, and development* (pp. 1–42). New York: Mc Graw Hill.

Rutter, M., Tizard, J., & Whitmore, K. (1970). *Education, health, and behavior*. London: Longmans.

Rutter, M. Yule, B., Quinton, D., Rowlands, O. Yule, W., & Berger, M. (1974). Attainment and adjustment in two geographic areas: Some factors accounting for area differences. *British Journal of Psychiatry, 125,* 520–533.

Saegert, S. (1982). Environment and children's mental health: Residential density and low income children. In A. Baum, & J. E. Singer (Eds.), *Handbook of psychology and health* (pp. 247–271). Hillsdale, NJ: Erlbaum.

Sameroff, A., Seiffer, R., Baldwin, A., & Baldwin, C. (1993). Stability of intelligence from preschool to adolescence: The influence of social and family risk factors. *Child Development, 64*, 80–97.

Sameroff, A., Seiffer, R., Barocas, R., Zax, M., & Greenspan, S. (1987). Intelligence quotient scores of 4 year old children: Social-environmental risk factors. *Pediatrics, 79*, 343–350.

Sampson, R. (1992). Family management and child development: Insights from social disorganization theory. In J. McCord (Ed.), *Facts, frameworks, and forecasts: Advances in criminology* (pp. 63–93). New Brunswick, NJ: Rutgers.

Sherrod, D. (1974). Crowding, perceived control, and behavioral aftereffects. *Journal of Applied Psychology, 4,* 171–186.

Taylor, S., Repetti, R., & Seeman, T. (1997). Health psychology: What is an unhealthy environment and how does it get under the skin? *Annual Review of Psychology, 48*, 411–447.

Tulkin, S. R. (1977). Social class differences in maternal and infant behavior. In P. Liederman, A. Rosenfeld, & S. R. Tulkin (Eds.), *Culture and infancy* (pp. 495–536). NY: Academic.

Wachs, T. D. (1989). The nature of the physical microenvironment: An expanded classification system. *Merrill Palmer Quarterly, 35*, 399–419.

Wachs, T. D., & Gruen, G. (1982). *Early experience and human development*. NY: Plenum.

Wandersman, A., & Nation, M. (1998). Urban neighborhoods and mental health. *American Psychologist, 53*, 647–656.

Wedge, P., & Petzing, J. (1970). Housing for children. *Housing Review, 19*, 165–166.

Welch, B. (1979). Extra-auditory health effects of industrial noise: A survey of foreign literature. *Aerospace Medical Research Laboratory. Wright Patterson Air Force Base*. AHRL-TR-79-41.

Wilson, J. (1987). *The truly disadvantaged*. Chicago: University of Chicago Press.

Woodall, K., & Matthews, K. A. (1989). Familial environment associated with Type A behaviors and psychophysiological responses to stress in children. *Health Psychology, 8,* 403–426.

Work, W., Cowen, E., Parker, G., & Wyman, P. (1990). Stress resilient children in an urban setting. *Journal of Primary Prevention, 11*, 3–17.

Wyman, P., Cowen, E., Work, W., & Parker, G. (1991). Developmental and family milieu correlates of resilience in urban children who have experienced major life stress. *American Journal of Community Psychology, 19*, 405–426.

21

THEORY DEVELOPMENT IN ENVIRONMENTAL PSYCHOLOGY

A Prospective View

Daniel Stokols

School of Social Ecology
Room 206, Social Ecology I Building
University of California, Irvine
Irvine, CA 92697

1. INTRODUCTION

Life in contemporary times confronts individuals with glaring disparities between the goals of achieving social cohesion and emotional well-being, on the one hand, and a host of political, economic, and technological circumstances that undermine those values, on the other. Efforts to cultivate cohesive families, neighborhoods, work organizations, and whole communities, for example, are hindered by trends in society toward: (1) dual-career parenting and job insecurities that create a sense of fatigue, work overload, and reduced opportunities for informal contacts among family members, colleagues, and friends; (2) a growing reliance on advanced telecommunications technologies (e.g., e-mail, voice mail, fax communications) that make it possible for individuals to work remotely and in isolation from each other, thereby decreasing opportunities for face-to-face contact in organizational settings and public places; (3) the growing prevalence of new telecommunications technologies and the rapidity of their development, which impose greater time constraints on individuals and higher levels of sensory overload as they struggle to keep up with new software, hardware, and other technological innovations, and the flow of information transmitted to them each day; (4) the trend toward managed health care that poses a threat to the quality of medical services provided to individuals, particularly those burdened by acute and chronic diseases more prevalent in an aging population; and (5) strident political discourse, incivility, and inter-ethnic conflict that are rampant in many regions of the world and further threaten individuals' efforts to achieve a sense

Theoretical Perspectives in Environment-Behavior Research, edited by Wapner *et al.*
Kluwer Academic / Plenum Publishers, New York, 2000.

of personal well-being and meaningful social ties within organizational and community settings.

These disparities and strains of modern life pose several directions for theory development in environmental psychology as we approach the 21st Century. First, we need to understand better the transactional processes that enable some families, work groups, and communities to achieve high levels of cohesion and to transcend the threats to social integration noted above. Second, the circumstances under which people's transactions with new technologies are supportive and satisfying rather than restrictive or oppressive warrant greater attention. Third, the ways in which individuals achieve a sense of self efficacy and security in a world of diminishing resources is an important topic for future study. This quest for greater security and sense of agency may be reflected in the growing popularity of alternative medical therapies in the United States and individuals' use of the Internet for obtaining medical information and assistance in an effort to manage their own care, rather than having it controlled by health maintenance organizations.

The present paper addresses these issues by examining the ways in which qualities of the physical and social environment jointly affect people's capacity to cope successfully with rapid changes in society and with contemporary threats to civility and cohesion in interpersonal, organizational and community settings. The qualities of sociophysical environments, identified below, are useful at a theoretical level in helping to understand the circumstances under which our encounters with work-related challenges, new telecommunications technologies, diminishing economic and natural resources, and global environmental changes either strengthen or undermine personal well-being and social cohesion.

2. THEORY DEVELOPMENT IN ENVIRONMENTAL PSYCHOLOGY: BASIC AND APPLIED GOALS

The field of environmental psychology and the broader, interdisciplinary domain of environment-behavior studies are fundamentally concerned with both the scientific goal of broadening our understanding of people's transactions with their surroundings, and the more practical goal of enhancing—even optimizing—people's relationships with their everyday environments (e.g., Craik, 1973; Moore, 1987; Stokols, 1978). The programmatic assumptions underlying environmental psychology research, and the very content of research foci in this field, exemplify the "action research" orientation espoused by Kurt Lewin (Lewin, 1936; Weisman, 1982), whereby the processes of theorizing, and the application of theories in the service of environmental improvement and community problem-solving, are inseparably intertwined.

Recognizing both the scientific and applied goals of theory development in environmental psychology, it is incumbent on theorists to evaluate whether their efforts to develop new conceptualizations of environment and behavior are, in fact, likely to result in significant scientific advances and/or practical benefits to society (e.g., the reduction of environmental problems and the enhancement of people's transactions with their surroundings). Several factors can be expected to influence the scientific and applied "yield" of our theory-development efforts but two, in particular, seem especially important: that is, the complexity of the environment-behavior relationships that are being modeled and the rate of change in the contextual circumstances that impinge on those relationships. To the extent that contextual factors are undergoing rapid and unpredictable transformation, it becomes increasingly difficult to make stable predictions about the environment-behavior relationships that are influenced by those factors. Moreover, rapid and dramatic changes in family, organizational, and community settings make it all the more difficult to anticipate both the benefits and costs of implementing theory-based environmental interventions within those settings.

3. MODELING PEOPLE-ENVIRONMENT RELATIONS IN AN ERA OF ACCELERATING SOCIETAL CHANGE

As we approach the 21st Century, the rate of societal change and the complexities of modern-day life appear to be increasing. The proliferation of desk-top computing in homes, schools, and workplaces since the 1970s, the expansion of the Internet, the aging of our population, corporate trends toward down-sizing, telecommuting and "managed" health care, and growing concerns about global environmental changes are examples of the societal transformations that are profoundly affecting people's transactions with their everyday environments (Craik, 1997; Myerowitz, 1985; Stern, 1992; Stokols, 1995).

The rapid pace of organizational and societal change poses major challenges for the development of scientifically valid and useful theories of environment and behavior. Most importantly, any effort to model people-environment relationships as they occur within residential, educational, occupational, health care, and recreational settings must take account of the contextual "turbulence" that is increasingly impinging on these categories of settings (Emery & Trist, 1965). That is, the complexity of our theoretical models must be commensurate with the increasing change and complexity that characterizes contemporary relationships between people and their sociophysical surroundings.

The remaining sections of this paper outline certain conceptual strategies that may facilitate the development of more robust and useful theories of environment and behavior, especially in the context of accelerating societal change and complexity.

4. STRATEGIES FOR DEVELOPING MORE ROBUST AND USEFUL THEORIES OF ENVIRONMENT AND BEHAVIOR

Early conceptualizations of environmental and ecological psychology emphasized certain core assumptions about the nature of people's transactions with their everyday environments. These core assumptions include the propositions that: (1) people-environment relationships are bi-directional and dynamic, rather than linear and static—that is, individuals and their environments mutually influence each other over time; (2) environment-behavior relationships often occur within the context of highly structured, systemically-organized situations and settings; (3) the physical and social features of our everyday environments are closely interrelated and jointly affect our behavior and well-being; and (4) the environmental contexts of behavior, encompassing individuals' immediate residential, work, and educational settings as well as more distant community and societal conditions, reflect an interdependent, nested structure in which multiple settings concurrently influence the behavior and health of individuals and groups (cf., Barker, 1968; Bronfenbrenner, 1979; Ittleson, Proshansky, Rivlin, & Winkel, 1974; Stokols, 1978; Wapner, Cohen, & Kaplan, 1976). These core assumptions provide a valuable starting point and suggest key strategies for developing more robust and useful theories of environment and behavior.

4.1. Target High-Impact Physical and Social Circumstances in Settings that Either Enhance or Constrain People's Efforts to Optimize Their Relationships with Their Environments

Certain physical and social features of individuals' everyday settings and activity systems have greater prominence and potential to influence their behavior and well being than do other, less influential conditions in those environments. These "high-impact" features of the sociophysical environment are theoretically significant in that they have the capacity to either enable or constrain people's efforts to optimize their relationships with their surroundings. Organizational policies that prohibit workers from decorating or personalizing their offices, for example, can weaken their involvement in and identification with the work environment (Wells, 1997). Also, work environments that fail to support employees' efforts to regulate their privacy during the workday may expose workers to chronic experiences of stress (Steele, 1986; Sundstrom, 1986). And conflict-prone individuals who occupy influential decision-making roles in organizations have the capacity to exacerbate the work demands and stress experiences of numerous co-workers (Stokols, 1992). Theories that attempt to explain people's goal-directed efforts to cope with complex and rapidly changing environments should identify (and account for) these high-impact features of the sociophysical milieu.

4.2. Give Greater Attention to the Ways in which Physical and Social Features of Environmental Settings Jointly Influence People's Behavior and Well-Being

The systemic organization of environmental settings suggests that the physical and social features of those environments are closely interrelated and exert joint influence on individuals' behavior and well-being (Minami & Tanaka, 1995). Yet, many theories and empirical studies in environmental psychology treat the physical and social features of behavior settings separately and in isolation from each other. Efforts to develop more robust and useful theories of environment and behavior should address the interdependencies that exist between the physical and social facets of human environments. Rather than thinking simply in terms of "physical" or "social" variables and design interventions, environmental psychologists should develop a nomenclature for describing qualities of the "sociophysical" environment and the ways in which physical and social features of settings jointly influence individuals' behavior and well-being. Certain qualities of the sociophysical environment that merit greater theoretical attention are noted below.

4.2.1. Symbolic Qualities of the Physical Environment. Environmental elements are often imbued with social meaning and symbolic value. Examples include historically significant landmarks in cities that have both physical and social imageability

(Lynch, 1960; Milgram & Jodelet, 1976; Stokols, 1981, 1990), and displays of team projects and photographs that are affixed to the walls of offices to strengthen employees' sense of identity with their company and work environment. These physical features of everyday environments have the potential to enhance not only the aesthetic quality of people's surroundings, but also the social cohesion and camaraderie that exist within organizational and community settings. The capacity of symbolically-laden objects and places to influence behavior and well-being in multiple ways (e.g., by enhancing individuals' aesthetic experiences, strengthening the perceived imageability of an urban area, or by reinforcing group members' identity with a workplace or neighborhood) suggests that they are especially significant and salient features of human environments and, as such, should be further examined in future research.

4.2.2. The Role of Intermediaries in Determining the Quality of People-Environment Transactions. Key individuals and intermediaries within organized settings can dramatically influence the perceived quality and healthfulness of environments shared by many people, and can encourage or thwart others' efforts to improve the quality of their surroundings (Stokols, 1996). For example, worksite health coordinators and facilities managers directly influence the well-being of their employees by providing (or withholding) workplace health promotion programs, or by purchasing office equipment and materials that are either toxic and injurious, or hygienic and safe (e.g., providing ergonomic vs. non-ergonomic office furniture to employees). Also, computing consultants employed by corporations, schools, and government agencies can enable people to become more proficient in using new telecommunications technologies (e.g., the Internet), thereby reducing their experiences of information overload and confusion. And, as noted earlier, high-impact individuals holding key decision-making roles can dramatically affect the healthfulness of the workplace by enacting conflict-promotive or conflict-resistant behavior (Stokols, 1992). In sum, the role of intermediaries, facilitators, and pivotal decision-makers warrants greater theoretical attention in environment-behavior research—especially, the ways in which these key individuals either undermine or enhance the quality of other people's transactions with their environments.

4.2.3. Processes by which Physical and Social Features of Environments Mediate or Moderate each others' Influence on Behavior and Health. The physical and social qualities of environmental settings can jointly influence individuals' behavior and well-being, both through mediational and moderating processes (Evans, Johansson, & Carrere, 1994; Stokols, 1996; Wachs, 1987). For example, high levels of ambient noise in a work environment can lead to personal feelings of annoyance, which make conflicts and hassles among co-workers more likely. These interpersonal experiences, in turn, can lead to elevated levels of emotional and physiological stress. Similarly, the physical separation of team members caused by poor space plans and adjacencies in offices can reduce informal social contacts and communications among co-workers which, in turn, create personal and group strains resulting from poor coordination and lack of cohesion. Alternatively, the health consequences of a non-supportive social environment may be moderated by the availability of certain physical resources (e.g., the availability of private work space and onsite fitness facilities), which enable employees to cope more effectively with interpersonal strains at work (e.g., by avoiding stressful interactions and maintaining a regular exercise regimen). These examples illustrate some of the ways in which the social and physical features of settings mutually influence participants' behavior and well-being. The mediational and moderating relationships among the physical and social features of behavior settings merit further study in future research.

4.3. Identify Those Settings within Individuals' Daily Activity Systems that Exert Greatest Influence on Their Coping Capacities and Resources

People's everyday activity systems are comprised of multiple environmental settings linked together, both spatially and psychologically, within a nested structure. Just as high-impact features of particular settings exert disproportionate influence on people's behavior and health, certain settings within individuals' daily activity systems exert a greater impact on their overall well-being than other, less influential environments. It is important for environment-behavior theorists to identify and understand better these high-impact settings as they have a disproportionate capacity to influence their participants in either positive or negative ways. For example, when individuals first relocate to a new area, they often establish "anchor points"—places within their daily activity settings that provide a base of security, familiarity, refuge, and a reference point for further exploration (Wapner, 1981). These anchor environments are particularly important sources of psychological security and respite from excessive stimulation, both among newcomers to an area and longer-term residents.

Similarly, designers of classrooms, child care facilities, and work environments often incorporate "stimulus shelters"—relatively secluded or private areas that afford experiences of solitude and relief from unwanted contacts with others (Steele, 1986; Wachs, 1979). At the community level, commemorative environments such as the Viet Nam Memorial in Washington, DC, "vest-pocket" parks, and public spaces that afford social interaction and contact with nature, such as Ghiradelli Square in San Francisco, provide opportunities for restoration and contemplation—experiences that enable people to cope more effectively with the demands and challenges of contemporary life (Carr, Francis, Rivlin, & Stone, 1992; Kaplan & Kaplan, 1989). And, with the advent of the Internet and the proliferation of "chat rooms", "listserve", and "use-net" support groups, electronic or "virtual" communities are becoming an increasingly important arena for information exchange, friendship formation, medical advice, and social support among individuals experiencing common problems and challenges (Rheingold, 1993; Schuler, 1996).

During times of rapid societal change and accelerating complexity, high-impact settings such as anchor places, stimulus shelters, restorative environments, and virtual communities assume greater psychological and theoretical significance, particularly in view of their capacity to expand individuals' repertoire of coping resources and to improve the overall quality of people's transactions with their everyday environments.

5. CONCLUSIONS

Physical objects, places, and behavior settings are not uniform in their capacity to influence human behavior and health. Certain features of the sociophysical environment are especially meaningful to people and exert a disproportionate influence their behavior and well-being. The development of more robust and useful theories of environment and behavior requires that these high-impact qualities and regions of the sociophysical environment be identified and given greater priority in our theoretical models. Moreover, the processes by which these pivotal environmental features and settings influence behavior and well-being—particularly in the context of rapid societal change, increasing technological complexity, and stimulation overload—warrant considerably greater attention in future theoretical and empirical research.

REFERENCES

Barker, R. G. (1968). *Ecological psychology: Concepts and methods for studying the environment of human behavior.* Stanford, CA: Stanford University Press.

Bronfenbrenner, U. (1979). *The ecology of human development.* Cambridge, MA: Harvard University Press.

Carr, S., Francis, M., Rivlin, L. G., & Stone, A. M. (1992). *Public space.* NY: Cambridge University Press.

Craik, K. H. (1973). Environmental psychology. In M. R. Rosenzweig & L. W. Porter (Eds.), *Annual Review of Psychology, 24,* 403–422. Palo Alto, CA: Annual Reviews.

Craik, K. H. (1997). Prospects for environmental psychology in the Third Millennium. In S. Wapner, J. Demick, T. Yamamoto, & T. Takahashi (Eds.), *Handbook of Japan-United States Environment-Behavior Research: Toward a transactional approach* (pp. 377–383). NY: Plenum Press.

Emery, F. E., & Trist, E. L. (1965). The causal texture of organizational environments. *Human Relations, 18,* 21–32.

Evans, G. W., Johansson, G., & Carrere, S. (1994). Psychosocial factors and the physical environment: Inter-relations in the workplace. In C. L. Cooper, & I. T. Robertson (Eds.), *International Review of Industrial and Organizational Psychology: Vol. 9* (pp. 1–29). Chichester, England: John Wiley, & Sons Ltd.

Ittelson, W. H., Proshansky, H. H., Rivlin, L. G., & Winkel, G. (1974). *An introduction to environmental psychology.* New York: Holt, Rinehart, & Winston.

Kaplan, R., & Kaplan, S. (1989). *The experience of nature: A psychological perspective.* New York: Cambridge University Press.

Lewin, K. (1936). *Principles of topological psychology.* (F., & G. Heider, Trans). NY: McGraw-Hill.

Lynch, K. (1960). *The image of the city.* Cambridge, MA: MIT Press.

Milgram, S., & Jodelet, D. (1976). Psychological maps of Paris. In H. M. Proshansky, W. H. Ittelson, & L. G. Rivlin (Eds.), *Environmental psychology* (2nd ed., pp. 104–124). New York: Holt, Rinehart, & Winston.

Minami, H., & Tanaka, K. (1995). Social and environmental psychology: Transaction between physical space and group-dynamic processes. *Environment and Behavior, 27,* 43–55.

Moore, G. T. (1987). Environment and behavior research in North America: History, developments, and unresolved issues. In D. Stokols, & I. Altman (Eds.), *Handbook of environmental psychology: Vol. 2* (pp. 1359–1410). New York: John Wiley, & Sons.

Myerowitz, J. (1985). *No sense of place: The impact of electronic media on social behavior.* NY: Oxford Univ. Press.

Rheingold, H. (1993). *The virtual community: Homesteading on the electronic frontier.* Reading, MA: Addison-Wesley Publishing Co.

Schuler, D. (1996). *New community networks.* Wired for change. Reading, MA: Addison-Wesley Publishing Co.

Steele, F. (1986). *Making and managing high-quality workplaces.* NY: Teachers College Press, Columbia University.

Stern, P. C. (1992). Psychological dimensions of global environmental change. In M. R. Rosenzweig, & L. W. Porter (Eds.), *Annual Review of Psychology, 43,* 296–302. Palo Alto, CA: Annual Reviews.

Stokols, D. (1978). Environmental psychology. In M. R. Rosenzweig & L. W. Porter (Eds.), *Annual Review of Psychology, 29,* 253–295. Palo Alto, CA: Annual Reviews.

Stokols, D. (1981). Group x place transactions: Some neglected issues in psychological research on settings. In D. Magnusson (Ed.), *Toward a psychology of situations: An interactional perspective* (pp. 292–415). Hillsdale, NJ: Lawrence Erlbaum.

Stokols, D. (1990). Instrumental and spiritual views of people-environment relations. *American Psychologist, 45,* 641–646.

Stokols, D. (1992). Conflict-prone and conflict-resistant organizations. In H. S. Friedman (Ed.), *Hostility, coping, & health.* Washington, DC: American Psychological Association, 65–76.

Stokols, D. (1995). The paradox of environmental psychology. *American Psychologist, 50,* 821–837.

Sundstrom, E. (1986). *Workplaces: The psychology of the physical environment in offices and factories.* New York: Cambridge University Press.

Wachs, T. D. (1979). Proximal experience and early cognitive-intellectual development: The physical environment. *Merrill-Palmer Quarterly, 25,* 3–41.

Wachs, T. D. (1987). Developmental perspectives on designing for development. In C. S. Weinstein, & T. G. David (Eds.), *Spaces for children: The built environment and child development* (pp. 291–318). NY: Plenum Press.

Wapner, S. (1981). Transactions of persons-in-environments: Some critical transitions. *Journal of Environmental Psychology, 1,* 223–239.

Wapner, S., Cohen, S., & Kaplan, B. (1976). *Experiencing the environment.* NY: Plenum Press.

Weisman, G. D. (1983). Environmental programming and action research. *Environment and Behavior, 15,* 381–408.

Wells, M. (1997). *Personalization of workspace and employee well-being.* Doctoral dissertation, School of Social Ecology, University of California, Irvine.

22

SPACE-FRAMES AND INTERCULTURAL STUDIES OF PERSON-ENVIRONMENT RELATIONS

George Rand

UCLA Department of Architecture and Urban Design
Los Angeles, CA 90024

1. INTRODUCTION

I think an interdisciplinary intercultural perspective is critical in the study of person-environment relations. It is important to complement understanding of cognitive maps and social cognition in relation to settings with the more psychoanalytic aspects of the built environment about which people have powerful, primitive, agonistic feelings, many of which that are passed from generation to generation.

There is a limit to how much can be learned when working within a definition of the problem by psychologists. It is important to include those practices that involve contact with contestations over space, for example, planning officials who mediate community conflicts, religious leaders who attempt to acquire new territories, politicians, and statesmen who mediate conflicts over national boundaries.

I find some guidance about the perception of this kind of space in the study of art and architectural history related to theories of perception and the concept of structure.

2. PROBLEM STATEMENT

Tarmo Pasto (1964) uses the concept of "space-frame"[1] to explain an aspect of figurative art that is otherwise difficult to capture in words. As a foundation for his theory Pasto relies on the research of Kai von Fieandt (1966) on perception of objects

---

[1] The Space Frame Experiment in Art, Tarmo Pasto, 1964.

Theoretical Perspectives in Environment-Behavior Research, edited by Wapner *et al.*
Kluwer Academic / Plenum Publishers, New York, 2000.

by the blind, and the empirical and theoretical work on "sensory-tonic theory" developed by Heinz Werner and Seymour Wapner (e.g., 1952). These theories hold that it is not valid to think of "seeing" as a purely "visual" affair. There are in fact *two* different modes of "seeing." The first is "geometric and technical" and is related to a way of seeing in which all of the senses are subordinated to vision as a model-sense. It is used in making the refined distinctions that are associated with the execution of logic and the formation of consciousness. The second is "expressive-physiognomic" and is related to harmonizing the senses through synesthesia and the use of intuition as a means of knowing that is associated with feeling and access to the unconscious (Werner & Wapner, 1954).

The issue for me has to do with the way in which pictures are made to come to life, as opposed to the way pictures are decoded or interpreted as representations. In the case of powerful figurative art, it involves the ability of the artist to present in the work the "ritual objectivity" he or she has witnessed. This is done using the "expressive-physiognomic" mode. The ritual objective aspect of experience is close to nature. It is this ritual aspect of the painting that allows us to assume the position of the artist and mentally enter the space-frame. The action-field of a sculpture or a painting has to be "imprinted" in our mind in a special way that is hard to understand if we rely solely on scientific modes of perception. The emotional relationship one establishes with the work under these conditions is different from mere "seeing" one based on rhythm and sequencing. It provides aural flow as well as optical data.

I define a "space frame" as follows: The experience of the person is based on more than using the forebrain as a tool for integrating information from the senses. The person is a vital, willful, vertical, moving object, that possesses momentum, has dynamic force, weight and thickness. This embodied self provides the background condition for our direct perception of the world. The embodied self creates "space-frames" for integrating our experiences. These "space-frames" are projected around the self. "Space frames" have two aspects: first, a positive (geometric-technical) space-frame is needed to ground us in the objective world where we experience ourselves as occupying a site in space defined in relation to the cardinal axes; second, a normative (expressive-physiognomic) space-frame is needed to provide a fluid medium which we move and can integrate new experiences. It is the medium within which we express ourselves. In both instances, the self is the dynamic and controlling center.

3. SPACE IN ART

According to critic and poet Kenneth Koch (1998), formal language meanings in poetry have to be augmented by sound and sense. Poetry is a special kind of language in which the sound of the words is raised to an importance equal to that of their meaning, and also to the importance of grammar and syntax. Repetition and variation of sounds, among other things, make it meditative, sad, and memorable. Take the sense of the word "sleep" and the impact of its phrasing, as in "SL EEP," a distinction made hearable by repetition of the sound pattern. Expressive-physiognomic uses of language create an awareness of the beauty and grace of words that can be achieved only by flexing them in different ways. Similarly, a powerful painting is more than a visual image ("stop-frame"). The moment has poetic depth partly because the image carries a sonic resonance that connects it with emotionally toned memories in the viewer's life-world.

Sound rather than vision is the source of access to these meanings and personal associations.

This process has historical roots in ancient or primitive art. Take for example, studies of Hindu cave-temples dating to about 700 AD. These magnificent panels depicting mythical stories can not be understood if approached as "stop-frames" representing part of an "event." The actions they picture are of struggles taking place among the Gods. In one case, the bodies of contesting enemies, Vishnu, one of the trinity of Gods, has come out of hiding inside a column and taken the form of a Man-Lion. He is locked in battle, legs knotted together with Hiranyakashipu who has committed something like the sin of hubris or arrogant misuse of powers that had been granted him, etc. The meaning of the piece can not be reduced to a metaphorical interpretation of the story it tells. That would imply that devotees could stand outside the reality that is depicted. The meaning is accessed through meditation and by achieving a state of one-ness with the panel. The focus is on the inner image of the deities (as opposed to their external form). This allows the observer to seek and arrive at a state of union or *samadhi*. Object and subject are one in this moment of "non-dual" awareness. This union resolves the conflict between opposite impulses represented in the physical object and the story that is portrayed Through the sacred geometry they participate in the agony of battle.

The panel preserves and protects the meaning as a social reality in the same way a fossil is captured inside a piece of clear amber. Discursive logic (geometric-technical) unpacks these emotions by breaking them into a series of discrete oppositions through language, in terms like jealousy, honor, pride. This allows for clear vision that is not "agonistically toned" as opposed to relying on metaphor to communicate the complex knots of feeling embodied in these images. Historically, where this clear language did not exist, one has to approach events in life using a likeness to other events, like a stalemate in war, or the association of personal courage and the image of a warrior's shield. Many Westerners were raised in this manner, where the figurative language of Homer's Iliad and Odyssey was used for guidance in life. The power of an "agonistic" image, once it is taken in, is that it captures an extended and multi-layered event in a single moment. The painting or sculpture that comes to life in this way uses concrete actions of bodies in conflict to create a powerful moment. This coalesces the narrative into a compressed "space-frame" which acts like a "magnetar" (the name given for a compressed star with a diameter of twelve or so miles that is full of potential magnetic energy).

Needless to say, these "agonistic" images do not function well if one is required to make critical distinctions between events. This is why a trauma (say of abuse in childhood) returns under conditions that have only a marginal connection to the original events. The mind that is captured by these images can not make differential comparisons between these events. It is also the reason that images, as in Biblical tales, remain through life as a frame used to understand other experiences of conflict or choice. The search for patterns related to ritual objectivity (expressive-physiognomic) and the discursive forms of objectivity (geometric-technical) are opposite sides of the waterfall of experience.

For a painting to accommodate this kind of embodied perception it has to provide the "space" for a viewer to mentally enter, move or walk around in it, to visually promenade among its objects, to assume the stance that the sculptor took as he/she created the object. Visual experience of an agonistic painting is processed with constant reference to background aural, kinesthetic and tactual information. Furthermore, if the viewer is involved in this manner, the eye is not operating under conditions of smooth movement associated with the normal "picture-plane" or two-dimensional surface. Imagine that the viewer is "walking around" in the painting. If the tableau does not support this kind of vision, there will be no involvement in the painting. We must remember that walking is an inherently unstable act, a form of falling forward in a rhythmic pattern. In addition to coordinating internal feedback from the muscles and tendons, vision while walking requires coordinating movement of the body with choppy and inconsistent feedback from the optical field. If the pattern of optical flow emitted by the walls is interrupted, even for a moment, the person's attempts to stay erect will fail and the body will fall to the floor unable to control against the pull of gravity. All these elements have to be present in the painting for it to have the agonistic qualities associated with a work of art that "moves us."

4. ROLE OF PERSPECTIVE

It is interesting to note in this respect that the notion of perspective (especially in painting) is associated with forms of consciousness that evolved in relation to geometric-technical perception. Leonardo daVinci called perspective the "rein and rudder and painting" because it was a tool steering the painter to a view of "objective" self-world relations in representations based on the shift from a geocentric to a heliocentric view of the universe.

Prior to discoveries of the renaissance (in many ways rediscoveries of Plato) there was no clear notion of a neutral space which acts as a container or holder for bodies or contents that move through it. The geocentric view of the cosmos prevailed. This view held that the core of the earth was the absolute center of the universe. The last celestial sphere that could be seen was its outermost limit. Infinity for the Scholastics meant a view of the cosmos as a closed system. No matter how far the divine omnipotence was able to reach, not spatial infinitude.

The new vision of the universe had to wait for the invention of instruments similar to the compass and sextant. Space became a continuous quantity with three physical dimensions capable of indifferently locating all objects using the cardinal axes. This entire process was later rationalized by Descartes and formalized by Kant. Rene Descartes (1596–1650) in his "meditations" developed the notion of "clear and distinct" ideas as a foundation for philosophy. He sought a philosophy that had the clarity of Newtonian physics and used calculus as a mathematical language to explain infinity and other mysteries. He viewed emotionally influenced ideas such as myth as pure superstition. He believed the "soul" had no extension or material realization.

He set the stage for modern psychology in which "bodies" (ideas) are represented independently of the space (mind) in which they are contained. By analogy to Newtonian space, he proposed a view of mind where the brain is a theater in which ideas are presented. Dennett[2] suspects that the "Cartesian Theater" was probably based on the idea of the Camera Obscura.

This model works well for scientific ideas, but has the effect of suppressing forms of experience that depend on direct, agonistically toned embodied perceptions. It was not until Freud's discovery of the unconscious in his studies of hypnosis and dreams

---

[2] Daniel Dennett, *Consciousness Observed*.

that the notion of "embodied" ideas laden with emotional content and symbolic associations could again be given the same status as subjects of scientific study as the "clear and distinct" ideas of Descartes.

All art (and by extension artistically conceived architecture) is based on seeing that is related to these embodied meanings. Great art maintains a balance between the visual symbols in aesthetic space and the logical form of theoretical space. In great art (as well as great ideas that are rooted in metaphors) we get both the objective clarity we expect from a calmer Cartesian mind as well as the depth of passion and subjective point of view we that is produced by hotter medieval or Gothic temperaments.

In other words, the relationship between object or idea and background is both geocentric and egocentric at the same time. The geocentric aspect allows us to articulate a modern systematic space based on quantitative measurement. The egocentric view allows us to enter the scene of the painting. When we look at the sculpture of "The Kiss" by Rodin, we are in the embrace that is formed by his gestures in stone.

What interests us here is the role of the media in the shaping the mind and its powers. In the history of art and architecture, there was a powerful role played by the invention of print. Print allowed the idea of perspective to dominate both the popular and the literate mind. It produced the idea, at least in the West, of a separate identity or ego capable of assuming an objective stance based on the notion of the vanishing point in space. Such a perspective allows one to see the relations among things in terms of their relative size and shape, and to distinguish them from the agonistic attraction and repulsion they may have in relation to the self.

This was convenient in a world that was attempting to "civilize" people by educating them in the modern manner, as individuals embedded in society, detached from their moorings in a tribal community. The transmission of culture in a medieval city was made through oral traditions associated with ritual and ceremony (cf. Carlo Ginzburg, 1992)[3] for examples of this relationship throughout history).

It is critical to recognize that adopting perspective as a norm was connected to the amazing differentiation among social and cultural systems made possible European culture including it excesses found in imperialist ambitions. The exploration of the inner world continued take place in Hindu and Buddhist traditions. These traditions continued the egocentric view of the universe. The impact that Western notions of perspective had on religion and metaphysics is powerful as well. It produced a new form of religion that eschewed any contact with the invisible mystical presence. Protestant religion, especially the Anglo-American Puritan version, demanded suppression of the embodied mind. This religion required reasoning based on the clear perspective offered by the printed word in the Biblical text. Meaning was found through silent reading of the text itself. Once reading occurred in silence, without the background voice of the community chant, poetic nuances that were carried in the voice of God were lost.

There are then two views of painting. First, the geometric-technical view associated with Protestant religion and the culture of silent reading suggests that the picture is something to look "through" as opposed to being a part of nature, another dimension of experience. Second, the expressive-physiognomic view suggests that the work bears similarities to the "phantom limb" phenomenon studied by Paul Schilder (1935/1972).[4] The body is projected into the painting. It flows into a tool for the purpose

---

[3] Carlo Ginzburg, *The Night Battles*.
[4] Paul Schilder, *Image and Appearance of the Human Body*.

of cutting or chipping a stone. It flows into the decoration of the handle of the tool with respect to the grip, or into the physiognomy of the rock transformed into a cutting devices.

5. SPACE FRAMES

The issue of the space-frame in relation to the agonistic experience objects is more complex than can be fully explained here. The "space-frame" is capable of taking the form of a cane, a car, or a property line, or a national boundary. It is absurd to try to derive these agonistic meanings based on geometric-technical understanding.

Suffice it to say that the model of the space-frame explains the intensity of the "relationship" the experient has with say a Rembrandt painting; it also explains the relationship of people to their personal environment. In the instance of a great work, the artist has produced an object wherein the notion of "fit" is deep, layered, and ever-changing. Each element of the work makes a profound contribution to the sense of complexity and richness of the work. Although it is contained within the frame of a canvas, it is experienced as allowing the experient to move through its components in order to heighten motoric as well as visual awareness.

A painting then becomes an "occasion map."[5] The experience is similar to the idea that a map (in two dimensions) of a country does not come to life until one has visited the territory and actually "walked" its landscape. Once there, an embodied sensation of its geographical orientation in relation to the sun is felt on the surface of the skin. This is the reverse of the historical phenomenon. The first maps were egocentric diagrams organized around journeys in which significant features were noted. The highest development of these "occasion maps" were the sea-charts used by navigators. With the compass and sextant they added a new dimension to the data. These lived, experiential maps had to be reconciled with "unlived" abstract conceptions related to the geographic totality.[6]

John Cage (and other artists) discovered that this ambient presence of information on the surface of the body (from sound and patterns of light and color) goes unnoticed because we are "tuned" to ignore the information. Visual absorption of the content of a painting involves a more extensive engagement to allow these qualities to be experienced as they are in music.

6. BRAIN PROCESSING OF SPACE

Perhaps there is another way to understand this conflict by following the lead offered by Roger Sperry (1983), Nobel prize winner for his research on split-brain

---

[5] Similarly, furniture is a map to the emotional space of a household. The body is merged with the object complement both in a literal and a symbolic sense.

[6] Coordinating these different kinds of theoretical space becomes a problem. First, "lived space" (the space in the minds of inhabitants including information from all the senses and emotional memories coming from the cartography of direct experience). Second, "conceived space" (abstract geometrical, frontal relations, based on maps, plans, numbers, and linear perspective). And finally, "perceived space" (space reflecting the appropriation and use of space to control or suppress behaviours, to maintain surveillance, to shape the character of work, private life, and leisure).

patients. In his last works he wrote of the brain which functions in a manner that bridges between the mystical and the scientific. It is capable of operating both in the temporal world (geometric-technical) and in the superearthly transcendent world. The brain develops new properties that are generated by interactions within the system as a whole as opposed to those produced by the acts of its individual constituents. New functions are generated based on "physical changes" in the brain due to experience.

To put it in a slightly different way, let us recall the fundamental observations made by Hughlings Jackson regarding the plasticity of the nervous system. As patients recovered from strokes he noted that they could re-access motor functions by relying on successively lower levels of the brain hierarchy. This was possible because the brain has such an intricately intertwined network of "organs" with overlapping and redundant functions.

The phenomenon of "blindsight" makes the point. These are patients who suffer the loss of the primary visual cortex on one side of the brain. These people are unable to "see" a target object (in the way we normally define "seeing" as for example, being able to discriminate an outline and color among other things) but they are able to "point" to a moving target. This is possible only when they are emotionally relaxed (not responding to the loss of function in their visual system). They are able to call upon a more primitive visual system that involves a lower level of visual brain functioning (e.g., a thalamic representation of retinal data rather than those normally sent by the retina to the—now defunct—occipital cortex via the lateral geniculate body).

This second system is designed for "emergency" use in relation to mating, feeding, hunting, avoidance of danger, adaptation of primordial settings, and territorial defense. Its biological function is to bypass thought and to respond without hesitation to a tree branch that looms ahead when we are walking through the forest. This type of vision is directly wired to motor and emotional receptors. It processes spatial data in a manner that is more autocentric (or egocentric) than allocentric (or objective), and involves a form of vision that processes spatial data unconsciously.

In this system, visual information is mixed with data from the other senses, hearing, touch etc. The cells are actually polymodal or synesthetic. The design of the secondary visual system has the role of "space tuning" the other senses. The auditory system is already tuned to search that same space. An auditory area is extended in the form of a "space-frame" in relation to the body. A barn owl, it is said, has a highly integrated visual and auditory map of its environment so it can locate a mouse accurately by the squeak it emits. It has a coordinated visual and aural space-frame.

7. HARMONIOUS SENSATIONS

The prospects of this system arose for me in thinking about the idea of developing a methodology for measuring "harmonious sensations," a combination of visual and aural sensations (this is an issue that came up in Japanese research on perception at the MERA conference in Tokyo, 1997). Much of the Japanese research in the studies on the experience of gardens requires a deep understanding of this kind of visual-aural system.

To extend this theory, imagine that the peak experience of the artist in responding to the emotional intensity of a particular color involves extending the body into the "space-frame" formed in relation to the color. The color has to be located in space in relation to other sensory data. Perhaps he dreams about some mud flecked on his boots.

The primary visual system may allow him to calculate how to mix the color as a visual memory. It does not help to recall the "feeling" of the color in relation to emotions and associations. The secondary visual system allows him to extend his body into a space frame that captures the qualities that captivated him when he first encountered the color on his boot? The paint color is now the experience of "tone" that fills a space that has been adjusted to include the grit of the mud. It is not a retinally apprehended "tint" or cover of the object surface.

New research on brain function suggests there is a far more complex architecture underlying these kinds of multi-modal spatial experiences, especially where diverse sources of stimulation are involved. The brain is now thought to be focused by a "conductor" function that modifies the relative role of different levels of the brain in relation to the environment as well as in relation to the pattern of will or concentration that is assumed by the experient. During waking hours the brain requires a high degree of orchestration due to the sheer abundance of external stimuli. In creative thinking or deep meditation, these "emergency" functions of the secondary visual system may be pressed into service to allow "emergent" qualities to come to the fore.

The key is to establish a viable relationship between the inner body schema and the environment. This requires defining a "space-frame" that allows for freedom of action, growth and change in the body image, provision of a sense of security, etc. If the environmental surfaces (rooms, buildings) are rigid or inadequate for accommodating this space-frame there will be an effort to change them. Alternatively, the ego can be forced to assume an alternate space-frame in response to environmental inadequacies. I spoke to a man recently who had sailed the Pacific alone about how he survived. He talked at length about the state of serenity he had simply looking at the night sky, and how during the day the boat required so much constant work that it became his universe.

8. GLOBAL APPLICATIONS OF SPACE FRAMES

These problems of space-frame apply on a global scale to territories and national boundaries. The Serbs will attempt to cleanse an historic area of Croats or Muslims, etc. The student in a dormitory room will add photographs of home to mollify his discomfort. Eventually, experience with new space-frames will supplant the old and cause the person (or group) to develop a new internal scheme (body image, self-concept). So for example, moving from the dormitory to a shared house will begin the process of extending the body-image of a young person to the boundaries of "home environment" as opposed to the fading memory of the loss of his or her childhood home. The settlement pattern of the West Bank and Jerusalem is an example of a complex arrangement of "States" with a mixed identity rather than complete autonomy. This requires a new formulation of the image of the body-politic of the Palestinian state. They will not be permitted the freedom to house foreign troops on their soil, for example.

The problem is that to understand these new types of individual or group person-environment relations requires a level of consciousness that goes beyond the geometric-technical. I believe it is necessary to look at person-environment relations associated with other means of concentration and focus, for example those associated with art, religion, mystical experiences, peak experiences, and deep experiences of meaning.

There are no person-environment studies that I know of in relation to homiletic interpretations of Biblical story space ("creationism") or the idea of Homeric space in Western education for citizenship. These are critical if one is to explore the differences that produce the underlying "fault lines" between Asian, Muslim, and Western views of space. There is no psychological examination of space in the context of cabalistic notions of "emanations of light." William James (1983) held that "consciousness of illumination" was the essential mark of the mystical state. Or the kind of lock in an endless space ("a block of ice 10,000 miles wide") that is commonly experienced as part of the process of meditation on a Koan by a Buddhist monk. Nor is there a comparative study of dream space related to mystic travel or "near death" experiences. Finally, there is no examination in psychological theories of mystical spaces as they appear in the context of different cultures as reported by Carlos Casteneda (1968).

The notion of the person in all these instances is different from that assumed in most studies. It includes "deep" powers of the personality that are not available in everyday life. It describes moments in life where people are required to attend to the "numinous" or invisible realm. Person-environment relations under these conditions are defined as an encounter with the sacred presence of a divinity, or as the result of the emptying of the mind common to Buddhist meditation. In the west the numinous takes the form of an external presence of the creative force outside the self. In the east the devotee has the sense of moving up toward the expression of a divine presence that already exists within the self and is immanent in everything else.

The "functional program" for a building (say a school) is simply to accommodate a space-frame for a small institution. The problem is to understand the emerging space-frame that will fit the mentality in a future. This is what Borromini accomplished in the Chapel at St. Ivo. The ecstatic spiral ceiling became an instance of a general space-frame that was adopted in counter-reformation Europe. Great art provides an intuitive connection to the unconscious in ways that takes hundreds of years to flower in rational consciousness.

To accept the split between the favored and so called "higher mental processes" and the "lower functions" is to perpetrate an undesirable stereotype. To challenge this split is to welcome the participation of lower levels of function to the mix. This not only means the contribution of the chair and floor surfacing to the office. It also involves the participation of "third world" languages, cultures and perspectives in the construction of "alter" realities.

The space-frame can be applied to the relationship of a person to physical models and metaphors as well as to environmental fit. There is often a contribution from "below." This depends on a more complex arrangement of brain function that "reasoning" related solely to the forebrain. This is confirmed in PET-scan studies of brain functioning under different kinds of concentration, discrimination, and problem solving. Some day it will be possible to perform mobile PET-scans in real environments to see how the space-frame affects people. One will measure the match between subjective accounts and brain activity. For example, if someone experiences a "harmonious" relationship between themselves and an environment, will this shift brain function from the occipital cortex to the thalamus and other centers of the brain involved with associations and emotional organization?

There are examples in the arts that offer suitable subjects for person-environment research involving specialized states of consciousness. Theater is a specialized autocentric domain where the viewer assumes the seat of the writer and director. The beholder must be able to extend himself into this space-frame. The experience of the folkish experience of the space of the tribal village are examples of opportunities to study the impact on "deeper" realms of the self as they are formed in relation to important space-frames. These inner modes of extending to culturally sanctioned space-frames may serve as a way of building an autocentic "ground" for the self. The language of space-frames is the same as the language by which we accumulate life's experiences.

One of the related areas I believe to be important is the study of field-dependence (e.g., Witkin & Goodenough, 1977) Cultures which rely on meditative practices tend to be more highly developed in the autocentric mode. If we assume that person-environment relations resolves the impact of the environment on personal development or cognitive processes, then it seems pretty clear that the autocentric and allocentric modes will have different ways of coping with environmental input. For example, say we study how the childhood home or growing up near a river effects the identity of an individual. From the point of view of the practical autocentric systems involved, these conditions have a major impact on the space-frames the individual can adjust to. But there is far more to it than that. We each carry with us a gendered and regionalized "personal" field. These patterns are also expressed in literature, for example in a New York City genre, or a Deep Southern mode. We learn to put ourselves in different space-frames, even the unfamiliar, when the lines are drawn clearly and the artist has created the space to make them accessible and safe to enter. In this way, the space-frame reflects the power of the unconscious to extend into domains such as science fiction, the new physics, deep ecology, distant religious practices, and poetry.

Critical experiences of person-environment relations are also "phatic," like a "call and response" relationship. In psychoanalysis, over a period of time, the analyst and the analysand co-create a space-frame that consists of specialized behaviors and interpretations. Transference and resistance are critical concepts for understanding the character of this small world. The analyst subordinates his concerns and maintains what Freud called a "hovering awareness." This enables him to help the analysand adjust to the space-frame in the office and then to adjust his space-frame to master the conditions of his life.

The brain has not changed much in the past 10–20,000 years. Most changes in function that occurred in Greece (changes in governance, the documentation of history and self-conscious philosophy) occurred as a result of the introduction of writing. Later, print and other forms of mediation between the individual and the environment, for example, money and other systems of regulation, required an adjustment of a similar magnitude. According to Tanto Pasto (as well as Werner, Wapner, 1952, etc.) this adjustment must have a motor component that is expressed in experiencing comfort in a new space-frame. With each major node point in history a new mentality is required. The Greeks experienced themselves as building a coherent culture within the boundaries of the state until Alexander's empire spread this culture (like a phantom limb) through his adventures into Persia and the Indus Valley where they mixed with and diluted by the space-frames of other cultures.

The problem is that there is a political connection between lower levels of function ("limbic brain") and the nation or state. In many instances the "suborn" can be related to "lower levels of functioning." As oral cultures, they are often not permitted to modify space to suit their needs. Like the American Indians whose mental space included large nomadic territories, they were unable to adjust to the confined conditions of a "reservation." This can be a source of tremendous internal distress if the space-frame of the reservation is inflexible and there is no room to make the needed motor adjustment. Any time a nomadic group is sedentarized there is a period of distress in forming a new motoric scheme.

During the period of European Enlightenment in the 17th and 18th centuries, part of the scientific justification for imperialism was achieved by placing exclusive focus on philosophy of the frontal lobe (epistemology). This was done at the expense of studies of the intuitive functions of the brain. This emphasis served European culture in establishing lines of trade and commerce based on an economic logic rooted in Newtonian physics. The new global map produced a mercantile economy that formed a space-frame for European nations and their colonies. But it also felt "natural" (like a phantom limb) as it gave them the freedom of action they needed to make motor adjustments in relation to world geography and resulted in imperialist ventures. Later, within its boundaries, and given the space to explore, the United States translated these motoric space-frame notions into concepts of "manifest destiny" and "rugged individualism." It pushed and shaped the frontier on the North American continent into a collective space-frame that served the interests of its Anglo-Teutonic majority.

From the beginning of the United States there were French and Spanish regions that were assimilated to the majority space frame. Interestingly, the Mexican "hacienda" had a powerful influence on the space-frame of European settlers in the Western United States, while it is only recently that the issue of conflicting comfort zones of Hispanics and Anglos has been focused on in Southern California. For example, the City of Southgate recently voted to ban the painting of private houses with the "tropical colors" (pink, light blue, mint green, etc.) that offer "comfort" in the Mexican and Central American space-frame and are considered to be "eye-sores" to the rest of the community.

9. SUMMARY

I take a long view of history of person-environment relations. This view is revealed in a serial understanding of history as a set of repeating and non-repeating patterns with regard to the impact of the deep psyche on the human environment. History is subject to radical phase-shifts due to the spread of culture through commerce, trade and war, the impacts of language and literacy (print culture), and their effects. New communication technologies with the potential for global "skywriting" (internet) will produce an evolution of the print medium with unknown future effects.

History is a reference plane for understanding the formation of global patterns and geopolitical lines of force. This global force field is associated with differences and similarities in cultural and environmental background. For example, the conflict in Bosnia and other ethically divided Balkan states, or Northern Ireland and the Middle East, are all "fault lines" wherein the impacts or more comprehensive shifting patterns have been intensified. These shifts reflect the search for a new orienting principle on a global scale.

Some cultures have developed in spatial isolation, as "tribal" cultures with local languages and habits restricted to adapting to their own geographical setting. Others have co-evolved in highly dense multicultural settings (such as within cities in the Ottoman Empire) where contrasting groups lived in close proximity. A geopolitical view of person-environment relations opposes the implicit idea that all cultures (based on the nation-state and ideology of the Enlightenment era) evolved under a common set of conditions:

- The notion of a person as bounded by the corporeal self as opposed to a collective definition of the self as Protean or made up of multiple personalities.
- A hegemonic view of the hierarchy of cultures as all developing in the direction of western models.
- The classical canonic view of culture based on ideas of a formal language based on logic underlying all vernacular languages.
- A stimulus centered notion of the object world as opposed to one involving a mystical perspective.
- A continuous and functionalist view of history as a coherent process as opposed to one capable of radical phase-shifts and changes of state.
- The idea that the world is organized according to lines of commerce and trade between independent nations.

International migration and resettlement will produce widespread patterns of multicultural society that will call to question the idea of the homogeneous state. National interests compete with concerns for the stability of the global economy. The world diagram has seismic fissures formed between competing cultures. As defined by Samuel Huntington, the world's cultural territories are Sinitic (Chinese), Japanese, Orthodox (Russian Steppes), African, Latin-American, Muslim, Hindu, American, European. Within each geopolitical territory there are competing religious and ethnic claims for centrality and acquisition of internal space-frames. These local claims need to be harmonized with pressures toward globalization around the new impacts of commerce and trade, transportation, information technology and services. CNN is a prime source of news for the Middle East and Russians are involved in Latin soap operas. There is a need to allow local identities to flourish (while instituting laws and practices against exclusionism) and accommodating the needs of global culture for common rules of disclosure, customs standards, airport agreements, etc. There is a danger that the market elite promoting the global market economy will take over for an elite concerned with the formation of democratic ethics and institutions that exhibit little awareness of the importance of space and person-environment relations.

REFERENCES

Casteneda, C. (1968). *Teachings of Don Juan: A Yaqui way of knowledge*. Berkeley, CA: University of California Press.

Ginzburg, C. (1992). *The night battles: Witchcraft and agrarian cults in the sixteenth and seventeenth centuries* (Anne Tedeschi and John Tedeschi, Trans.). New York: Penguin USA.

James, W. (1983). *The varieties of religious experience: A study in human nature*. NY: Penguin USA. (Original work published in 1902)

Koch, K. (1998). *Making your own days: The pleasures of reading and writing poetry*. NY: Scribner.

Pasto, T. (1964). *The space frame experiment in art*. NY: A. S. Barnes.

Schilder, P. (1935/1972). *Image and Appearance of the human body*. London: Kegan Paul, Trench, Trubner & Co., Ltd.

Sperry, R. W. (1983). *Science and moral priority: Merging mind, brain, and human values*. New York: Columbia University Press.

von Fieandt, K.(1966). *The world of perception*. Homewood, IL: Dorsey Press.

Werner, H., & Wapner, S. (1952). Toward a general theory of perception. *Psychological Review, 59*, 324–338.

Werner, H., & Wapner, S. (1954). Studies in physiognomic perception: I. Effect of configurational dynamics and meaning-induced sets on the position of the apparent median plane. *Journal of Psychology, 38*, 51–65.

Witkin, H., & Goodenough, D. R. (1977). Field-dependence and interpersonal behavior. *Psychological Bulletin, 84*, 661–689.

23

EPILOGUE

Similarities and Differences across Theories of Environment-Behavior Relations

Seymour Wapner,* Jack Demick,** Takiji Yamamoto,***
and Hirofumi Minami****

*Heinz Werner Institute for Developmental Analysis
Clark University
Worcester, Massachusetts 01610-1477
**Department of Psychology
Suffolk University
Boston, Massachusetts 02114
and Center for Adoption Research and Policy
University of Massachusetts
Worcester, Massachusetts 01605
***Research Institute of Human Health Science
10-51-215, Ichibaue-machi
Tsurumi-ku, Yokohama, 230 Japan
****Department of Urban Design
Planning and Disaster Management
Graduate School of Human-Environment Studies
Kyushu University
6-19-1, Hakozaki, Fukuoka, 812 Japan

1. INTRODUCTION

The papers in this volume take a major step in articulating and elaborating many of the assumptions underlying the work of contemporary environment-behavior researchers. This is particularly important for a variety of reasons. First, as we have asserted elsewhere (e.g., Wapner & Demick, 1998), there is a need for researchers to acknowledge that inquiry and knowledge are always biased and that there is no process of "neutral" observation, inquiry, or conclusion in any science. This assertion

Theoretical Perspectives in Environment-Behavior Research, edited by Wapner *et al.*
Kluwer Academic / Plenum Publishers, New York, 2000.

is based on the notion of perspectivism, which—in its most general form—assumes that any object, event, or phenomenon is always mentally viewed from a particular standpoint, or world view, which is capable of definition (cf. Lavine, 1950a, 1950b, on interpretationism).

Second, strongly believing in the interrelations among problem, theory, and method in science, we have also maintained that there is considerable value—for both the scientist and the practitioner (cf. Wapner & Demick, 1999)—in uncovering the ways in which one's theoretical orientation determines, at least in part, what one studies (problem) and how one studies it (method).

Third, as Werner and Altman (chapter 3; cf. Altman, 1997) have stated:

". . . we have found articulating research assumptions to be quite liberating. By putting traditional assumptions in perspective, we recognize they are just one of several ways of doing research. We feel comfortable trying out alternative approaches and exploring new ways of thinking about and studying phenomenon (e.g., Oxley, Haggard, Werner, & Altman, 1985). Indeed, being aware of alternative ways of knowing has helped us see limitations in traditional psychological approaches. In our own work, it helps us see where we have been, where we could go, as well as enabling us to see what we have overlooked." (p. 22)

Fourth, uncovering assumptions is helpful not only for individual scientists but also for their potential collaborators. Again in the words of Werner and Altman (chapter 3):

"Emerging opportunities for environment-behavior researchers also argue for a more careful articulation of research assumptions. More and more funding agencies expect research proposals to adopt a multidisciplinary approach . . . In order for researchers to communicate effectively across disciplinary boundaries, it is essential that we be aware of our fundamental research assumptions, know how to select methodologies most appropriate for different problems." (p. 36)

2. COMPARISON OF THEORIES AGAINST THE HOLISTIC, DEVELOPMENTAL, SYSTEMS-ORIENTED APPROACH

Thus, against this backdrop, the remainder of this chapter takes the following form. As described in chapters 2 and 16, our elaborated perspective is currently comprised of a set of interrelated assumptions about human action and experience in the complex everyday life environment—that is, what we believe to be the fundamental problem of environment-behavior research. In Table 1, these assumptions are delineated and, where relevant, compared across the contributing authors to this volume. Further, the major of these assumptions are described in more detail below and then used as categories for comparing our approach and those represented herein.

Before proceeding with the following analysis that is based on our perspective, we emphasize that this choice was clearly an arbitrary and convenient one. Parallel comparisons can be readily made using other perspectives presented in this volume as the basis for comparison. Indeed, we would encourage such undertakings because, in keeping with the basic goal of this volume, they would lead to better understanding among the contributors of the large variety of perspectives represented here.

2.1. World View

There is a major difference between the underlying "world view" (Altman & Rogoff, 1987) or "world hypothesis" (Pepper, 1942, 1967) of our approach and some of the others. Specifically, *our approach adopts elements of both organismic (organicist) and transactional (contextual) world views*. The organismic world view is embodied in an attempt to understand the world through the use of synthesis, that is, by putting its parts together into a unified whole. Such a view highlights the relations among parts, but the relations are viewed as part of an integrated process rather than as unidirectional chains of cause-effect relationships. The major feature of a transactional world view is that the person and the environment are considered parts of a whole so that one cannot deal with one aspect of the whole without treating the other (cf. Cantril, 1950; Ittelson, 1973; Lewin, 1935; Sameroff, 1983; Wapner, 1987). Specifically, the transactional view treats the "... person's behaving, including his most advanced knowings as activities not of himself alone, nor even primarily his, but as processes as the full situation of organism-environment" (Dewey & Bentley, 1949, p. 104).

These world views have figured within our approach as follows. First, they have impacted our choice of paradigmatic problems—for example, person-in-environment transitions across the life span—as well as of methods, that is, methodological flexibility depending on the nature of the problem coupled with the use of multiple methods for as complete a characterization of action and experience as possible. Second, they have suggested that holistic, ecologically-oriented research is a necessary complement to more traditional laboratory work and that is might be conducted through reducing the number of focal individuals studied rather than the number and kind of interrelationships among aspects of the person, of the environment, and of the systems to which they belong. This helps us conceptualize problems that are more in line with the complex character of everyday life and that cut across various aspects of persons and various aspects of their environments (cf. Demick & Wapner, 1988a).

With respect to the other theories represented in these volumes that differ from ours, those like Rapoport's (chapter 10) that is restricted to natural science and those that treat the biological and/or psychological levels (e.g., Michelson, chapter 11; Ohno, chapter 12; Evans & Saegert, chapter 20) would predominantly be termed "interactional" or "mechanistic," emphasizing a sensorial analysis of the effects of isolated independent variables within the person (e.g., perceptual processes) and/or the environment (e.g., stressors such as crowding and noise) on dependent variables (e.g., psychological functioning of the person). In contrast, our underlying transactional world view appears to be shared by many authors in this volume, perhaps signifying their tendency to embrace openly this relatively recent theoretical development in light of the typical complexity of environment-behavior research problems.

As two extremely clear examples of approaches that are most compatible with our underlying transactional world view, consider the relatively different views of Werner and Altman (chapter 3) versus Seamon (chapter 13) and Childress (chapter 14). Assuming that individuals and their psychological processes are embedded in and inseparable from their physical and social contexts, Werner and Altman have provided the example of how a transactional view might frame the study of landscaping around the home:

Table 1. Assumptions of Holistic, Developmental, Systems-oriented Perspective and Comparison with Other Perspectives

Category of Comparison	Holistic, Developmental, Systems-oriented Approach (Wapner & Demick)	Werner/Altman	Kobayashi/Miura	Rivlin	Bechtel	Bonaiuto/Bonnes	Little	Churchman	Rapoport	Michelson	Ohno	Seamon	Childress	Canter	Demick/Wapner/ Yamamoto, Minami	Nagasawa	Takahashi	Minami/Yamamoto	Evans/Saegert	Stokols	Rand
World View	Organismic	X			X										X			X			
	Transactional	X	X	X	X	X	X	X				X	X	X	X	X	X	X		X	X
Philosophical Underpinnings	Constructivism (Interpretationism)		X	X		X	X			X		X	X	X	X	X	X	X		X	X
Levels of Integration	Functioning at different though related levels of integration (biological, psychological, sociocultural)	X		X											X						X
Unit of Analysis	Person-in-environment as system state	X		X	X		X					X	X	X	X	X		X		X	X
Holism/Aspects of Experience/ Equilibration Tendencies	Person-in-environment system operates as a unified whole in dynamic equilibrium	X							X					X	X						
Concept of Person	Defined with respect to levels of integration: Physical/biological, psychological (cognition, affect, values), and sociocultural (e.g., roles) as mutually defining aspects of person; multiple intentionality						X							X	X					X	
Concept of Environment	Defined with respect to levels of integration: Physical (things), interpersonal (people), and sociocultural (rules, mores, customs) as mutually defining aspects of environment			X			X						X			X	X	X	X	X	X

Category	Feature / Description													
Structural and Dynamic Analysis	Focus on structural (part-whole) and/or dynamic (means-ends) analyses								X					
Concepts, Principles, and Endpoints of Development	Mode of analysis of person-in-environment contextual functioning across the life span; in addition to ontogenesis concerned with microgenesis, phylogenesis, etc.; homeostasis works in accordance with orthogenetic principle toward optimal self-world relations (e.g., microgenetic, mobility, freedom, self-mastery)	X							X			X		X
Temporal Features	General change	X	X	X		X	X	X	X	X	X	X	X	X
	Developmental change	X	X						X			X	X	X
Adaptation	Congruent person-in-environment system state: Optimal relations between person and environment	X							X		X	X	X	X
Individual Differences	Differential developmental psychology complementary to general developmental psychology								X					
Multiple Worlds	Different experiential worlds or spheres of activity		X					X	X					X
Theory of action	Relations of experience (intention) and action	X	X	X	X	X	X	X	X		X	X	X	X
General Perspective	Natural Science	X	X			X	X	X	X		X	X	X	X
	Human Science	X	X			X	X	X	X	X	X	X	X	X
Problem Formulation and Methodology	Methodological flexibility depending on level of organization and nature of problem under scrutiny; draws from experimental, naturalistic, observational, and phenomenological methods toward formal theory developmental; qualitative understanding of context-specific psychological events; praxis	X			X				X					X

"... the transactional view emphasizes the dynamic unity between people and setting. In this approach, one focuses on psychological processes that can be used to define both the yard and the residents, and one assumes that both yard and residents are changed through their transactions. For example, 'identity expression' is a concept that can be used equally well to describe a setting and the people who live there."

"People use their homes and yards to express their identities as unique individuals as well as their identities as members of groups and the broader society. Thus the yard and family are inseparable and mutually defining. Research questions based on these processes include: How does the yard reflect the family's self-expression processes? Do families with different individual or communal identities select different kinds of yards and change the landscaping in ways that support their self-expressive goals? How do individual and collective styles change across the lifespan, and how do yards mirror these changing styles? How do different cultures express identity, and how do yards reflect these expressive styles? ..." (pp. 23 and 26)

In a related vein, the phenomenological studies of Seamon and Childress are alternative, though no less compelling, examples of transactional views of environment-behavior research. Stated succinctly, these researchers believe that individuals and the environment compose an indivisible whole and that the wholeness of person and place is best conveyed through the use of one of several phenomenological (e.g., first-person, existential, hermeneutic, narrative) methodologies. Thus, at this point in time, transactional views of environment-behavior research have become manifest in many different theoretical and methodological forms. Both Nasagawa (chapter 17) and Takahashi (chapter 18), for example, point to the shift from a deterministic toward a transactionlist view in the field of architecture.

2.2. Philosophical Underpinnings

Our perspective assumes a *constructivist* view of knowledge. Specifically, we assume that *cognitive processes involve the person's active construction of objects of perception and thought.* Such an approach rejects all "copy" theories of perception and asserts that reality is relative to a person's interpretation or construction (cf. Lavine, 1950a, 1950b). In line with this, human beings are regarded as striving agents capable of creating, constructing, and structuring their environments in various ways and of acting in terms of their own experience. Wapner, Kaplan, and Cohen (1973) have characterized such striving in terms of Kuntz's (1968) notion that the individual, functioning at the sociocultural level, exhibits a "rage for order." These notions also lead to consideration of the distinction between the *physical* versus *experienced* environment, which has also been referred to as the *behavioral environment* (Koffka, 1935), *umwelt*, *phenomenal world, self-world* (von Uexkull, 1957), and *psychological environment* (Lewin, 1935).

This constructivist assumption is readily apparent in a study by Dandonoli, Demick, and Wapner (1990), which demonstrated that the cognitive-developmental status of the individual penetrates and plays a relatively powerful role in the way in which he or she organizes a new environment to which he or she is exposed. That is, adults experienced and represented part-quality room arrangements (e.g., furniture stacked as in a storeroom) as socially relevant meaningful wholes (e.g., sitting area, mail area).

The constructivist assumption is readily shared by many contributors to this volume. For example, Bonaiuto and Bonnes (chapter 7)—stressing the nature and features of environmental issues constructed by different social actors through different discursive selections and framings—have reported a study on environmental pollution in which ". . . the stronger the local identity of residents, the less polluted they perceived the beaches of their own towns to be . . . the stronger the national identity of residents, the less polluted they perceived the beaches of their own nation to be" (p. 72). In a related vein, Rand (chapter 22) has drawn on Werner's (1940/1957) distinction between geometric-technical (objective) and physiognomic (expressive/dynamic/affective) perception to develop Pasto's (1964) notion of space frame (pervasive spatial scheme) as a unifying theme in environment-behavior studies. What distinguishes elaborated Wernerian theory and research from that of other constructivists, however, is that, for us, the constructivist assumption reinforces the notion that there is the need to complement assessments of, for example, the geographic environment with those of the experienced environment.

2.3. Levels of Integration

This assumption states that *organism-in-environment processes may be categorized in terms of levels of integration* (Feibelman, 1954; Herrick, 1949; Novikoff, 1945a, 1945b; Schneirla, 1949), *that is, biological (e.g., breathing), psychological (e.g., thinking), and sociocultural (e.g., living by a moral code)*. Thus, there is a contingency relationship: functioning at the sociocultural level requires functioning at the psychological and biological levels, and functioning at the psychological level requires functioning at the biological level. The levels differ qualitatively and functioning at one level is not reducible to functioning on the prior, less complex level since we assume that higher level functioning does not substitute for, but rather integrates and transforms, lower level functioning (Wapner & Demick, 1990; Werner, 1940/1957).

This assumption has played a very important role within our research and highlights a significant driving force, namely, *rejection of reductionism*. Though most theoretical perspectives recognize that various relationships obtain between biological and psychological functioning and between psychological and sociocultural functioning, the contingency relationship for defining various levels of integration is interpreted for use in different ways. One alternative that we strongly reject is biological reductionism, which assumes that functioning is determined completely by the biological structure and state of functioning of the organism. Such a reductionist approach usually means an attempt to understand psychological functioning by translating its principles into those involving only biological terms (e.g., genetic bases of personality). In contrast, our approach has taken the position that levels of integration must be considered in any analysis of psychological functioning. Focus on a particular level depends on the specific question or issue posed or confronted together with the recognition that impact on one level affects all other levels of functioning, that is, the functioning of the whole. In general, there is no single way of analyzing or "explaining" action and experience independent of the goals of the analysis. This position permits exploration of the same phenomena that concern both biological and sociocultural determinists without excluding either biology or culture (cf. Wapner & Demick, 1998).

Thus, from our perspective, those formulations that treat individual functioning primarily at the biological (e.g., Ohno, chapter 12), psychological (e.g., Churchman, chapter 9), or sociocultural (e.g., Minami & Yamamoto, chapter 19) level of organization are erroneous insofar as they logically permit the possibility of biological, psychological, or sociocultural factors playing no role at all. The assumption that the individual functions at multiple levels of integration means that biological, psychological, and sociocultural factors must *always* be taken into account. Thus, our position fits most closely with those expressed by Werner and Altman (chapter 3), Rivlin (chapter 5), and Rand (chapter 22). For example, as Rivlin has stated about her current research on homelessness:

"There are multiple levels of analysis . . . of homelessness that offer a complex view beyond the first generalization of the trauma faced by people. This seems to transcend location and points to specific details that can be used to examine homelessness, across sites. For example, research on homeless children in New York City has identified their irregular attendance in school and their misplacement in classes . . . For some children, certainly not the majority, who are able to keep going to school, it can be the most 'normal' component of the child's life. The findings offer a view of the resources that are needed to buffer the impacts of homelessness on children, and school stability is one of them. This could apply to any kind of community and transcends the specific location . . . " (p. 6)

2.4. Unit of Analysis

This basic assumption holds that *the person-in-environment system is the unit of analysis with transactional (experience and action) and mutually defining aspects of person and environment.* Treating the person-in-environment as the unit of analysis has the advantages that it corresponds to and represents the complexity of the real life situation, that it suggests analysis of the individual's behavior and experience in a variety of contexts (thus, environmental context is built into and an essential part of the unit of analysis), and that it is both comprehensive and flexible in uncovering sources of variation underlying behavior. This assumption has figured prominently in our research for some time. Specifically, it has led to an unwavering commitment to conceptualizing the individual (as a system at the various levels of integration), individual-in-group, individual-in-organization, and organization-in-environment systems (see Mayo, Pastor, & Wapner, 1995).

Relevant to this conceptualization, for example, is Bechtel's (chapter 6) work in the field of ecological psychology, which centers around the behavior setting (". . . a standing pattern of behavior and a part of the milieu which are synomorphic and in which the milieu is circumjacent to the behavior"; Barker & Wright, 1955, p. 45) as the fundamental unit of analysis of human behavior. While Bechtel's unit of analysis (behavior setting) shares some basic similarity with ours (person-in-environment system) insofar as human behavior and the environment are conceptualized as inseparable, there is also a fundamental difference. In Bechtel's own words:

"The fundamental difference is the broader view that these two views take compared to the setting specificity of ecological psychology. Ecological psychology would not deny the larger context, but it would insist that all influences are mediated within and through the behavior setting boundaries." (p. 62)

By implication, Bechtel's approach negates the notions of individual differences and development—notions that figure prominently within our approach—since, for him, the context (i.e., behavior setting) always defines and prescribes the behavior independent of individual differences. Nonetheless, we agree with his conclusion that:

"The evolution of the transactional and organismic views tend to incorporate much of the assumptions and methods of ecological psychology and to keep many of the assumptions of conventional psychology also. It will be interesting to see what the future brings from such a mix." (p. 66)

2.5. Holism, Aspects of Experience, and Equilibration Tendencies

We also assume that *the person-in-environment system operates as a unified whole so that a disturbance in one part affects other parts and the totality.* This holistic assumption holds not only for functioning among levels of integration (biological, psychological, sociocultural), but also for functioning within a given level. For example, on the psychological level, such part-processes as the *cognitive* aspects of experience (including sensory-motor functioning, perceiving, thinking, imagining, symbolizing) as well as the *affective* and *valuative* aspects of experience and *action* operate contemporaneously and in an integrated fashion in the normal functioning adult (cf. Wapner & Demick, 1990). Finally, related to this notion, we also assume that the tendency toward equilibration is a basic end that operates at all levels of organization (e.g., homeostatic mechanisms at the biological level, perceptual adaptation at the psychological level, sociocultural adaptation to a new environment following relocation). Thus, *person-in-environment systems are also assumed to operate in a dynamic equilibrium.*

These holistic assumptions are consonant with the work of several of the other contributors to this volume. For example, Rapoport (chapter 3) has begun by posing what he considers the three most basic questions for the field of environment-behavior studies, viz.: (a) what biosocial, psychological, and cultural characteristics of human beings influence which characteristics of built environments?; (b) what effects do which aspects of which environments have on which groups of people under what circumstances and why?; and (c) what are the mechanisms that link person and environment? He has then attempted to answer these questions (and to integrate the field) by demonstrating holistically how culture (e.g., world views, values, rules) is interrelated with the individual (e.g., schemata, lifestyle, activity systems) and the built environment (e.g., organization of space/time/meaning/communication, system of settings, cultural landscape, fixed/semi-fixed/non-fixed features). In a related manner, Canter (chapter 15) has discussed the interrelations between the meanings and actions that people associate with places that lead to place preferences. While both these authors imply that the person-in-environment system operates in a state of dynamic equilibrium, our approach is different insofar as we state these notions explicitly and attempt to incorporate them into our empirical research.

2.6. Concept of Person

We define the person aspect of the person-in-environment system with respect to levels of integration and so assume that *the person is comprised of mutually defining physical/biological (e.g., health), intrapersonal/psychological (e.g., self-esteem), and sociocultural (e.g., role as worker, family member) aspects.* Further, in line with our constructivist assumption, we regard individuals as active, striving, purposeful, goal-oriented agents capable of spontaneously structuring, shaping, and construing their environments in various ways and acting in terms of their own experience.

We also assume that human beings are characterized by a "rage for order" (Kuntz, 1968) and by *multiple intentionality*. That is, the person transacting with his or her environment has the capacity to focus on different objects of experience such as the self, an object out-there, and the relations between both. Moreover, the person has the capacity to *plan*, which involves plotting future courses of action that move the person-in-environment system to some end state (Wapner & Cirillo, 1973).

Many of the authors in this volume share a multifaceted conception of the person. For example, Bonaiuto and Bonnes (chapter 7) have considered the individual not ". . . as an isolated entity, but as a part embedded in a social network or system, that is, an entity materially and symbolically connected to other individuals and social entities, to be studied and understood with respect to the social processes such as those of communications, groups, institutions, etc." (p. 68). Thus, like many other authors, they have integrated two levels in their concept of person (here, psychological and sociocultural), but have not holistically employed all three levels of integration. Of the authors represented here, Little's (chapter 8) work implies the most similar concept of person to our own. That is, through his focus on personal projects (extended sets of personally relevant action) as the unit of analysis, he has detailed concrete ways that individuals act on environments. These ways, or projects, may typically involve (among other things) physical (e.g., going on a diet), intrapersonal (e.g., doing well at school), and sociocultural (e.g., being a good father) aspects of self. However, relations among personal projects at the differing levels of integration remain an open empirical problem.

2.7. Concept of Environment

Analogous to our conceptualization of person, we assume that *the environment aspect of the person-in-environment system is comprised of mutually defining physical (e.g., natural and built objects), living organism (e.g., spouse, friend, pet), and sociocultural (e.g., rules and mores of the home, community, and other cultural contexts) aspects* (cf. Demick & Wapner, 1988b). Again, we do not focus on the person or on the environment *per se* but rather consider the person and the environment relationally as parts of one whole.

While several contributors (e.g., Rivlin, chapter 5; Little, chapter 8; Childress, chapter 14; Nagasawa, chapter 17; Takahashi, chapter 18; Evans & Saegert, chapter 20; Stokols, chapter 21) share our multidimensional view of the environment, the work of Rand (chapter 22) is particularly noteworthy. That is, in his discussion of the space frame as a notion that accounts for many aspects of person-environment relations in a manner compatible with art and architectural history, he has attempted to explicate all forms of contextual input (including the historical) within a unifying framework.

2.8. Structural and Dynamic Analyses

We view the person and the environment as structural components of the person-in-environment system. Drawing on the theme of self-other differentiation, our *structural, or part-whole, analyses* focus on the characteristic structure of person-in-environment systems with an eye toward discerning whether the parts of subsystems (e.g., person, environment) are more or less differentiated and/or integrated with one another in specifiable ways. We also view *dynamic, or means-ends, analyses* as complementary aspects of a formal description of a person-in-environment system. Focusing on the dynamics of a system entails a determination of the means (e.g., rational vs.

irrational) by which a characteristic structure or goal (e.g., preparation for a hurricane) is achieved or maintained. For some time now, we have, for example, focused on the cognitive process of planning, that is, the verbalized plotting of a future course of action, as one of a number of the means by which the person-in-environment system moves from some initial state of functioning to some end state.

While some of the other contributors have capitalized on either structural (e.g., Bechtel, chapter 6, on the size of the social unit) or dynamic (e.g., Kobayashi & Miura, chapter 4, on positive and negative means following a natural disaster) analyses, no other theory of person-environment functioning except our own attempts to integrate structural and dynamic analyses as complementary aspects of a formal description of person-in-environment system states. Moreover, the relations between structural and dynamic aspects of systems is an area that appears worthy of further empirical investigation.

2.9. Concepts, Principles, and Endpoints of Development

Our view of development transcends the boundaries within which the concept of development is ordinarily applied and not restricted to child growth, to ontogenesis. We, in contrast, view *development more broadly as a mode of analysis of diverse aspects of person-in-environment functioning*. This mode of analysis encompasses not only ontogenesis, but also microgenesis (e.g., development of an idea or percept), pathogenesis (e.g., development of neuro- and psycho-pathology), phylogenesis (development of a species), and ethnogenesis (e.g., development of a culture) (cf. Evans & Saegert, chapter 20).

Components (person, environment), relations among components (e.g., means-ends), and part-processes (e.g., cognition) of person-in-environment systems are assumed to be developmentally orderable in terms of the orthogenetic principle (Kaplan, 1966, 1967; Werner, 1940/1957, 1957). *The orthogenetic principle defines development in terms of the degree of organization attained by a system. The more differentiated and hierarchically integrated a system is, in terms of its parts and of its means and ends, the more highly developed it is said to be.* Optimal development entails a differentiated and hierarchically integrated person-in-environment system with flexibility, freedom, self-mastery, and the capacity to shift from one mode of person-in-environment relationship to another as required by goals, by demands of the situation, and by the instrumentalities available (e.g., Kaplan, 1959, 1966; Wapner, 1987; Wapner & Demick, 1990).

The orthogenetic principle has also been specified with respect to a number of polarities, which at one extreme (left) represent developmentally less advanced and at the other (right) more advanced functioning (cf. Kaplan, 1966, 1967; Werner, 1940/1957; Werner & Kaplan, 1956). These polarities are synoptically described as follows: (a) *integrated to subordinated* (in the former, ends or goals are not sharply differentiated; in the latter, functions are differentiated and hierarchized with drives and momentary states subordinated to more long-term goals); (b) *syncretic to discrete* (syncretic refers to the merging of several mental phenomena, whereas discrete refers to functions, acts, and meanings that represent something specific and unambiguous); (c) *diffuse to articulate* (diffuse represents a relatively uniform, homogeneous structure with little differentiation of parts, whereas articulate refers to a structure where differentiated parts make up the whole); (d) *rigid to flexible* (rigid refers to behavior that is fixed and not readily changeable; flexible refers to behavior that is readily changeable or plastic); and (e) *labile to stable* (labile refers to the fluidity and inconsistency that go along with changeability; stable refers to the consistency or unambiguity that occurs with fixed properties).

As noted above, while several contributors have attempted to introduce developmental notions within their conceptualization (e.g., Rivlin, chapter 5; Childress, chapter 14), few have done so utilizing systematic developmental principles. As an exception, Kobayashi and Miura (chapter 4) have posited three developmental stages in victims' responses following natural disaster, namely, the short-term response, the long-term effects, and the recovery process. Further, we argue that significant advances in the theory and method of environment-behavior studies may obtain were investigators to incorporate developmental notions such as those posited here into their theorizing and research.

2.10. Temporal Features

Time and temporal qualities are properly considered as a complicated, largely ignored but central issue by Werner and Altman (chapter 11). They raised many interesting questions about temporality but have not considered that both temporal qualities and spatial qualities represent abstractions from ongoing human functioning that is spatio-temporal in character. As we see it, human beings can structure their experience of spatial features as if independent of temporality as well as structure their experience of temporal features as if independent of space. Moreover, we believe a distinction should be made between general change and developmental change. Developmental change is directed toward some end state (e.g., as we have specified, the orthogenetic principle speaks to an end state of differentiation and hierarchic integration); in contrast, no particular end state is specified for general change.

2.11. Adaptation

We conceptualize *adaptation as a congruent person-in-environment system state consisting of optimal relations between the person and his or her environment.* This stands in marked contrast to those approaches that conceptualize adaptation as either the general adaptation level of the person (Helson, 1948) or adaptation of the individual to a particular sociocultural context such as the family (e.g., Minuchin, 1974) or society at large (e.g., Vygotsky, 1978).

Conceptualizations of adaptation represented in this volume range from the general adaptation level of the person (e.g., Evans & Saegert, chapter 20) to one somewhat similar to our own (e.g., Werner & Altman, chapter 3; Stokols, chapter 21). However, what distinguishes our concept of adaptation from that of the others lies in our underlying world view and corresponding analytic strategy. That is, whereas other authors have often advocated matching characteristics of the individual with characteristics of the environment, our transactional approach—which has advocated that the parts, viz., person and environment, must be treated relationally as parts of one whole—has alternatively led to the assessment of adaptation through examination of the structural characteristics of the person-in-environment system as a whole (see below).

2.12. Individual Differences

There is a major difference in the concept of individual differences between our approach and all of the others. That is, several other authors either interpret individual differences as a source of error or as manifest in different personality dimensions (e.g., Churchman, chapter 9) or pathological states (e.g., Nagasawa, chapter 17). In contrast, we see individual differences as contributing to a differential developmental psychology that is complementary to a general developmental psychology.

Our developmental analysis of self-world relations utilizing the orthogenetic principle may be applied to describe individual differences in a broad variety of content areas and modes of coping. One approach to formulating individual differences according to this developmental scheme was presented by Apter (1976). Specifically, she studied some relations between the individual's formulation of plans for leaving an environment and his or her coping with discrepancies between the expectation and the actuality of environmental transactions. Planning groups included: college seniors with clearly articulated plans as to what they would do after graduation; and those with no such plans.

In terms that are developmentally orderable with respect to the orthogenetic principle, status of plans was related to modes of coping as follows:

1. *Dedifferentiated P-in-e state*: The senior without plans coped with transactional conflict by *accommodation*, where a student who expected the facilities to work properly took no action, conforming outwardly to "fit in" with the environmental conflict between expectation and actuality;
2. *Differentiated and isolated P-in-e state*: The senior in the process of making plans, somewhat less invested in and more differentiated from the environment, coped by *disengagement*, distancing himself or herself from the environment by mocking it;
3. *Differentiated and in conflict P-in-e state*: The senior, again without completely articulated plans, kicks a washing machine that is the source of the conflict and copes with frustration, anger, and disappointment through general usage of the strategy of *nonconstructive ventilation*; and
4. *Differentiated and hierarchically integrated P-in-e state*: The senior with securely established and highly articulated plans handles transactional conflict by *constructive assertion*, that is, with coping by planned action where he or she is less dominated by emotions and engaged in a hierarchically integrated self-world relationship.

Among the authors contained herein, our approach to individual differences is unique insofar as no other contributor has identified individual differences within the framework of a general developmental psychology. The above example also highlights the potentially important role of the individual's planning processes (that may be conceptualized developmentally) and suggests that such processes need to be brought to more significant prominence within environment-behavior theorizing.

2.13. Multiple Worlds

Another approach to individual differences has concerned our assumption of multiple worlds. In our culture, different yet related experiential worlds (Schutz, 1971) or spheres of activity consist of the *multiple worlds of family, of work, of school, of recreation, of community, etc.* Each of these worlds usually occupies distinct spatiotemporal regions, involves different sets of people, and operates according to different sets of rules. For us, empirical issues concern the relative centrality of these worlds in the individual's life, the relation of these worlds to one another, and the manner of movement from one world to another. In line with the orthogenetic principle, we ask whether the worlds are relatively fused, relatively isolated, in conflict, or integrated with one another. For example, critical person-in-environment transitions are often initiated when there is acquisition of a new world (e.g., beginning a new job), replacement of one world by another (e.g., migration to a new country), or excision or deletion of one world (e.g., retirement).

We also distinguish *mytho-poetic*, *ordinary-pragmatic*, and *scientific-conceptual* attitudes as well as the experienced worlds corresponding to these attitudes. Consider the experience of thunder. As an aspect of the mytho-poetic world, thunder has expressive qualities: it is personally threatening, bodes ill, and is ominous. As part of the ordinary pragmatically viewed world, thunder may signal impending rain and remind the individual to take a course of action to avoid being struck by lightning. As a phenomenon in the scientific world, thunder is viewed as occurring in specifiable, objective circumstances and has general features such as its intensity that may be defined, measured, and investigated. These worlds may be analyzed similarly to multiple worlds. For example, we might examine: the *patterns* of an individual's experienced worlds (e.g., dedifferentiated, differentiated and isolated, differentiated and in conflict, differentiated and integrated); the *centrality* of a particular reality in his or her total experience (e.g., a scientifically-oriented person with an artistic side); individual differences in the *dominant*, overall character of the experienced world at a particular time and the relative *salience* of each mytho-poetic, practical, and scientific aspect; and the individual's relative *control* over shifts among these general attitudes.

Our assumption of multiple worlds is most compatible with the work of Rivlin (chapter 5), who in her research on children's experience of homelessness, has ascertained that the multiple worlds of these children is quite important. As she has stated, "The findings offer a view of the resources that are needed to buffer the impacts of homelessness on children, and school stability is one of them" (p. 57). As our approach advocates, further advances in theory and method might be brought about through more refined structural analyses aimed at assessing the relations among these children's worlds.

2.14. Theory of Action

We believe that an important problem related to the transactions of the person with the environment concerns the relation between experience and action. This problem has entered into our paradigmatic studies of critical person-in-environment transitions where we have attempted to study the change from "wanting to do something" (i.e., intentionality) to "doing it" (i.e., carrying out the action). Here, based on the musical instrument metaphor (tuning vs. activating inputs) that Turvey (1977) employed in analyzing neurophysiological processes, we have examined: usage of automobile safety belts prior to and following mandatory legislation (e.g., see Demick *et al.*, chapter 16); initiating a diet regime; abstention from the use of tobacco; and abstention from the use of alcohol.

For example, tuning corresponds to the category we have described as *general factors*, which may be necessary but not sufficient for action (e.g., in the case of automobile safety belt usage, a basic belief in the need for safety belts). Activating corresponds to our second category of *precipitating events* (*specific precursors* or *triggers*), which serve as the initiators of action (e.g., wanting to serve as a role model for children in the car). Thus, the translation of experience to action requires both general factors and specific precipitating events.

Further, since the goals and instrumentalities governing action differ at the various levels of integration (biological, psychological, sociocultural), we have attempted to develop systematic, holistic constructs accounting for relations between and among levels. Utilizing Wapner's (1969) earlier application of these categories to relations among cognitive processes, we have documented that the relations between any two levels may be *supportive* (e.g., personal opinion about safety belt usage and public policy coincide), *antagonistic* (e.g., personal opinion and public policy are antithetical), or *substitutive/vicarious* (e.g., substituting the good of society for one's own desires). Such conceptualization has the potential to shed light on such important questions as: What are the mechanisms that underlie the ways in which rules and regulations at the sociocultural level become translated into individual functioning at the psychological level? Conversely, how does the need for the quality of human functioning at the biological and psychological levels become actualized at the sociocultural levels (cf. Demick & Wapner, 1990)?

Numerous contributors to this volume have treated the problem of the relations between experience and action (See Nagasawa, chapter 17, on wayfinding). On the one hand, Bechtel (chapter 6) has asserted that:

"People willingly join settings, in fact, are born into them, learn the proper pattern of behavior, and graduate to other settings. A person's life is defined by the series of settings entered. There is more choice about entering a setting then there is about how one will behave after entering. One enters a setting because one wants to do the behaviors in it. While the behavior may be experienced as originating from self, the demand of the setting is seen as the cause." (p. 66)

On the other, Stokols (chapter 21) has struggled with theorizing about the complex ways in which the current sociocultural conditions of accelerating societal change (e.g., informational and technological complexity) become translated into individual experience and action. Little's (chapter 7) work on personal projects has also been—though somewhat less explicitly—concerned with the relations between experience and action. However, in contradistinction to our approach that attempts to assess the ways in which intentionality and experience become translated into action, he has emphasized the assessment of intentional action in context and, in so doing, has been concerned with the end product of the translation in and of itself.

2.15. Problem Formulation and Methodology

From our point of view, environment-behavior research should be open to a range of methodologies from both the "natural science" and the "human science" perspectives. For the most part, the former is concerned about prediction, which is linked to: a focus on observable behavior and explanation in terms of cause-effect relations; an analytic analysis that begins with parts and assumes that the whole cannot be understood through addition of those elements; and the adoption of scientific experimentation as the appropriate methodology. In contrast, the latter (e.g., Giorgi, 1970):

"... specifies as its goal the understanding of experience or the explication of structural relationships, pattern, or organization that specifies meaning ... adopts the descriptive method ... seeks detailed analysis of limited numbers of cases that are presumably prototypic of larger classes of events ... carries out qualitative analysis through naturalistic observation, empirical, and phenomenological methods." (Wapner, 1987, p. 1434)

Thus, as we see it, since explication makes focal the meaning of a phenomenon or what is explained and causal explanation focuses on underlying process (e.g., the question of "how" and under what conditions developmental transformation is reversed, arrested, or advanced), both approaches have advantages and limitations. While the "natural science" approach may be characterized by precision and reliability, it may also suffer from lack of validity. In contrast, while the "human science" approach may be characterized by validity, it may suffer from lack of precision and reliability. Accordingly, both of these approaches to understanding complement each other and should be fostered.

A central issue for us then concerns when these methods should be used. This depends on the level of complexity of the phenomenon under investigation. Phenomena on the level of complex features of human experience—where manipulation and control of conditions are not possible—may more appropriately be analyzed by "human science" methods; in contrast, less complex and more simplified phenomena where conditions can be manipulated and controlled may be appropriately addressed by the methods of "natural science."

Several contributors to these volumes share our methodological eclecticism. In this regard, most notable is the work of: Werner and Altman (chapter 3), who have illustrated the need to select assumptions and methodologies most appropriate for different problems (cf. Maslow, 1946); Rivlin (chapter 5), who has advocated the customary use of more than one method or "triangulation," which she believes ". . . represents an approach to validate findings by using multiple 'imperfect measures'" (p. 55); and Canter (chapter 15), who has spoken to the need for understanding the everyday experience of people in naturally occurring contexts, which require new and old methodologies that accurately reflect such experience. In fact, methodological eclecticism is one of the most common themes that unifies the work of the contributors to this volume.

3. SUMMARY AND CONCLUSIONS

The above comparison of similarities and differences in the theoretical assumptions of leading researchers in the field of environment-behavior studies has suggested that this field has come a long way since its inception about 40 years ago. That is, in its infancy, environment-behavior studies consisted largely of a disparate group of researchers, problems, theories, and methodologies. Currently, there appear to be more similarities than dissimilarities in the theoretical orientations at least among this current cohort of researchers. That is, common assumptions that unify this group revolve around the notions of: transactionalism; constructivism; person-in-environment system as the unit of analysis; a multidimensional conception of the environment; the relations between experience and action; and methodological eclecticism. Less common assumptions among this group included: the relations among levels of integration; the equilibration tendencies of the person-in-environment system; a multifaceted conception of person; a broader concept of development; and multiple worlds.

It should be emphasized, as mentioned earlier, that the commonalities and differences reported here are based on the assumptions ingredient in our holistic, developmental, systems-oriented perspective. We strongly encourage others, both contributors and readers of this volume, to make similar comparisons utilizing their own perspective as its basis. Such endeavors will bring us ever closer to understanding similarities and differences among researchers and thereby serve to advance the study of environment-behavior relationships.

In closing, we note that this volume was completed during the last year of the 20th Century, more than four decades since environment-behavior studies began to become an area of interest to such diverse fields as anthropology, architecture, geography, psychology, sociology, and urban planning. We hope that the uncovering of similarities and differences of assumptions, research problems, and methodologies of the broad variety of investigators represented in this volume will serve as a basis for even further accelerated growth of environment-behavior research in the 21st Century. Such research, it is anticipated, will have powerful theoretical and practical value and thereby serve at least two functions: optimizing the transactions (experience and action) of human beings with their diverse physical, interpersonal, and sociocultural environments; and integrating their relations with people living in different physical and/or cultural contexts.

REFERENCES

Altman, I. (1997). Environment and behavior studies? A discipline? Not a discipline? Becoming a discipline. In S. Wapner, J. Demick, T. Yamamoto, & T. Takahashi (Eds.), *Handbook of Japan-U.S. environment behavior research: Toward a transactional approach* (pp. 423–434). New York: Plenum Press.

Altman, I., & Rogoff. (1987). World views in psychology: Trait, interactional, organismic and transactional perspectives. In D. Stokols & I. Altman (Eds.), *Handbook of environmental psychology* (pp. 7–40). New York: Wiley.

Apter, D. (1976). *Modes of coping with conflict in the presently inhabited environment as a function of variation in plans to move to a new environment.* Unpublished Master's thesis, Clark University.

Barker, R., & Wright, H. (1955). *Midwest and its children.* New York: Row Peterson.

Cantril, H. (Ed.). (1950). *The why of man's experience.* New York: Macmillan.

Dandonoli, P., Demick, J., & Wapner, S. (1990). Physical arrangement and age as determinants of environmental representation. *Children's Environments Quarterly, 7*(1), 26–36.

Demick, J., & Wapner, S. (1988a). Children-in-environments: Physical, interpersonal, and sociocultural aspects. *Children's Environments Quarterly, 5*(3), 54–62.

Demick, J., & Wapner, W. (1988b). Open and closed adoption: A developmental conceptualization. *Family Process, 27,* 229–249.

Demick, J., & Wapner, S. (1990). Role of psychological science in promoting environmental quality: Introduction. *American Pychologist, 45*(5),631–632.

Dewey, J., & Bentley, A. F. (1949). *Knowing and the known.* Boston: Beacon Press.

Feibelman, J. K. (1954). Theory of integrative levels. *British Journal of Philosophy of Science, 5,* 59–66.

Giorgi, A. (1970). Towards phenomenologically based research in psychology. *Journal of Phenomenological Psychology, 1,* 75–98.

Helson, H. (1948). Adaptation level as basis for a quantitative theory of frame of reference. *Psychological Review, 55,* 297–313.

Herrick, C. J. (1949). A biological survey of integrative levels. In R. W. Sellars, V. J. McGill, & M. Farber (Eds.), *Philosophy for the future* (pp. 222–242). New York: Macmillan.

Ittelson, W. H. (1973). Environmental perception and contemporary perceptual theory. In W. H. ittelson (Ed.), *Environment and cognition* (pp. 1–19). New York: Seminar Press.

Kaplan, B. (1959). The study of language in psychiatry. In S. Arieti (Ed.), *American Handbook of Psychiatry:* Vol. 3 (pp. 659–668). New York: Basic Books.

Kaplan, B. (1967). Meditations on genesis. *Human Development, 10,* 65–87.

Koffka, K. (1935). *Principles of gestalt psychology.* New York: Harcourt Brace.

Kuntz, P. G. (1968). *The concept of order.* Seattle: University of Washington Press.

Lavine, T. (1950a). Knowledge as interpretation: An historical survey. *Philosophy and Phenomenological Research, 10,* 526–540.

Lavine, T. (1950b). Knowledge as interpretation: An historical survey. *Philosophy and Phenomenological Research, 11,* 80–103.

Lewin, K. (1935). *A dynamic theory of personality.* New York: McGraw-Hill.

Maslow, A. H. (1946). Problem-centering vs. means-centering in science. *Philosophy of Science, 13,* 326–331.

Mayo, M., Pastor, J. C., & Wapner, S. (1995). Linking organizational behavior and environmental psychology. Relations between environmental psychology and allied fields. *Environment and Behavior, 27*(1), 73–89.

Minuchin, S. (1974). *Families and family therapy.* Cambridge, MA: Harvard University Press.

Novikoff, A. B. (1945a). The concept of integrative levels and biology. *Science, 101,* 209–215.

Novikoff, A. B. (1945b). Continuity and discontinuity in evolution. *Science, 101,* 405–406.

Oxley, D., Haggard, L. M., Werner, C. M., & Altman, I. (1985). Transactional qualities of neighborhood social networks: A case study of "Christmas Street." *Environment and Behavior, 18,* 640–677.

Pasto, T. (1964). *The space frame experiment in art.* NY: A. S. Barnes.

Pepper, S. C. (1942). *World hypotheses.* Berkeley: University of California Press.

Pepper, S. C. (1967). *Concept and quality: A world hypothesis.* LaSalle, IL: Open Court.

Sameroff, D. J. (1983). Developmental systems: Contexts and evolution. In P. H. Mussen (Ed.), *Handbook of child psychology Vol. 1,* (pp. 237–294). New York: Wiley.

Schneirla, T. C. (1949). Levels in the psychological capacities of animals, In R. W. Sellars, V. J. McGill, & M. Farber (Eds.), *Philosophy for the future.* New York: Macmillan.

Schutz, A. (1971). In M. Natanson (Ed.), *Collected papers: Vols. 1–3.* The Hague, Netherlands: Nijhoff.

Turvey, M. T. (1977). Preliminaries to a theory of action with reference to vision. In R. Shaw & J. Bransford (Eds.), *Perceiving, acting, and knowing.* Hillsdale, NJ: Erlbaum.

Von Uexküll, J. (1957). A stroll through the world of animals and men. In C. H. Schiller (Ed.), *Instinctive behavior.* New York: International Universities Press.

Vygotsky, L. S. (1978). *Mind in society.* Cambridge, MA: Harvard University Press.

Wapner, S. (1969). Organismic-developmental theory: Some applications to cognition. In J. Langer, P. Mussen, & N. Covington (Eds.), Trends and issues in developmental theory (pp. 35–67). New York: Holt, Rinehart and Winston.

Wapner, S. (1987). A holistic, developmental, systems-oriented environmental psychology: Some beginnings. In D. Stokols & I. Altman (Eds.), *Handbook of environmental psychology* (pp. 1433–1465). NY: Wiley.

Wapner, S., & Cirillo, L. (1973). *Development of planning* (Public Health Service Grant Application). Worcester, MA: Clark University.

Wapner, S., & Demick, J. (1990). Development of experience and action: Levels of integration in human functioning. In G. Greenberg & E. Tobach (Eds.), *Theories of the evolution of knowing. The T. C. Schneirla Conference Series, Vol. 4* (pp. 47–68). Hillsdale, NJ: Erlbaum.

Wapner, S., & Demick, J. (1998). Developmental analysis: A holistic, developmental, systems-oriented perspective. In R. M. Lerner (Ed.), *Theoretical models of human development. Vol. 4 Handbook of child psychology* (5th ed., Editor-in-chief: William Damon). New York: Wiley.

Wapner, S., & Demick, J. (1999). Developmental theory and clinical child psychology: A holistic, developmental, system-oriented approach. In W. K. Silverman & T. H. Ollendick (Eds.), *Developmental issues in the clinical treatment of children and adolescents* (pp. 3–30). Boston: Allyn and Bacon.

Wapner, S., Kaplan, B., & Cohen, S. (1973). An organismic-developmental perspective for understanding transactions of men in environments. *Environment and Behavior, 5,* 255–289. (Reprinted in: G. Broadbent, R. Bunt, & T. Llorens, Eds., (1980). *Meaning and behavior in the built environment.* New York: Wiley).

Werner, H., & Kaplan, B. (1956). The developmental approach to cognition: Its relevance to the psychological interpretation of anthropological and ethnolinguistic data. *American Anthropologist, 58,* 866–880.

Werner, H. (1940/1957). Comparative psychology of mental development. New York: International Universities Press. (Originally published in German, 1926 and in English, 1940).

Werner, H. (1957). The concept of development from a comparative and organismic point of view. In D. B. Harris (Ed.), *The concept of development: An issue in the study of human behavior.* Minneapolis: University of Minnesota Press.

NAME INDEX

Pages in *italics* refer to citations in references.

SUBJECT INDEX

"Action research," 192
Action(s)
  as providing meaning, 204
  as requiring interpretation, 204
  theory of, 293, 302–303
Adaptation, 293, 300
  to new sociocultural environments, 211–212
"Affordance," 153
"Alternative constructionism," 192
Analysis, modes of, 10, 64
Anthropology, 119
Applicable measurement, 86
Architectural design, 235; see also Housing design
  modern assumptions of, 230
Architectural education, 231–234
  moving beyond the stereotypical in, 234
Architectural model of environmental–behavior situation, 231
Architecture, universal language of, 167–168
Art, 279–282
Assumptions, 51–53, 79; see also under specific topics
  articulating and uncovering, 290
Averages, 202–203

Behavior, see also Environment and behavior; Environment–behavior relations
  linear vs. cyclical, 32
Behavior change, 32
Behavior setting theory, 142
Behavior settings, 62–64, 234–235, 296; see also Setting(s)
  analysis of, 126
Behavioral functioning, level of, 64
Beliefs, personal, 180–187; see also Assumptions
Biophilia, 35
"Blindsight," 283
Body and self experience, in Japan vs. U.S., 209–210
"Body subject," 162
Boring tasks, finding ways to make them interesting, 28–34

"Camping," 235
Categorization process, 69

Causality, 111
  efficient vs. formal, 27, 31–32
Classical approach/perspective, 195, 202
Collectivism, 240
Communities, paradigmatic research problems in, 65–66
Compressibility, 117
Conativity, 83–85
"Conceptual diffusion," 186
Conjoint measurement, 85
Consiliency, 85–86
Constructivism/interpretationism, 7–9, 80–82, 91, 115, 205, 292
Constructivist assumption, 294–295
Context, 64
  as providing meaning, 199
  transactions between person and, 199
Contextualism, 82–83, 91
Continuity, principle of, 117
Correlations, 202–203
Cross-cultural environment–behavior research, 208–213, 280–281, 288
  implications for interdisciplinary environment–behavior research, 213–214
  space-frames and, 284–287
Cross-cultural psychology, 208
Crowding, 239–240, 250–251
  and inner city stress, 253–262
  residential, in the context of inner city poverty, 247–253
    cumulative risk and, 253
    multiple stressors and, 253–254
Cultural assumptions underlying E-B research, 238–244
Cultural landscape, environment as, 121–123
Culture(s), 121, 208
  conditions under which they all evolve, 287–288
  environment and, 128, 130

Data, as insufficient, 196–197
Deconstruction, 173
Determinism, 33–34, 230
  forms of, 27
Development, 64, 299
  concepts, principles, and endpoints of, 293, 299–300